Water Scarcity Management

Water Scarcity Management

Enabling Technologies

Edited by

Kanchan D. Bahukhandi
Sustainability Cluster, School of Advanced Engineering
UPES, Dehradun, Uttarakhand, India

Manish Kumar
Escuela de Ingeniería y Ciencias, Tecnológico de Monterrey
Monterrey, Nuevo León, Mexico

Sustainability Cluster, School of Advanced Engineering
UPES, Dehradun, Uttarakhand, India

Research and Development Initiative, Chuo University
Kasuga, Bunkyo-ku, Tokyo, Japan

Durga P. Panday
Sustainability Cluster, School of Advanced Engineering
UPES, Dehradun, Uttarakhand, India

Escuela de Ingeniería y Ciencias, Tecnologico de Monterrey
Monterrey, Nuevo León, Mexico

Tushara G.G. Chaminda
University of Ruhuna
Matara, Sri Lanka

WILEY

This edition first published 2025

Registered Offices
John Wiley & Sons, Inc., 111 River Street, Hoboken, NJ 07030, USA
John Wiley & Sons Ltd, New Era House, 8 Oldlands Way, Bognor Regis, West Sussex, PO22 9NQ, UK

For details of our global editorial offices, customer services, and more information about Wiley products visit us at www.wiley.com.

The manufacturer's authorized representative according to the EU General Product Safety Regulation is Wiley-VCH GmbH, Boschstr. 12, 69469 Weinheim, Germany, e-mail: Product_Safety@wiley.com.

Wiley also publishes its books in a variety of electronic formats and by print-on-demand. Some content that appears in standard print versions of this book may not be available in other formats.

Library of Congress Cataloging-in-Publication Data Applied for:

Hardback ISBN: 9781394176717

Cover Design: Wiley
Cover Image: Courtesy of Manish Kumar

Set in 9.5/12.5pt STIXTwoText by Straive, Pondicherry, India
Printing and Binding: CPI Group (UK) Ltd, Croydon, CR0 4YY

C9781394176717_310325

Contents

About the Editors

Professor Kanchan D. Bahukhandi

Dr. Kanchan D. Bahukhandi is a Senior Associate Professor in the Sustainability Cluster at the School of Advanced Engineering, UPES, with over 22 years of experience in environmental science, sustainability, water pollution, hydrogeochemistry, and solid waste management. She has published more than 50 research papers in international journals, contributed 30 book chapters, and edited 3 books. Dr. Bahukhandi has delivered invited lectures at premier institutions and organizations and presented around 50 research papers at national and international conferences.

Professor Manish Kumar

Dr. Manish Kumar received his PhD in Environmental Engineering from the University of Tokyo. He is a Distinguished Professor at Tecnológico de Monterrey, Monterrey, Mexico; UPES, Dehradun; and Chuo University, Hachioji, Japan. A fellow of FRSC, JSPS, and WARI, he has supervised 10 PhD students, edited over 15 books, and published more than 250 SCI/SCIE papers with over 14,000 citations. Featured among the world's top 2% researchers, he has led major global projects on water sustainability and wastewater epidemiology while serving on editorial boards of top journals including *npj Clean Water*.

Professor Durga P. Panday

Dr. Durga P. Panday, an Associate Professor at UPES and investigator postdoc at Tecnologico de Moneterrey, specializes in Water Resources Engineering and Management, focusing on water quality, hydroclimatic extremes, and transboundary water sharing using game theory. With a decade of teaching experience, he is dedicated to sustainable water management through high-impact research and pragmatic strategies in premier journals.

Professor Tushara G.G. Chaminda

Dr. Tushara G.G. Chaminda is a full Professor in Civil and Environmental Engineering at the Faculty of Engineering, University of Ruhuna, Sri Lanka. He also serves as a Director Board member of the National Water Supply and Drainage Board, Sri Lanka. With over 200 scientific publications, his primary research interests include emerging pollutants in urban waters, antibiotic-resistant bacteria, virus indicators in water and soil, sustainable water usage, and green building technology.

List of Contributors

Sonali Aggarwal
Department of Management Studies
Graphic Era (Deemed to be University)
Dehradun, Uttarakhand, India

Apoorva
Sustainbaility Cluster, School of Advanced Engineering, UPES, Dehradun
Uttarakhand, India

Urvashi Arya
Department of Management Studies
Graphic Era (Deemed to be University)
Dehradun, Uttarakhand, India

Sushila Arya
Department of Agriculture, Dev Bhoomi Uttarakhand University, Dehradun
Uttarakhand, India

Durgesh Bahuguna
Graduate School of Business
Tulas Institute, Dehradun
Uttarakhand, India

Kanchan D. Bahukhandi
Sustainability Cluster, School of Advanced Engineering, UPES
Dehradun, Uttarakhand, India

Vishal Bhardwaj
Department of Zoology and Environmental Science, Gurukul Kangri (Deemed to be University)
Haridwar, Uttarakhand, India

Ganesh Datt Bhatt
College of Agriculture Sciences
An ICAR Accredited College
Teerthanker Mahaveer University
Moradabad, Uttar Pradesh, India

Aditi Bisht
Department of Zoology and Environmental Science, Gurukul Kangri (Deemed to be University)
Haridwar, Uttarakhand, India

Uttamasha B. Borah
Department of Petroleum Engineering and Earth Sciences (Energy Cluster)
School of Advanced Engineering
UPES, Dehradun, Uttarakhand, India

Khem Chand
MM Institute of Management
Maharishi Markandeshwar
(Deemed to be University)
Ambala, Bharat, India

S.S.K. Chandrasekara
Department of Agricultural Engineering
Faculty of Agriculture, University of
Peradeniya, Peradeniya, Sri Lanka

Divya Chaudhary
Department of Genetics and Plant
Breeding, G.B. Pant University of
Agriculture and Technology
Udham Singh Nagar
Uttarakhand, India

Kanika Dogra
Sustainability Cluster, School of Advanced
Engineering, UPES, Dehradun
Uttarakhand, India

Navjot Hothi
Department of Physics, School of
Engineering, UPES, Dehradun
Uttarakhand, India

Kishor Joshi
CTVS, Mira Heart Centre, Dehradun
Uttarakhand, India

Nitin Kamboj
Department of Zoology and Environmental
Science, Gurukul Kangri (Deemed to be
University), Haridwar
Uttarakhand, India

Amrit Kumar
Department of Zoology and Environmental
Science, Gurukul Kangri (Deemed to be
University), Haridwar
Uttarakhand, India

Avinash Kumar
Department of Zoology and Environmental
Science, Gurukul Kangri (Deemed to be
University), Haridwar
Uttarakhand, India

Manish Kumar
Escuela de Ingeniería y Ciencias
Tecnológico de Monterrey, Monterrey
Nuevo León, México

Sustainability Cluster, School of Advanced
Engineering, UPES, Dehradun
Uttarakhand, India

Research and Development Initiative
Chuo University, Kasuga, Bunkyo-ku
Tokyo, Japan

Pawan Kumar
Department of Zoology and Environmental
Science, Gurukul Kangri (Deemed to be
University), Haridwar, Uttarakhand, India

Sunil Kumar
Gurukula Kangri (Deemed to be University)
Haridwar, Uttarakhand, India

Aanchal Kumari
Sustainability Cluster, School of Advanced
Engineering, UPES
Dehradun, Uttarakhand, India

Gagan Matta
Department of Zoology and Environmental
Science, Gurukul Kangri (Deemed to be
University), Haridwar
Uttarakhand, India

Ruchi Mehrotra
Department of Management Studies,
Graphic Era (Deemed to be University)
Dehradun, Uttarakhand, India

Anjali Nayak
Department of Zoology and Environmental
Science, Gurukul Kangri (Deemed to be
University), Haridwar
Uttarakhand, India

Nirlipta P. Nayak
Energy Cluster, School of Advanced
Engineering, UPES, Dehradun
Uttarakhand, India

Durga P. Panday
Sustainability Cluster, School of Advanced Engineering, UPES, Dehradun
Uttarakhand, India

Escuela de Ingeniería y Ciencias
Tecnológico de Monterrey, Monterrey
Nuevo León, Mexico

Minakshi Pandey
Department of Chemistry, School of Science, IFTM University, Moradabad
Uttar Pradesh, India

Neeraj Pandey
Department of Zoology and Environmental Science, Gurukul Kangri (Deemed to be University), Haridwar, Uttarakhand, India

Gaurav Pant
Department of Zoology and Environmental Science, Gurukul Kangri (Deemed to be University), Haridwar
Uttarakhand, India

Saumya Patel
Department of Petroleum Engineering and Earth Sciences (Energy Cluster), School of Advanced Engineering, UPES, Dehradun
Uttarakhand, India

M.P. Perera
Department of Geography
Faculty of Arts, University of Peradeniya
Peradeniya, Sri Lanka

Deepali Rana
Department of Zoology, Dolphin (PG)
Institute of Biomedical & Natural Sciences
Dehradun, Uttarakhand, India

Saumyaranjan Sahoo
Department of Civil Engineering, School of Engineering, UPES, Dehradun
Uttarakhand, India

Himanshu Saini
Department of Zoology and Environmental Science, Gurukul Kangri (Deemed to be University), Haridwar
Uttarakhand, India

Aseem Saxena
Sustainability Cluster, School of Advanced Engineering, UPES
Dehradun, Uttarakhand, India

Madhu Sharma
Centre of Excellence – Energy and Eco-sustainability Research, Uttaranchal Institute of Technology, Uttaranchal University, Dehradun
Uttarakhand, India

Pushpa Sharma
Department of Petroleum Engineering and Earth Sciences (Energy Cluster)
School of Advanced Engineering, UPES
Dehradun, Uttarakhand, India

Rajesh Tiwari
Department of Management Studies
Graphic Era (Deemed to be University)
Dehradun, Uttarakhand, India

Sachin Tripathi
Sustainability Cluster, School of Advanced Engineering, UPES, Dehradun
Uttarakhand, India

Escuela de Ingeniería y Ciencias
Tecnológico de Monterrey, Monterrey
Nuevo León, Mexico

M. Vithanage
Ecosphere Resilience Research Center
Faculty of Applied Sciences, University of Sri Jayewardenepura, Nugegoda
Sri Lanka

National Institute of Fundamental Studies
Kandy, Sri Lanka

Foreword

Water scarcity is emerging as one of the most critical challenges facing humanity in the 21st century. As the global population increases, industrial activities expand, energy generation rises, the standard of living improves, and climate change continues to reshape ecosystems, the demand for fresh, clean water is growing at an unprecedented rate. Addressing this challenge requires not only technological innovation but also a deep understanding of traditional and indigenous knowledge, which has led to sustainable management of water resources for centuries.

This book, *Water Scarcity Management: Enabling Technologies*, offers a comprehensive examination of water scarcity through both modern and historical lenses. Divided into four sections, it provides a thorough exploration of current water contaminants, conservation practices, and security trends across different regions of the world.

The book's first section sets the stage by outlining current water contaminants and conservation practices. It serves as a reminder that the roots of water scarcity are complex, often involving a web of factors, such as pollution, mismanagement, and overexploitation. The second section turns our attention to the status of water security and availability across continents, offering detailed analyses of regions that are facing some of the most severe water crises, from Asia and Africa to the Americas, Australia, and Europe.

One of the most innovative aspects of this book is its third section, which highlights the significance of indigenous technologies and traditional water management practices. From Asia to the islands, these practices offer a wealth of knowledge, often overlooked, that could be vital in developing sustainable solutions for modern water challenges.

The final section focuses on actionable solutions – both traditional and modern – that could mitigate the growing water crisis. It not only discusses the technological advances in water treatment but also dives into the critical question of sustainability. Chapters on traditional water-related practices, modern technological solutions, and the challenges in achieving long-term water security shed light on how integrated approaches are essential for securing our global water future.

In compiling this volume, the editors have made a significant contribution to both academic discourse and practical applications. The diverse perspectives and interdisciplinary approaches presented here are crucial, as we navigate the complex relationship between water, technology, and society in the Anthropocene.

This book comes at a critical time, and I am confident it will provide valuable insights to researchers, practitioners, and policymakers who are working tirelessly to solve one of the defining issues of our time.

Professor Vijay P. Singh, Ph.D., D.Sc., D. Eng. (HC, Waterloo),
D.Sc. (HC, Guelph), Ph.D. (HC, Basalicata), D.Sc. (HC, McGill),
P.E., P.H., BC WRE (Hon.), D. EWRA (Hon.), Dist.M. ASCE,
Hon.Dist. M. IWRA, Dist.F. AGGS, Hon.M. AWRA,
F.AGU, EASA, Academician (GFA), NAE
Distinguished Professor and Regents Professor
Caroline and William N. Lehrer Distinguished Chair in
Water Engineering
Department of Biological and Agricultural Engineering
Zachry Department of Civil & Environmental Engineering
Texas A&M University
College Station, TX, USA

Preface

With the world facing an acute water crisis in terms of both quantity and quality, our focus shifts towards traditional knowledge and indigenous technologies to ensure water security. This book extensively traversed continents, observing the interconnections between stakeholders, diverse water laws, and evolving water security concerns. As we delve into the past, we find the ever-changing landscape of water pollutants, from early industrialization to modern-day contaminants. This historical context prepares the background to address the complexities of present-day pollution.

The book has revealed the traditional wisdom and indigenous technologies that have long played a role in managing water resources through its chapters. We have brought out innovative practices of tribal, rural, and urban communities and modern water treatment methods that align with ecological balance.

The last phase of the book leads to sustainable solutions. We analyzed the long-term viability and environmental impacts of nature-based and modern technologies. This book aims at providing tangible, sustainable solutions to the world's water challenges. This book represents our collective effort to deepen our knowledge of the current issues surrounding water security and sustainability, offering a roadmap for safeguarding this precious resource on a global scale. We believe the solutions to the present problems can be solved through historical traditional practices and indigenous knowledge but with modern technological implementation.

Editors:
Kanchan D. Bahukhandi
Manish Kumar
Durga P. Panday
Tushara G.G. Chaminda

1

Emerging Contaminants and Water Conservation

Challenges, Strategies, and Solutions for Sustainable Water Management

Sachin Tripathi[1,2], *Durga P. Panday*[1,2], *and Manish Kumar*[1,2,3]

[1] *Sustainability Cluster, School of Advanced Engineering, UPES, Dehradun, Uttarakhand, India*
[2] *Escuela de Ingeniería y Ciencias, Tecnológico de Monterrey, Monterrey, Nuevo León, Mexico*
[3] *Research and Development Initiative, Chuo University, Korakuen Campus 1-13-27, Kasuga, Bunkyo-ku, Tokyo, Japan*

1.1 Introduction

As the number of people living on Earth continues to increase and there is a corresponding decrease in the amount of available water, water conservation has emerged as an increasingly pressing issue (Alotaibi et al. 2023). Emerging pollutants, on the other hand, pose a substantial danger to both the quality of the water supply and human health (Tripathi et al. 2023). The purpose of this chapter is to give a complete analysis of contemporary developing pollutants as well as water conservation practises, covering the sources, impacts, and management measures associated with these contaminants.

Emerging pollutants may have their origins in a wide number of sources, such as agricultural practises, industrial operations, wastewater treatment plants (WWTPs), or consumer products. Emerging contaminants (ECs) may come from industrial, municipal (household), agricultural, hospital, or laboratory effluent (Figure 1.1). Surface water, groundwater, drinking water, and WWTP effluent include environmental pollutants (Tripathi et al. 2023). Municipal wastewater is known to emit novel contaminants into the environment. These contaminants come from non-point and point sources, industrial activities, storm water runoff, home wastewater, and water treatment facilities (Pradhan et al. 2023). Due to high EC values in sludge, management is becoming more concerned (Das et al. 2022, Kumar et al. 2023a, b).

The use of agricultural practises such as pesticides and fertilisers, among other things, can contribute to the presence of newly discovered pollutants in both surface water and groundwater. WWTPs, which are designed to eliminate conventional pollutants but are less successful at removing ECs, are a substantial source of ECs (Dubey et al. 2023). This is because of the way that they are designed. Manufacturing and mining are two examples of industrial activity that might contribute to the discharge of chemicals into rivers. Last but not least, consumer goods, such as flame retardants and plasticizers, have the potential to make their way into water sources (Macklin et al. 2023).

Water Scarcity Management: Enabling Technologies, First Edition. Edited by Kanchan D. Bahukhandi, Manish Kumar, Durga P. Panday, and Tushara G.G. Chaminda.

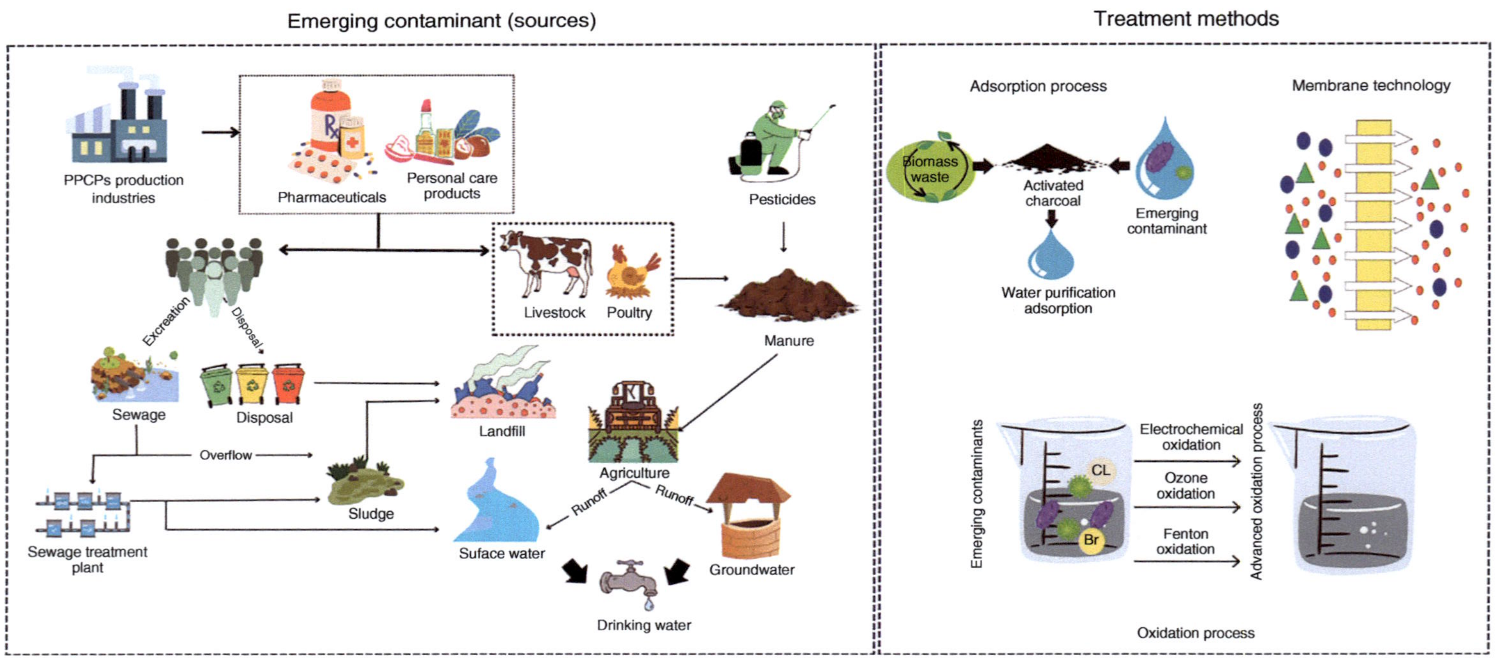

Figure 1.1 Classification of emerging contaminants that impact on soil, water, plants, and treatment processes.

It has been discovered that ECs have a wide range of effects on both human health and the environment (Neog et al. 2024). These effects can be broken down into several categories. Pharmaceuticals and personal care items, for instance, have been connected to endocrine disruption, which can have an effect on both the function and development of the reproductive system (Kumar et al. 2023d). There is evidence that flame retardants cause neurotoxicity, and certain per- and polyfluoroalkyl substances (PFAS) have been related to cancer and dysfunction in the immune system. Microplastics, which are small plastic particulates that can be discovered in water sources, have the potential to have both a physical and a chemical impact on the creatures that live in water (Farooq et al. 2023).

Emerging pollutants, in addition to having these effects, can also have ecological repercussions, such as changing the composition of microbial communities and disrupting the regular functioning of ecosystems (Li et al. 2023). These consequences can have cascade repercussions throughout food webs, which can have an effect on the health of organisms living in both aquatic and terrestrial environments. The Figure 1.1 represents a classification of emerging contaminants and their varied impacts on soil, water, and plants, highlighting the complex challenges they pose to environmental health. Additionally, it illustrates how these contaminants influence treatment processes, often requiring advanced or modified remediation strategies due to their persistence and resistance to conventional methods.

1.1.1 Sources of Heavy Metal

Approximately 40% of the world's lakes and rivers have been contaminated by heavy metals (Zhou et al. 2020), stemming from both natural and human activities. Natural sources involve interactions with metal-containing rocks and volcanic eruptions (Ali et al. 2019). Volcanic emissions, including geothermal activity and degassing, contribute sporadically (Naggar et al. 2018). Anthropogenic sources encompass industrial processes, agriculture, and domestic practices (Gautam et al. 2014); as shown in Figure 1.2.

Mining, pivotal for many economies, releases heavy metals into water bodies, impacting groundwater, soil erosion, and health. Urbanisation and industrialisation exacerbate pollution levels, as evidenced by arsenic in India's drinking water and various heavy metals in Nigeria's mining communities. Latin America faces chronic exposure issues, with millions affected by arsenic-contaminated water exceeding WHO limits. China grapples with high metal concentrations in coastal rivers (Xu et al. 2017), while mercury contamination plagues Venezuela's artisan gold mining areas. Turkey also battles heavy metal contamination. Mitigating heavy metal pollution is crucial globally, with economic challenges hindering remediation efforts in developing nations.

1.1.2 Irrigation Water Quality

1.1.2.1 Sodium Percent

Sodium, when interacting with soil, diminishes its permeability. Higher levels of sodium prompt a cation exchange process, leading to a reduction in water and air movement within

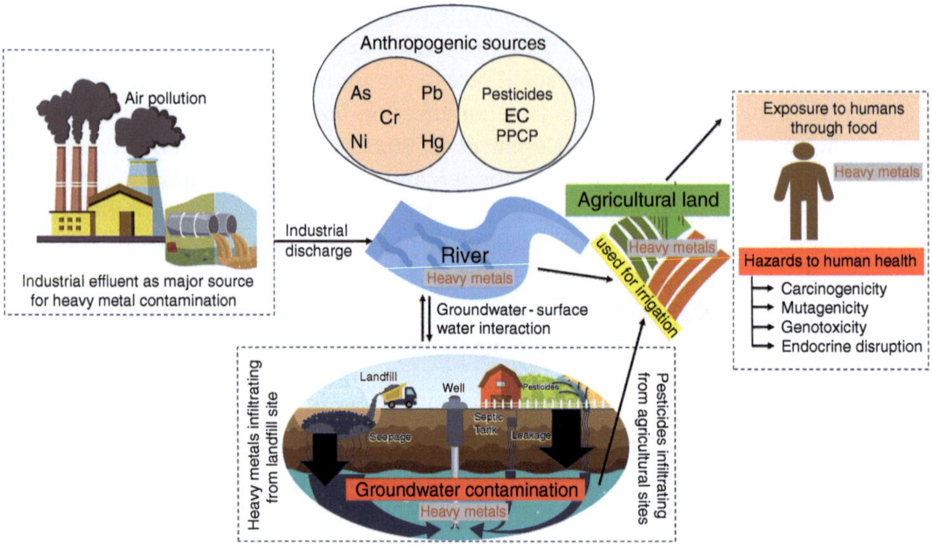

Figure 1.2 Sources of heavy metal concentration in food plants and trophic transfer to human

the soil, particularly under moist conditions (Hopkins et al. 2007). The term "sodium percent" is defined as follows:

$$Na^{+}(\%) = \left[\left(Na^{+}\right) \times 100 / \left(Ca^{2+} + Mg^{2+} + Na^{+} + K^{+}\right)\right]$$

Wilcox diagram is used to classify irrigation water based on sodium percent.

1.1.2.2 Chloride

Chloride levels in irrigation water contribute to its overall salinity and can pose toxicity risks to plants when concentrations are excessively high. Elevated chloride levels can lead to foliar burns when deposited on leaves. Some plant species are more vulnerable to chloride damage than others. To mitigate the harm caused by high chloride levels in irrigation water, options include selecting less sensitive crop varieties, utilising irrigation methods such as furrow, flood, or drip irrigation to minimise foliar contact, and rinsing plants at the conclusion of each irrigation cycle if a source of high-quality water is accessible. Excessive chlorine in plants can lead to leaf tissue accumulation, resulting in a burnt appearance, despite chlorine being a micronutrient essential for plant growth (Hopkins et al. 2007).

1.1.2.3 Salinity

Irrigation water with an electrical conductivity below 0.2 μS/cm, as discussed earlier, can lead to issues with soil permeability. When water salinity is very low, it can leach out calcium and cause soil particles to become more prone to breaking apart, resulting in difficulties with water infiltration. To prevent these infiltration problems, it is suggested to add a calcium salt such as gypsum or calcium chloride to the irrigation water, increasing the salinity to 0.2–0.3 μS/cm (Hopkins et al. 2007).

1.2 Emerging Contaminants

1.2.1 Pharmaceuticals

Due to the widespread use of pharmaceuticals in the medical treatment of human and animal diseases, accidents, and illnesses, this industry is considered to be one of the largest contributors to the world's most critical environmental problems. Pharmaceutically active compounds are complex molecules that contain a variety of diverse functions and physical properties in order to engage in specific biological activities (Kumar et al. 2023c). Pharmaceuticals and medications contain these compounds. The vast quantity of these compounds necessitates their subdivision into a variety of sub-classes.

Non-steroidal anti-inflammatory medications (NSAIDs) are the compounds most frequently used to relieve pain which are frequently detected in surface water and are especially resistant to conventional wastewater treatment methods (Khumalo et al. 2023). Hormones and oestrogens are considered to be significant emerging pollutants due to their ability to persist in living organisms for extended periods of time and also disrupt the endocrine systems (Almazrouei et al. 2023). Compared to other oestrogens, 17-ethinylestradiol is a prevalent oral contraceptive for women that possesses endogenous activity (Thacharodi et al. 2023). Typically discovered in wastewater, it is notoriously challenging to eradicate. Antibiotics are complex chemical compounds used to inhibit or eliminate pathogenic microorganisms. Most frequently detected in the environment are the antibiotics amoxicillin, ciprofloxacin, erythromycin, and penicillin. Table 1.1 shows the concentration of various emerging contaminants reported globally.

1.2.2 Personal Care Products

There are a variety of over-the-counter health, cosmetic, and drug products referred to as PCP (personal care product) (Benny et al. 2024). Organic ultraviolet (UV) filters, preservatives, and fragrances are the three categories of contaminants of emerging concerns (CECs) that are most prevalent in PCPs (Chaturvedi et al. 2021a). In general, the compounds that comprise PCPs are conjugated with other medications or pollutants present in pharmaceutical products, such as antidandruff cleansers (Krishnan et al. 2021). As preservatives in cosmetics and pharmaceuticals, parabens are compounds. Triclosan is another compound extensively used as an antimicrobial agent in soaps, deodorants, skin creams, toothpaste, and plastics, and it is frequently detected in surface water. Among the most frequently detected PCPs in surface water, parabens must be considered; these compounds are used in cosmetics and pharmaceuticals. Due to their lipophilic nature, a number of these pollutants can bioaccumulate in organisms after being ingested or absorbed through the epidermis. They are then excreted through urine or merely removed by washing, and eventually, they make their way into aquatic and soil environments.

1.2.3 Plasticizers

Plasticizers are clear, colourless liquids with a viscous consistency. They have a low-molecular weight and are commonly used as additives to increase the flexibility or distensibility of a

Table 1.1 Emerging contaminants (ECs) in various water sources, their concentrations, health and environmental impacts, and detection methods (liquid chromatography-mass spectrometry/mass spectrometry [LC-MS/MS], gas chromatography-mass spectrometry [GC-MS], high-performance liquid chromatography [HPLC]).

Category of ECs	ECs	Source	Max. conc. (ng/L)	Detection method	Potential health/ environmental impact	Regulatory limit (if applicable)	Country	References
Pesticides	Atrazine	Surface water	171	LC-MS/MS	Endocrine disruption, carcinogenic	3 μg/L (EU)	Brazil	Abubakar et al. (2020)
	Hexazinone	Surface water	44.1	GC-MS	Potential groundwater contamination	N/A		
	2-hydroxy atrazine	Surface water	130	HPLC	Toxic to aquatic life	N/A		
	Hexazinone	Surface water	100	GC-MS	Soil leaching, toxic to plants	N/A	Australia	Almazrouei et al. (2023)
	Simazine	Surface water	100	LC-MS/MS	Possible endocrine disruptor	0.1 μg/L (EU)		
Pharmaceutical products	Triclosan	Surface water	8.6	LC-MS/MS	Antibiotic resistance, aquatic toxicity	N/A	Brazil	Benny et al. (2024)
	Ibuprofen	Groundwater and surface water	49.4	HPLC	Disrupts aquatic ecosystems, possible kidney effects in humans	N/A	India	Chaturvedi et al. (2021b)
	Paracetamol	Surface water	106,900	LC-MS/MS	Toxic to aquatic species	N/A	Kenya	Kesari et al. (2021)
	Ibuprofen	Surface water	106,900	LC-MS/MS	Aquatic toxicity, persistent pollutant	N/A		
	Carbamazepine	Surface water	660	GC-MS	Persistent, harmful to aquatic life	N/A		

Personal care products	Benzophenone-3	Drinking water	5.7	HPLC	Skin and eye irritation, environmental toxicity	N/A	Italy	Krishnan et al. (2021)
	Diethyltoluamide	Drinking water	98	GC-MS	Neurotoxic, environmental hazards	N/A	United States	Kumar et al. (2023a)
	Galaxolide	Drinking water	110	HPLC	Bioaccumulative, disrupts aquatic life	N/A		
	Propyl paraben	Surface water	3.61	LC-MS/MS	Endocrine disruptor, possible carcinogen	N/A	Spain	Kumar et al. (2023b)
Surfactants and plasticizers	Bisphenol A	Surface water	4800	HPLC	Endocrine disruption, reproductive harm	0.1 μg/L (EU)	Spain	Kumar et al. (2023b)
	4-n-nonylphenol (4-n-NP)	Surface water	1030	LC-MS/MS	Persistent, bioaccumulative, aquatic toxicity	N/A		
	Bis(2-ethylhexyl) phthalate	Surface water	2050	GC-MS	Endocrine disruptor, possible carcinogen	6 μg/L (EPA)	Taiwan	Gou et al. (2016)
	Dibutyl phthalate	Surface water	163	LC-MS/MS	Endocrine disruption, developmental effects	N/A		
Anthropogenic markers	Bromoform	Source water	3300	GC-MS	Possible carcinogen, harmful to aquatic life	80 μg/L (EPA)	United States	Glassmeyer et al. (2017)
	Nicotine	Groundwater	28.3	HPLC	Neurotoxic, affects aquatic organisms	N/A	Italy	Riva et al. (2018)
Fluorinated compounds	Perfluorobutanoic acid	Surface water	725	LC-MS/MS	Persistent, bioaccumulative, toxic to humans and wildlife	N/A	Spain	Pico et al. (2019)
	Perfluorooctane sulfonate	Source water	48.3	LC-MS/MS	Persistent, bioaccumulative, toxic to aquatic life	70 ng/L (EPA)	United States	Glassmeyer et al. (2017)
	Sulfluramid	Surface water	2	LC-MS/MS	Highly toxic to aquatic organisms	N/A	Australia	Sardina et al. (2019)

material. This makes working with the material simpler during the manufacturing process and during the formulation of final products. Plasticizers can be categorised into a variety of groups based on the molecular structure of their molecules; among these, phthalates have garnered a great deal of attention due to the frequency with which they are employed and the quantity of phthalates that are discharged into water sources. They are esters of phthalic acid and have a benzene ring with two functional ester groups, as in diethylhexyl phthalate (DEHP), diethyl phthalate (DEP), and dibutyl phthalate (DBP) (Mehraie et al. 2022). In addition, there are DEP and DBP. Phthalates can readily migrate and disperse throughout the environment due to their low-molecular weight. In addition, the lipophilic nature of phthalates allows them to pass through natural barriers such as the skin, lungs, and gut tissue of humans, which can contribute to bioaccumulation and biomagnification.

1.2.4 Pesticides

Pesticides are any mixture of substances that can prevent, eliminate, repel, or mitigate the effects of an insect or other pest (Abubakar et al. 2020). The most prevalent varieties of pesticides are fungicides, herbicides, bactericides, and insecticides. Two of the most prevalent sources of pesticide contamination are surface runoff from agricultural areas and urban wastewater. It is possible to recognise various sub-classes based on chemical structure, including organochlorines (which include aldrin and DDT), organophosphates (which include diazinon and malathion), carbamates (which include carbaryl and propoxur), triazines (which include atrazine), and chloroacetamides (which include metolachlor and alachlor) (Parlapiano et al. 2021). Unfortunately, because pesticides are persistent in the environment, such as organochlorines, which have a lengthy half-life in the environment, they bioaccumulate in the food chain. Moreover, the extensive use of these compounds can readily induce poisoning via a variety of distinct toxicity pathways (Savoca and Pave 2021). When it comes to environmental contamination, the use of these substances in liquid form is far more detrimental than when they are administered in powder form.

1.3 Best Management Practises for Water Conservation

The demand for freshwater resources can be reduced by the implementation of water conservation practises, which in turn lowers the risk that newly discovered toxins will make their way into water supplies (Banaduc et al. 2022). Reusing and recycling water, collecting rainwater, and using technologies that are more water-efficient are just a few of the many methods that can be utilised in order to save water (Varma 2022).

The process of processing effluent in order to reuse it for non-potable applications, such as irrigation or industrial operations, is referred to as water reuse and recycling. This practise has the potential to lower the demand for freshwater resources and also to assist in lowering the amount of toxins that are discharged into water sources (Kesari et al. 2021). The process of collecting and storing rainwater for later use, often for irrigation or other inside applications that do not require potable water, is known as rainwater harvesting. This practise has the potential to lower the demand for freshwater resources, which can be especially advantageous in locations that receive a low amount of precipitation. Finally, water-efficient technology, such as low-flow fixtures and appliances, can help cut down on the quantity of water needed for day-to-day activities like taking showers and washing clothes. These technologies can lower the amount of water required for these tasks (Abu-bakar et al. 2021).

1.3.1 Strategies for the Management of Newly Emerging Contaminants

The management of ECs calls for an approach that incorporates a number of different strategies, including source control, remediation technology, monitoring programmes, and regulatory strategies (Ritter et al. 2002). Eliminating or minimising the discharge of newly discovered pollutants at their point of origin is an essential component of source control. For instance, best management practises in agriculture can assist reduce the amount of pesticides and fertilisers that are used, and industrial processes can be adjusted to limit the amount of hazardous chemicals that are used (Leong et al. 2020). Emerging pollutants can be eradicated from water sources through the utilisation of treatment technology. For instance, it has been demonstrated that highly developed oxidation techniques, which make use of chemical reactions to break down contaminants, are successful at removing pharmaceuticals and personal care products from effluent (Xu et al. 2017). Last but not least, monitoring programmes can assist in locating newly discovered pollutants in water sources and tracking the quantities of these contaminants over time (Panday et al. 2021a, b; Panday and Kumar 2023).

In addition, regulatory strategies are being developed in order to handle newly discovered pollutants (Geissen et al. 2015). For instance, in the United States, the Environmental Protection Agency (EPA) has compiled a Contaminant Candidate List (CCL) of chemicals that are currently being researched for the possibility of being regulated by the Safe Drinking Water Act. In a similar manner, the European Union has compiled a list called the Priority Substances list, which identifies newly discovered toxins that call for additional research and monitoring (Lapworth et al. 2019).

1.3.2 The Integration of Water Conservation with the Management of Emerging Contaminants

The incorporation of water conservation practises and the management of ECs might potentially yield major benefits for both the health of humans and the water resources themselves. For instance, the implementation of water-saving technologies can cut down on the need for available freshwater resources, which in turn lowers the risk that newly discovered toxins will be able to make their way into water supplies (Talib and Randhir 2017). In a similar vein, the recycling of effluent after it has been treated has the potential to cut down on the discharge of newly discovered toxins into surface water and groundwater (Panday and Kumar 2022). In addition, the incorporation of EC management into water conservation programmes can assist in the process of increasing knowledge regarding the possible effects that emerging pollutants may have on both the health of humans and the water resources (Gogoi et al. 2018). Table 1.2 provides an overview of various contaminant classes and their associated contaminants, emphasizing the potential health hazards linked to each. It also outlines water conservation practices that can mitigate pollution risk, offering insights into sustainable approaches for reducing contamination and safeguarding public health.

Table 1.2 Overview of contaminant classes, associated contaminants, health hazards, and water conservation practices for mitigating pollution risk.

Contaminants class	Contaminants	Major health hazards	Water conservation practices	Environmental impact	Regulatory standards (if applicable)	Source of contamination	Remediation techniques
Pharmaceuticals	Acetaminophen, amoxicillin, caffeine, carbamazepine, ciprofloxacin, diclofenac, estradiol, ibuprofen, naproxen, sulfamethoxazole, trimethoprim, triclosan	Endocrine disruption, antibiotic resistance, cancer, developmental effects, reproductive effects, neurological effects, hormone disruption, and immune system effects	• Proper disposal of unused medications • Implementing effective wastewater treatment technologies • Educating the public about the risks of improper disposal of pharmaceuticals	Persistent in aquatic ecosystems, promotes antibiotic-resistant bacteria	N/A (varies by country)	Wastewater discharge, improper disposal of pharmaceuticals	Advanced oxidation processes, activated carbon filtration
Personal care products	Benzophenone-1, benzophenone-3, octylphenol, octyl phenol ethoxylates, phthalates, triclocarban, triclosan	Endocrine disruption, cancer, developmental effects, reproductive effects, neurological effects, hormone disruption, and immune system effects	• Avoiding the use of personal care products containing emerging contaminants • Using natural and eco-friendly personal care products • Proper disposal	Bioaccumulation in wildlife, toxic to aquatic life	N/A (some limits in EU for phthalates)	Domestic wastewater, runoff from personal care product usage	Biodegradation, eco-friendly alternatives, membrane filtration
Flame retardants	Brominated diphenyl ethers (BDEs), hexabromocyclododecane (HBCD)	Endocrine disruption, neurotoxicity, developmental effects, reproductive effects, and cancer	• Reducing the use of products containing flame retardants • Proper disposal of products containing flame retardants • Implementing effective waste management and recycling programs	Persistent organic pollutants, bioaccumulation, toxic to marine organisms	EU Restriction on BDEs, Stockholm Convention bans	Industrial waste, household waste	Adsorption, incineration, waste recycling programs
Per- and polyfluoroalkyl substances (PFAS)	PFAS	Cancer, thyroid hormone disruption, immune system effects, developmental effects, reproductive effects, and liver damage	• Avoiding the use of products containing PFAS. Implementing effective wastewater treatment technologies	Persistent, bioaccumulative, toxic to aquatic life and humans	70 ng/L (EPA limit in drinking water for some PFAS)	Industrial effluents, firefighting foams	Granular activated carbon, ion exchange, reverse osmosis
Microplastics	Microbeads, microfibers	Ingestion of microplastics by organisms in food chain, transfer of contaminants from microplastics to organisms	• Avoiding the use of products containing microbeads and using natural and eco-friendly cleaning products	Accumulation in marine environments	Microbeads banned in cosmetics (US, EU)	Domestic runoff, textile washing, improper plastic disposal	Waste management programs, bioremediation

1.3.3 Removal of CECs

In a particular instance that takes place in the real world, the adsorption process is affected by a variety of elements, including surface area, the kind of analytes being absorbed and their starting concentration, temperature, the pH of the solution, the kind of adsorbate being absorbed and how much of it there is, etc. It is necessary to carry out tertiary treatments that make use of a wide variety of sorbent materials in order to get rid of CECs, which are generally found in wastewater (Joseph et al. 2019).

Activated carbon, often known as AC, is the type of adsorbent that is used most frequently and extensively in WWTPs. AC is a form of charcoal that has undergone both physical and chemical modification in order to enhance the adsorbent qualities it possesses (Mestre et al. 2022). The production of AC is possible using a wide variety of raw materials, some of which include coal, coconut shells, lignite, and wood, among others. Following the pyrolysis of organic matter with oxidising gases, the inorganic elements of the carbonised material are effectively eliminated, leaving behind either powder activated carbon (PAC) or granular activated carbon particles (GAC) with a significant surface area (Yu et al. 2022). There are two different ways to accomplish this goal. There has been some consideration given to the possibility of utilising AC as an efficient adsorbent for the removal of persistent or non-biodegradable organic molecules such as CECs. Although many writers feel that it is difficult to differentiate biochar (BC) from AC, BC is a stable carbon source that may be created from biomass via thermal or hydrothermal methods at high temperatures with limited or zero oxygen environment (Panwar and Pawar 2022). BC can be used in a wide variety of applications, including the production of fuels and chemicals, as well as the treatment of water and soil. A wide number of applications are attainable with the use of BC. BC is capable of receiving additional chemical activation in a manner analogous to that of AC, and this can take place either through an acidic or a basic process. This ultimately leads to the creation of novel functionalities on the surface of the sorbent, which, in turn, improves the interactions with the contaminants and increases the adsorption of those pollutants (Veeramalai et al. 2022).

Carbon nanotubes (CNTs) are an allotrope of carbon with a structure similar to graphite (Tiwari et al. 2016). These carbon allotropes display varied degrees of adsorption depending on the degree of coil, the formation of the original sheet, the diameter, the internal geometry, their physicochemical properties, and the treatment procedure that was used to synthesise them.

Clay minerals are well-known naturally occurring materials that display plastic capabilities, and they can be found in a variety of different forms. Clay minerals are sub-micrometer-sized particles that are mostly made up of hydrous layer silicates of aluminium, but they may also contain magnesium and iron. Clay minerals can be found in a variety of different environments. Clays such as montmorillonite, mica, kaolinite, pyrophyllites (talc), bentonite, and diatomite are good candidates for use as adsorbents. Clays like these are inexpensive, have a high porosity, and can be found in large quantities in nature. In the body of scientific literature, it has been demonstrated that a number of different sorbents, such as zeolites, metal oxide, graphene oxide, and polymeric resins, are capable of efficiently extracting CECs from water matrices. Because of their high selectivity towards bound target molecules and their adaptability to couple with a wide variety of materials in order to take advantage of synergistic features (Surana et al. 2022), molecularly imprinted polymers (MIPs) have the potential to be an effective alternative sorbent in this context. MIPs with a core-shell of TiO_2 particles, which were generated through surface polymerisation, are used in applications such as the removal and photocatalytic degradation of the fungicide *ortho*-phenylphenol (Motamedi et al. 2022).

The removal of ECs through biodegradation might be thought of as one of the most effective and kindest to the environment techniques available. The value of the kinetics constant is generally defined by the rate of biodegradation (compounds emerging/organic). Caffeine, estradiol, acetaminophen, and ibuprofen are examples of pollutants that are simple to biodegrade due to their high biodegradation constants. Carbamazepine and iopamidol, on the other hand, have a low biodegradation constant, which makes it difficult for them to be broken down. Over the course of many years, a number of different efforts have been made to enhance the biodegradation of persistent medication (Taoufik et al. 2020). This section will provide a brief summary of the different approaches, both conventional and unconventional, that have been employed to get rid of EC oxidoreductase, nitrifying agent, and fungal culture. In order for pollutants to be degraded, it is necessary to take into account a wide range of biological and non-biological factors, including temperature, molecular features, redox potential, electron acceptor availability, main substrates, and physical and chemical qualities. The biodegradation of EC is believed to be the formation of metabolites and co-metabolites, as stated by the relevant EC. Previous research has shown that the organism known as "*Pseudomonas aeruginosa*" is capable of transferring estradiol by way of its metabolism. Similarly, the degradation of estradiol as the only source of carbon by degrading (bacterial cultures with enhanced estradiol) is one example of this. In the event that bacteria have developed resistance to antibiotics and other antibacterial agents, ammonium and carbonate must be employed to stimulate enzymes that are responsible for the biotransformation of such ECs. There are primarily two categories of bacteria that are responsible for biodegradation: autotrophic oxidants and heterotrophic bacteria. Among these, autotrophic bacteria that oxidise ammonia (such as ibuprofen, acetaminophen, and bisphenol) play the most important role in the breakdown of various ECs. The amount of ammonium in the environment is directly proportional to the rate of EC biodegradation. The evaluation of the biological mechanism that gets rid of EC has been going on for a good number of years. Some ECs have been successfully eliminated through the use of anaerobic membranes, aerobic membranes, and anoxic membrane reactors, whereas the untreated EC is treated through the utilisation of traditional and unconventional approaches. For example, the highest rate of xenoestrogens removed is 92% by traditional treatment through activated sludge, 80% with two oxidation ditch methods, 70% with various bioreactors, and 64% of which have 10 lagoons connected in series. This is the highest rate of xenoestrogens removed. In a similar vein, the removal of certain of the ECs necessitates the use of methods that are effective but costly due to the environmental persistence of the substance.

1.3.4 Irrigation and Water Conservation Practices

The main water source for irrigation was groundwater (Jara-Rojas et al. 2012). Surface water from streams and bayous also contributed to irrigation. Additionally, some growers constructed on-farm water storage and tailwater recovery systems capable of fully meeting their irrigation needs for certain fields. Growers using groundwater from depletable aquifers were expected to be more open to adopting water conservation practices. The most common irrigation method for row crops was furrow irrigation, Practices such as deep tilling, computerised hole selection (CHS), and surge irrigation were employed by a significant portion of growers to enhance furrow irrigation performance and conserve water.

Effective irrigation scheduling is essential for water management, with the use of soil moisture sensors (SMS) having the potential to save up to 50% of total water applied (Hassanli et al. 2009). Studies have shown that SMS can improve water use efficiency in furrow-irrigated soybeans and corn without compromising yields compared to traditional farmer-managed scheduling (Bryant et al. 2017). Flow meters along with pump timers which allow for automated starting and stopping of irrigation events at irrigation wells are crucial management tools.

1.3.4.1 Reducing Salts in Irrigation Water

Controlling salinisation involves leaching salt from the root zone, altering farm practices, and using salt-tolerant plants. Improved irrigation methods such as partial rootzone drying and drip irrigation sustain irrigated farming (Shrivastava and Kumar 2015). To combat dryland salinity, deep-rooted perennials are reintroduced to prevent excess water movement. Integrating perennials in farming systems enhance resilience to salinity stress, though widespread adoption is hindered by cost and water availability. Research focuses on developing strategies such as breeding resilient crop varieties and refining resource management practices (Shrivastava and Kumar 2015). Amendments like sulphur and gypsum can address soil salinity, but excess water is still needed for leaching. However, these treatments are effective only for sodium-affected soils and may worsen salinity issues in saline soils (Fipps 2003).

1.4 Conclusion

Water conservation practices and EC management are both critical components of ensuring the sustainability of water resources and protecting human health. The integration of these two approaches can provide significant benefits for both, reducing the demand for freshwater resources while also managing the potential risks posed by ECs. Continued research and monitoring of ECs, along with the development and implementation of effective management strategies, will be crucial for ensuring the long-term health and sustainability of our water resources.

References

Abubakar, Y., Tijjani, H., Egbuna, C. et al. (2020). Pesticides, history, and classification. In: *Natural Remedies for Pest, Disease and Weed Control* (ed. C. Egbuna and B. Sawicka), 29–42. Academic Press.

Abu-Bakar, H., Williams, L., and Hallett, S.H. (2021). A review of household water demand management and consumption measurement. *Journal of Cleaner Production* 292: 125872.

Ali, H., Khan, E., and Ilahi, I. (2019). Environmental chemistry and ecotoxicology of hazardous heavy metals: environmental persistence, toxicity, and bioaccumulation. *Journal of Chemistry.* https://doi.org/10.1155/2019/6730305 2019.

Almazrouei, B., Islayem, D., Alskafi, F. et al. (2023). Steroid hormones in wastewater: sources, treatments, environmental risks, and regulations. *Emerging Contaminants* 9 (2): 100210.

Alotaibi, B.A., Baig, M.B., Najim, M.M. et al. (2023). Water scarcity management to ensure food scarcity through sustainable water resources management in Saudi Arabia. *Sustainability* 15 (13): 10648.

Bănăduc, D., Simić, V., Cianfaglione, K. et al. (2022). Freshwater as a sustainable resource and generator of secondary resources in the 21st century: stressors, threats, risks, management and protection strategies, and conservation approaches. *International Journal of Environmental Research and Public Health* 19 (24): 16570.

Benny, S.M., Gupta, S.D., Ismail, S.P., and Francis, D. (2024). Pharmaceutical pollution of water bodies: sources, impacts, and mitigation. In: *Handbook of Water Pollution* (ed. T.A. Inamuddin and A. Alrooqi), 371–416. Wiley.

Bryant, C.J., Krutz, L.J., Falconer, L. et al. (2017). Irrigation water management practices that reduce water requirements for Mid-South furrow-irrigated soybean. *Crop, Forage & Turfgrass Management* 3 (1): 1–7.

Chaturvedi, P., Shukla, P., Giri, B.S. et al. (2021a). Prevalence and hazardous impact of pharmaceutical and personal care products and antibiotics in environment: a review on emerging contaminants. *Environmental Research* 194: 110664.

Chaturvedi, M., Mishra, A., Sharma, K. et al. (2021b). Emerging contaminants in wastewater: sources of contamination, toxicity, and removal approaches. In: *Emerging Treatment Technologies for Waste Management* (ed. I. Haq and A.S. Kalamdhad), 103–132. Springer.

Das, A., Kumar, M., Jha, P.K. et al. (2022). Isotopic and hydrogeochemical tracking of dissolved nutrient dynamics in the Brahmaputra River System: a source delineation perspective. *Chemosphere* 307: 135757. https://doi.org/10.1016/j.chemosphere.2022.135757.

Dubey, M., Vellanki, B.P., and Kazmi, A.A. (2023). Removal of emerging contaminants in conventional and advanced biological wastewater treatment plants in India – a comparison of treatment technologies. *Environmental Research* 218: 115012.

Farooq, M., Nisa, F.U., Manzoor, Z. et al. (2023). Abundance and characteristics of microplastics in a freshwater river in northwestern Himalayas, India-Scenario of riverbank solid waste disposal sites. *Science of the Total Environment* 886: 164027.

Fipps, G. (2003). Irrigation water quality standards and salinity management. In: *Texas Cooperative Extension*, vol. B-1667 (ed. G. Fipps), 4–03. Texas A&M University.

Gautam, R.K., Sharma, S.K., Mahiya, S., and Chattopadhyaya, M.C. (2014). Contamination of heavy metals in aquatic media: transport, toxicity and technologies for remediation. In: *Heavy Metals in Water: Presence, Removal and Safety* (ed. S. Sharma), 1–24. Royal Society of Chemistry. http://dx.doi.org/10.1039/9781782620174-00001.

Geissen, V., Mol, H., Klumpp, E. et al. (2015). Emerging pollutants in the environment: a challenge for water resource management. *International Soil and Water Conservation Research* 3 (1): 57–65.

Glassmeyer, S.T., Furlong, E.T., Kolpin, D.W. et al. (2017). Nationwide reconnaissance of contaminants of emerging concern in source and treated drinking waters of the United States. *Science of the Total Environment* 581: 909–922.

Gogoi, A., Mazumder, P., Tyagi, V.K. et al. (2018). Occurrence and fate of emerging contaminants in water environment: a review. *Groundwater for Sustainable Development* 6: 169–180.

Gou, Y.Y., Lin, S., Que, D.E. et al. (2016). Estrogenic effects in the influents and effluents of the drinking water treatment plants. *Environmental Science and Pollution Research* 23: 8518–8528.

Hassanli, A.M., Ebrahimizadeh, M.A., and Beecham, S. (2009). The effects of irrigation methods with effluent and irrigation scheduling on water use efficiency and corn yields in an arid region. *Agricultural Water Management* 96 (1): 93–99.

Hopkins, B. G., Horneck, D. A., Stevens, R. G., Ellsworth, J. W., and Sullivan, D. M. (2007). Managing irrigation water quality for crop production in the Pacific Northwest.

Jara-Rojas, R., Bravo-Ureta, B.E., and Díaz, J. (2012). Adoption of water conservation practices: a socioeconomic analysis of small-scale farmers in Central Chile. *Agricultural Systems* 110: 54–62.

Joseph, L., Jun, B.M., Jang, M. et al. (2019). Removal of contaminants of emerging concern by metal-organic framework nanoadsorbents: a review. *Chemical Engineering Journal* 369: 928–946.

Kesari, K.K., Soni, R., Jamal, Q.M.S. et al. (2021). Wastewater treatment and reuse: a review of its applications and health implications. *Water, Air, & Soil Pollution* 232: 1–28.

Khumalo, S.M., Makhathini, T.P., Bwapwa, J.K. et al. (2023). The occurrence and fate of antibiotics and nonsteroidal anti-inflammatory drugs in water treatment processes: a review. *Journal of Hazardous Materials Advances* 10: 100330.

Krishnan, R.Y., Manikandan, S., Subbaiya, R. et al. (2021). Removal of emerging micropollutants originating from pharmaceuticals and personal care products (PPCPs) in water and wastewater by advanced oxidation processes: a review. *Environmental Technology & Innovation* 23: 101757.

Kumar, M., Panday, D.P., Bhagat, C. et al. (2023a). Demystifying the decadal shift in the extent of groundwater in the coastal aquifers of Gujarat, India: a case of reduced extent but increased magnitude of seawater intrusion. *Science of the Total Environment* 165451. https://doi.org/10.1016/j.scitotenv.2023.165451.

Kumar, M., Mazumder, P., Silori, R. et al. (2023b). Prevalence of PPCPs, microplastics and co-infecting microbes in the post-COVID-19 era and its implications on antimicrobial resistance and potential endocrine disruptive effects. *Science of the Total Environment* 166419. https://doi.org/10.1016/j.scitotenv.2023.166419.

Kumar, M., Silori, R., Mazumder, P., and Tauseef, S.M. (2023c). Screening of pharmaceutical and personal care products (PPCPs) along wastewater treatment system equipped with root zone treatment: a potential model for domestic waste leachate management. *Journal of Environmental Management* 335: 117494.

Kumar, M., Silori, R., Mazumder, P. et al. (2023d). Wars and pandemics: AMR accelerators of the 21st century? *Environmental Science & Technology Letters* 10 (4): 289–291.

Lapworth, D.J., Lopez, B., Laabs, V. et al. (2019). Developing a groundwater watch list for substances of emerging concern: a European perspective. *Environmental Research Letters* 14 (3): 035004.

Leong, W.H., Teh, S.Y., Hossain, M.M. et al. (2020). Application, monitoring and adverse effects in pesticide use: the importance of reinforcement of Good Agricultural Practices (GAPs). *Journal of Environmental Management* 260: 109987.

Li, Y., Zhang, C., Wang, X. et al. (2023). Pollutant impacts on bacteria in surface water and sediment: conventional versus emerging pollutants in Taihu Lake, China. *Environmental Pollution* 323: 121334.

Macklin, M.G., Thomas, C.J., Mudbhatkal, A. et al. (2023). Impacts of metal mining on river systems: a global assessment. *Science* 381 (6664): 1345–1350.

Mehraie, A., Shariatifar, N., Arabameri, M. et al. (2022). Determination of phthalate acid esters (PAEs) in bottled water distributed in tehran: a health risk assessment study. *International Journal of Environmental Analytical Chemistry* 1–15.

Mestre, A.S., Campinas, M., Viegas, R.M. et al. (2022). Activated carbons in full-scale advanced wastewater treatment. *Advanced Materials for Sustainable Environmental Remediation* 433–475.

Motamedi, M., Yerushalmi, L., Haghighat, F., and Chen, Z. (2022). Recent developments in photocatalysis of industrial effluents: a review and example of phenolic compounds degradation. *Chemosphere* 296: 133688.

Naggar, Y.A., Khalil, M.S., and Ghorab, M.A. (2018). Environmental pollution by heavy metals in the aquatic ecosystems of Egypt. *Open Access Journal of Toxicology* 3. https://doi.org/10.19080/OAJT.2018.03.555603.

Neog, N., Goswami, R., Panday, D.P. et al. (2024). Arsenic contamination in the groundwater of Northeastern India: critical understandings on geotectonic controls and the need for intervention. *Current Opinion in Environmental Science & Health* 100539. https://doi.org/10.1016/j.coesh.2024.100539.

Panday, D.P. and Kumar, M. (2022). Climate indices and hydrological extremes: deciphering the best-fit model. *Environmental Research* 215: 114301. https://doi.org/10.1016/j.envres.2022.114301.

Panday, D.P. and Kumar, M. (2023). Application of remote sensing techniques to deal with scale aspects of GRACE data to quantify groundwater levels. *MethodsX* 102108. https://doi.org/10.1016/j.mex.2023.102108.

Panday, D.P., Khosa, R., Maheswaran, R. et al. (2021a). Game-theoretic-based modelling of Krishna waters dispute: equilibrium solutions by metagame analysis. *The European Physical Journal B* 94: 101. https://doi.org/10.1140/epjb/s10051-021-00107-w.

Panday, D.P., Khosa, R., Maheswaran, R. et al. (2021b). Game theoretic-based modelling of Krishna waters dispute: equilibrium solutions by hypergame analysis. *The European Physical Journal B* 94: 131. https://doi.org/10.1140/epjb/s10051-021-00135-6.

Panday, D.P., Singh, A., Siddhant et al. (2023). Probable maximum precipitation analysis of high rainfall regions in India. *Groundwater for Sustainable Development* 100893. https://doi.org/10.1016/j.gsd.2022.100893.

Panwar, N.L. and Pawar, A. (2022). Influence of activation conditions on the physicochemical properties of activated biochar: a review. *Biomass Conversion and Biorefinery* 12: 925–947.

Parlapiano, M., Akyol, Ç., Foglia, A. et al. (2021). Selective removal of contaminants of emerging concern (CECs) from urban water cycle via Molecularly Imprinted Polymers (MIPs): potential of upscaling and enabling reclaimed water reuse. *Journal of Environmental Chemical Engineering* 9 (1): 105051.

Pradhan, B., Chand, S., Chand, S. et al. (2023). Emerging groundwater contaminants: a comprehensive review on their health hazards and remediation technologies. *Groundwater for Sustainable Development* 20: 100868.

Ritter, K.S., Sibley, P., Hall, K. et al. (2002). Sources, pathways, and relative risks of contaminants in surface water and groundwater: a perspective prepared for the Walkerton inquiry. *Journal of Toxicology and Environmental Health Part A* 65 (1): 1–142.

Riva, F., Castiglioni, S., Fattore, E. et al. (2018). Monitoring emerging contaminants in the drinking water of Milan and assessment of the human risk. *International Journal of Hygiene and Environmental Health* 221 (3): 451–457.

Sardiña, P., Leahy, P., Metzeling, L. et al. (2019). Emerging and legacy contaminants across land-use gradients and the risk to aquatic ecosystems. *Science of the Total Environment* 695: 133842.

Savoca, D. and Pace, A. (2021). Bioaccumulation, biodistribution, toxicology and biomonitoring of organofluorine compounds in aquatic organisms. *International Journal of Molecular Sciences* 22 (12): 6276.

Shrivastava, P. and Kumar, R. (2015). Soil salinity: a serious environmental issue and plant growth promoting bacteria as one of the tools for its alleviation. *Saudi Journal of Biological Sciences* 22 (2): 123–131.

Surana, D., Gupta, J., Sharma, S. et al. (2022). A review on advances in removal of endocrine disrupting compounds from aquatic matrices: future perspectives on utilization of agri-waste based adsorbents. *Science of the Total Environment* 826: 154129.

Talib, A. and Randhir, T.O. (2017). Managing emerging contaminants in watersheds: need for comprehensive, systems-based strategies. *Sustainability of Water Quality and Ecology* 9: 1–8.

Taoufik, N., Boumya, W., Janani, F.Z. et al. (2020). Removal of emerging pharmaceutical pollutants: a systematic mapping study review. *Journal of Environmental Chemical Engineering* 8 (5): 104251.

Thacharodi, A., Hassan, S., Hegde, T.A. et al. (2023). Water a major source of endocrine-disrupting chemicals: an overview on the occurrence, implications on human health and bioremediation strategies. *Environmental Research* 116097.

Tiwari, S.K., Kumar, V., Huczko, A. et al. (2016). Magical allotropes of carbon: prospects and applications. *Critical Reviews in Solid State and Materials Sciences* 41 (4): 257–317.

Tripathi, S., Kumar, R., Mondal, P. et al. (2023). Urban fingerprinting on the River Ganges: a perspective of emerging contaminants status. *Total Environment Research Themes* 5: 100028.

Varma, V.G. (2022). Water-efficient technologies for sustainable development. In: *Current Directions in Water Scarcity Research*, vol. 6 (ed. A.L. Srivastav, S. Madhav, A.K. Bhardwaj, and E. Valsami-Jones), 101–128. Elsevier.

Veeramalai, S., Ramlee, N.N., Mahdi, H.I. et al. (2022). Development of organic porous material from pineapple waste as a support for enzyme and dye adsorption. *Industrial Crops and Products* 181: 114823.

Xu, Y., Liu, T., Zhang, Y. et al. (2017). Advances in technologies for pharmaceuticals and personal care products removal. *Journal of Materials Chemistry A* 5 (24): 12001–12014.

Yu, S., Wang, J., Zhao, Z., and Cai, W. (2022). Simultaneous coupling of fluidized granular activated carbon (GAC) and powdered activated carbon (PAC) with ultrafiltration process: a promising synergistic alternative for water treatment. *Separation and Purification Technology* 282: 120085.

Zhou, Q., Yang, N., Li, Y. et al. (2020). Total concentrations and sources of heavy metal pollution in global river and lake water bodies from 1972 to 2017. *Global Ecology and Conservation* 22: e00925. https://doi.org/10.1016/j.gecco.2020.e00925.

2

Water Security in Asia

Kanika Dogra[1]*, Aseem Saxena*[1]*, Kanchan D. Bahukhandi*[1]*,*
and Manish Kumar[1,2,3]

[1] *Sustainability Cluster, School of Advanced Engineering, UPES, Dehradun, Uttarakhand, India*
[2] *Escuela de Ingeniería y Ciencias, Tecnológico de Monterrey, Monterrey, Nuevo León, México*
[3] *Research and Development Initiative, Chuo University, Kasuga, Bunkyo-ku, Tokyo, Japan*

2.1 Introduction

Water is extremely important for all living creatures and delicate political problem. Attention must be given right away to the weaknesses in the current infrastructure, overuse of groundwater resources, and deterioration of water quality brought on by pollution. Due to a complex and quickly shifting geographic pattern of water supply, demand, and use, it is challenging to predict how adequate freshwater resources will be utilized in the future on a worldwide scale. Governmental and nongovernmental organizations, the commercial sector, and academics have to take an interest in the term "water security" and its underlying ideas in both policy and practice (UNESCO 2019). Water security is persuaded by a number of monetary and public factors as well as the substantial accessibility of freshwater resources about the urge for water (such as effective water planning and management techniques, institutional capacity to provide water services, and economic policies for sustainable development) (Das et al. 2022; Kumar et al. 2023a, b).

UN defines water security as "the capacity of a population to safeguard sustainable access to adequate quantities of acceptable quality water for sustaining livelihoods, human well-being, and socioeconomic development; for ensuring protection against water-borne pollution and water-related disasters; and for preserving ecosystems in a climate of peace and political stability." The ability of a population to protect sustainable access to sufficient amounts of water of acceptable quality for sustaining livelihoods, human well-being, and socioeconomic development, for ensuring protection against waterborne pollution and water-related disasters, and for preserving ecosystems in an environment of peace and political stability was defined as water security globally (Moumen et al. 2019).

Water Scarcity Management: Enabling Technologies, First Edition. Edited by Kanchan D. Bahukhandi, Manish Kumar, Durga P. Panday, and Tushara G.G. Chaminda.

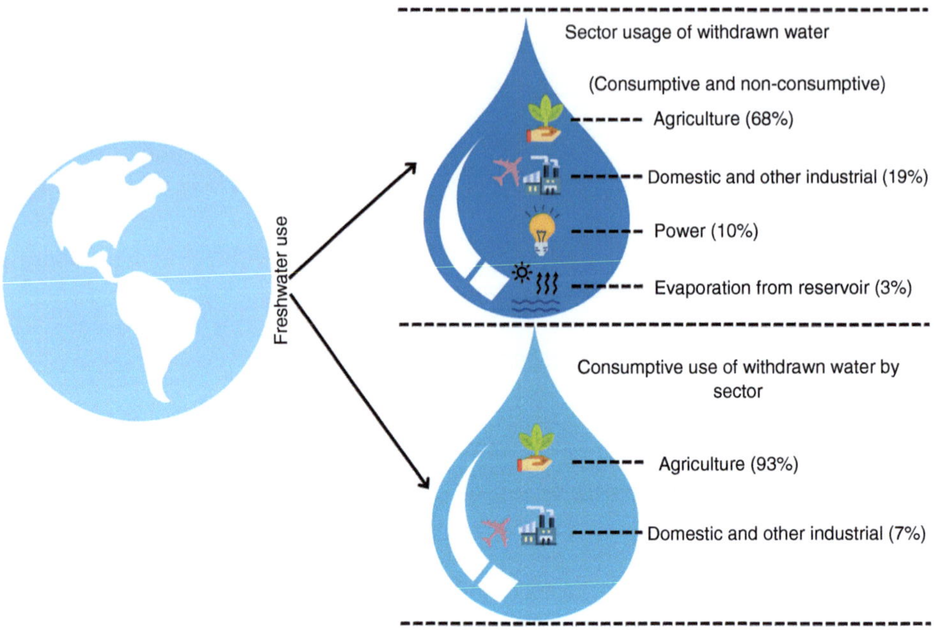

Figure 2.1 Earth's freshwater distribution highlighting the sectoral consumption of water. Agriculture dominates the consumptive use of freshwater.

Whereas, nationally water security is defined as an acceptable degree of water-related threat to the public, environment, and economies, as well as the approachability of water in an adequate abundance and quality for well-being, livelihoods, nature, and production. In present, almost 2 billion people reside in nations where freshwater withdrawals surpass 25% of the total amount of sustainable freshwater resources. According to the FAO (2016), agriculture accounts for approx. 68% of the world's freshwater withdrawals and an even larger portion of its consumptive freshwater usage as shown in Figure 2.1. In addition to increasing demand on water resources, this increase is also having an influence on the water cycle due to climate change, which might reduce the amount of freshwater available. Therefore, keeping an eye on the strain on water supplies is crucial as a precursor to future shortages (Gun 2012). Therefore, the objective of this paper is to highlight the present scenario of water security and its availability in Asia and also to discuss how the key dimension (KD) of water security helps in indexing the rankings of different countries (Neog et al. 2024).

2.2 Water Security in Asia

2.2.1 Southeast Asia

Southeast Asia is home to nine nations: Malaysia, Laos, Cambodia, Myanmar, Vietnam, Thailand, Philippines, Indonesia, and Singapore. Southeast Asia is home to over 670 million people, or 9% of the worldwide population, who mostly depend on reclaimed water and groundwater (Frederick and Leinbach 2020). Due to urbanization and population increase, this region is under extreme pressure. Several countries in Southeast Asia recorded a high proportion of household groundwater use in rural areas such as East Timor (81%), Indonesia (90%), and Myanmar (78%) (Carrard et al. 2019). Due to the proximity of Southeast Asian nations to the equator, these nations' water supplies are occasionally affected by extended periods of dry weather. Water shortages arise when demand and usage outweigh supply, placing the nation under stress (Burek et al. 2016).

Southeast Asia is experiencing a surge in freshwater demand due to factors like fast urbanization, industrialization, population expansion, and an increase in cultivated land to meet the necessity for the production of basic foods. The pattern of water withdrawal in Southeast Asian countries depends on socioeconomic factors. Indonesia consumes the most agricultural and industrial water (92.76 billion $m^3$and 24.65 billion m^3) in Southeast Asia, followed by Vietnam and the Philippines. Similarly, Indonesia led the list of countries with the biggest population for household water withdrawal, followed by the Philippines and Malaysia as shown in Figure 2.2 (Ritchieand Roser 2017; Maniam et al. 2021). According to the report of the World Bank, In Jakarta, 4.5 million people have access to piped water, while 35% of residents use water from artesian wells, 15% drink bottled water, 8% still rely on rainwater and about 80% of Indonesians lack access to clean water.

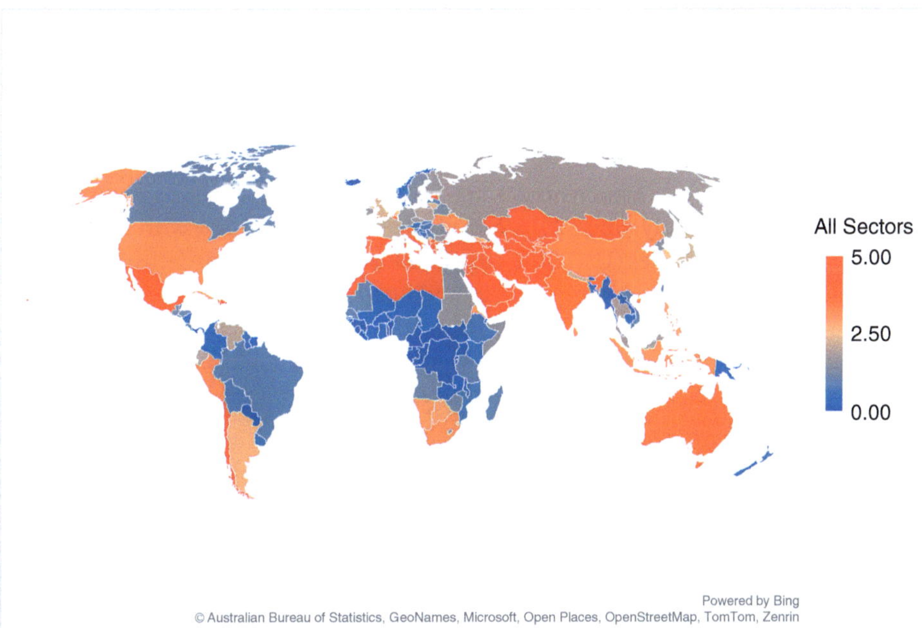

Figure 2.2 Future country-level water stress for 2040 under pessimistic scenarios that include different sectors (agricultural, domestic, and industrial). *Source:* Adapted from Bing Australian Bureau of Statistics, GeoNames, Microsoft, Open Places, OpenStreetMap, Tom, Zenrin.

2.2.2 South Asia

The geologic history of the area has a significant impact on the security of the water supply in South Asia. When the Indian and Eurasian tectonic plates gently collide, the Himalayan Mountain range is being created. The Indus, Ganges, and Brahmaputra River systems are based on the basins that were created by the erosion of silt from that range. The comparatively poor performance of rural household water security (HWS) and urban water security has a significant impact on South Asia's water security.

2.2.3 Central and West Asia

For countries in the former Soviet Union, scores on water security in Central and West Asia are respectable, but Afghanistan and Pakistan have rather poor outcomes. Water security in the region has improved between 2013 and 2020. Water security for rural households is expanding rapidly. Economic water security increased between 2013 and 2016, but it declined between 2016 and 2020. Urban water security is still low and requires improvement. Environmental water security and disaster security related to water are both low when compared to those of the other regions. Managing trade-offs, maintaining integrity, and involving stakeholders are priority tasks for improving governance in Central and West Asia (Panday et al. 2021a, b; Panday and Kumar 2023).

2.2.4 East Asia

East Asia during 2013–2020 has shown a robust increase in water security. Different countries in East Asia such as the People's Republic of China, Taipei is experiencing the fastest growth and has the highest level of water security. Mongolia has also made tremendous improvements. Water security for rural households and the environment both have room for improvement. Advanced economies have achieved a high level of economic water security (Panday and Kumar 2022). Water-related disaster security has reformed considerably since 2013 but needs additional progress to achieve the level of advanced economies (UNESCO 2019). In East Asia, some initiatives have been undertaken to improve governance that include policy consistency, managing trade-offs, and stakeholder involvement. Water stress assesses the total water usage (municipal, industrial, and agricultural), expressed as a ratio of the total yearly available blue water. Higher values signify high competition among users shown in Table 2.1.

2.3 UNSDG Goals and Water Security Index

One of the main focuses of the Sustainable Development Goals is water because it is essential for social and economic development and is also needed for the production of energy and food, the maintenance of healthy ecosystems, and human existence. The Sustainable Development Goal (SDGs) offer a unprecedented potential for addressing the water security challenge by improving coordination among stakeholders, and private and government sectors. The availability of water security, therefore, entails achieving the three

Table 2.1 Various industrial activities and their impact on water quality.

Type of industry	Usage of water	Discharge water characteristics	Impact on water bodies
Chemical manufacturing	Cooling, solvent, cleaning	Contaminated with chemicals	Alters aquatic ecosystems
Mining operations	Dust suppression, ore	Contains heavy metals, sediments	Increases sedimentation
Power generation	Cooling, steam production	Elevated temperature, chemicals	Thermal pollution
Food processing	Cleaning, cooking, processing	Organic matter, nutrients	Nutrient enrichment
Textile manufacturing	Dyeing, cleaning, finishing	Dyes, chemicals	Alters aquatic ecosystems
Petroleum refining	Cooling, processing	Hydrocarbons, heavy metals	Oil spills, contamination

Source: Kanu and Achi (2011) and Ewa et al. (2011).

key facets of sustainability: social sustainability (by securing access to clean water sufficient for domestic use and small agriculture for the poor majority), environmental sustainability (by ensuring the protection of surface water and groundwater), and economic sustainability (by ensuring a reliable supply and improving the productivity of water use in agricultural, industrial, urban, and rural development) as shown in Figure 2.3 (Bengtsson and Shivakoti 2015; Moumen et al. 2019). According to the United Nations Sustainable Development Goals report (2021), about 2.3 billion people lived in countries with significant water stress in 2018 (Maniam et al. 2021). Most Sub-Saharan African, Asian, and Oceanian nations struggle with either a physical shortage of water or a financial inability to access it.

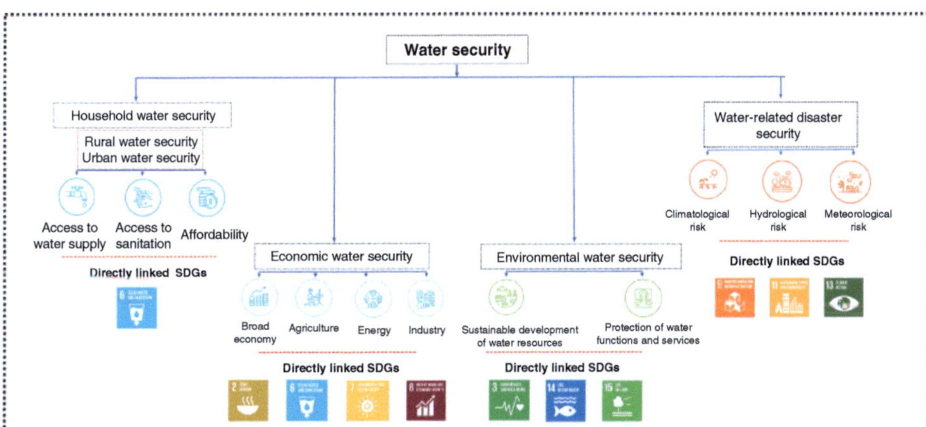

Figure 2.3 Linkage between United Nations Sustainable Development (UN SDG) Goals and Water Security.

Access to enough water to meet fundamental human and environmental demands would be compromised by water scarcity. Improvements in health, education, gender equality, and eventual efforts to counteract global warming would be made possible with continued success in improving drinking water (Asok Raja et al. 2019; Panday and Kumar 2023).

2.3.1 Water Security Index

The water security index (WSI) is a composite index that is formed based on five KDs and their integral subindices. The Report develops a composite index called the "water security index", which quantifies the access of the population across countries/regions to the various sub-indicators of the five KDs of water security on a scale of 1–5. As a result, a WSI of 1 denotes a country in the first phase where the national water situation is perilous and there is a significant gap between the present status and the appropriate standards of water security. At WSI Stage 5, the nation may be regarded as a model for its management of water resources and services, and the nation is as water secure as it can be given the existing situation. From an analytical standpoint, a significant report titled Asian Water Development Outlook attempts to explore the idea of national water security through five essential components, including urban water security, HWS, environmental water security, economic water security, and catastrophe preparedness relating to water as shown in Figure 2.4.

Figure 2.5 shows the WSI of different regions: SE Asia (2.4), Central and West Asia (2.2), East Asia (2.6), the Pacific (2.5), and mature Asian countries (3.3). According to the United Nations, Department of Economic and Social Affairs (2019), Population Division in South Asia, Pakistan has the lowest WSI on a regional scale, at 1.6, which means the national water situation is dangerous and there is a significant gap between the present condition and the appropriate standards of water security. This demonstrates how inadequate water security has a negative impact on the people living in the SA region. In terms of home water security (including sanitation), urban water security, environmental water security, and

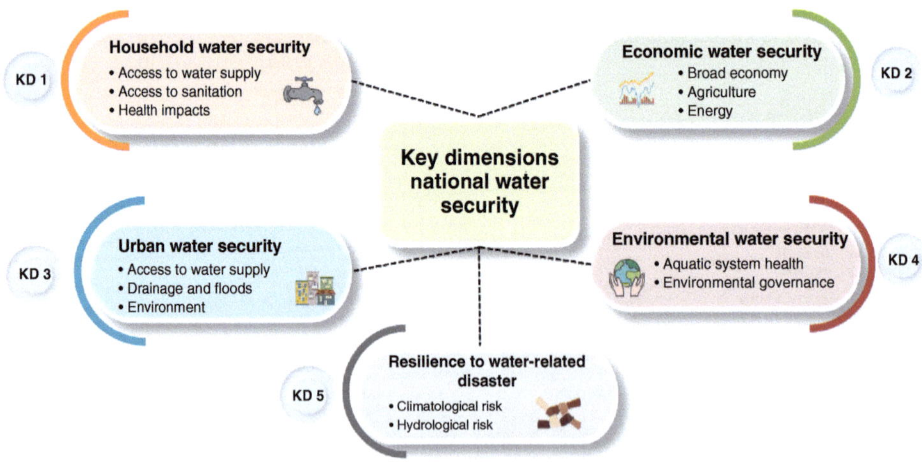

Figure 2.4 Five Pillars of National Water Security (NSW): Household (KD1), Economic (KD2), Urban (KD3), Environmental (KD4) and Resilience to water-related disaster (KD5).

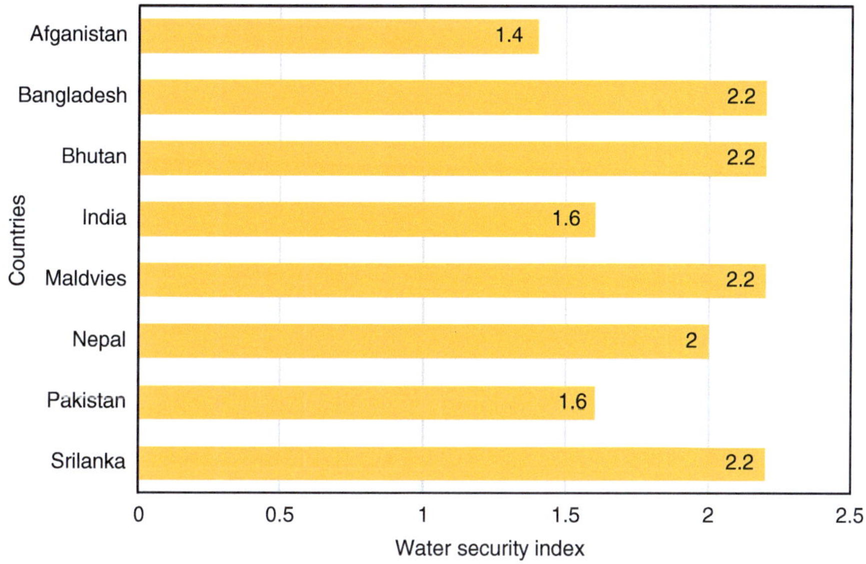

Figure 2.5 Ranking of countries according to the Water Security Index that includes five key dimensions.

resilience to water-related calamities, South Asia is less secure than all other regions. In terms of economic water security, South Asia is marginally more secure than Central and West Asia. When it comes to overall regional water security, Central and West Asia is ranked second to last (Akinyemi et al. 2022).

2.4 Key Dimensions (KD) of Water Security

2.4.1 Household Water Security (KD 1)

HWS includes a variety of water-related issues that may jeopardize the health and well-being of family members as well as households' ability to get an acceptable amount of drinkable water to meet their basic needs (Akinyemi et al. 2022). The fundamental elements of HWS include indices of water quality, per-person water use, accessibility, dependability, resilience (seasonal change in water delivery services), and affordability. More than 60% of households lack access to better hygiene and secure, piped water in Asia and the Pacific. Asia and the Pacific must address crucial water security challenges because they have the lowest freshwater resources per capita in the globe (Lohani and Ait-Kadi 2013).

2.4.2 Economic Water Security (KD 2)

Economic water security is the second crucial component of water security for a country. It includes mainly three elements: agriculture water security, energy water security, and industrial water security as shown in Figure 2.6. Economic water security refers to the efficient use of water that enables the food, production, and energy sectors of the economy

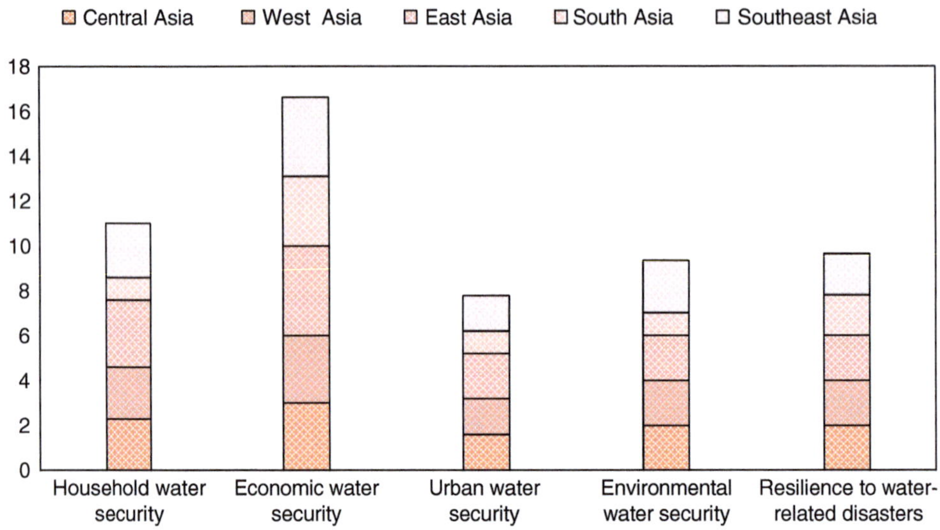

Figure 2.6 Key dimension of water security in different regions of Asia.

to continue growing economically. In Asia and the Pacific, 79% of the annual average extractions of water account for agriculture (Lohani and Ait-Kadi 2013; Akinyemi et al. 2022).

2.4.3 Urban Water Security (KD 3)

Urban water security is a complex idea with connections to other ideas and frameworks on a larger scale, including ecological security, the web of water-energy-food securities, integrated urban water management risk management, resilient and adaptive water management, and water-sensitive cities (Maheshwari et al. 2016; Artioli et al. 2017; Scanlon et al. 2017; Aboelnga et al. 2018; Aboelnga et al. 2019). In Asia and the Pacific, the proportion of people who live in urban areas has increased by 29% during the previous 20 years, accounting for over 43% of the population. A hotspot for the development of habitable cities in South Asia, where as little as 22% of wastewater discharges are treated.

2.4.4 Environmental Water Security (KD 4)

Deals with reestablishing a healthy environment and waterways. In the last few years, there is been a rapid expansion of the economy in Asia that is impacting the environment and natural resources negatively. The major cause for this condition is that governments across the globe prioritize economic security before environmental security (Panella et al. 2020). Eighty percent of Asia's rivers are in poor condition, endangering economies and quality of life. A total of $1.75 trillion in environmental services is threatened annually.

2.4.5 Resilience to Water-Related Disaster (KD 5)

Refers to build resilient communities that can adapt to change or recover due to the hazards related to water. To improve the continuity of vital services like energy, transportation, water, wastewater, waste, and digital communications, which support the efficient operation of health, education, and other sectors, a set of guiding principles, important steps, and guidelines are described in principles for resilient infrastructure (Vorosmarty et al. 2010). Rapid changes have occurred in economic activity, urbanization, cuisine, trade, culture, and communication as a result of the region's expanding socioeconomic growth. To lessen the effects of upcoming disasters, rapid progress must be made in creating resilient communities that can adapt to change and lower their vulnerability to water-related natural disasters. 90% of people affected by water-related disasters live in Asia (Manish Kumar et al. 2020).

2.5 Factors Affecting Water Security

The Asian zone is a very populated area that is rapidly advancing in terms of energy production, urbanization, agricultural development, and industrialization (Vorosmarty et al. 2010). These are all crucial elements for the region's social and economic development (Benneyworth 2016). However, these factors could function as a catalyst for adverse water stress circumstances and a decline in the region's ability to maintain a proper balance in socioeconomic water development needs (Cashman 2014).

2.5.1 Climate Change

Most environmental change dynamics in nations and ecosystems are conveyed through the water. Water supply and consumption habits are both predicted to change due to the change in climatic conditions. Evaporation will increase with a rise in temperature, affecting groundwater recharge and groundwater-surface water interactions, and reducing the availability of grey and blue water (Asok Raja et al. 2023). In many parts of Asia, alpine snowfall is the principal water source for sustaining productive agriculture and other residential and commercial operations (IPCC 2007). The accumulation of snow will decrease due to climate change, diminishing water availability. The Himalayan River systems of Asia, which give food and fiber to the majority of South Asia and China, will be badly affected by this change in snow hydrology. As temperatures rise and evaporation from land and the sea occurs more quickly, the hydrological cycle is anticipated to speed up. Several significant flood incidents that have already occurred in Asia this decade was the Himalayan Tsunami of 2013, the Thailand floods of 2011, and the Pakistan flood of 2011 (Ali 2013). However, low-lying deltas like the Mekong Delta and the Ganga-Brahmaputra Delta, as well as many Asian coastal zones, will experience degradation of the land and water environment due to sea level rise, increased incidences of cyclones, inundation and flooding, saltwater intrusion, and coastal erosion, leading to loss of agriculture land and biodiversity (Nelson et al. 2010).

2.5.2 Non-Revenue Water

In the cities of the region, "Non-Revenue Water" (NRW), which is defined as the gap between the amount of water supplied into the distribution system and the amount of water paid to customers, averages 35% and can go much higher. In Asia, there are an estimated 29 billion cubic meters (m^3) of NRW produced annually by metropolitan water utilities (Frauendorfer and Liemberger 2010). Using a $0.30 per m^3 water value as an estimate, Asia's water utilities lose close to $9 billion annually. In addition to being crucial for operating sustainably and contributing to the achievement of the UN SDG 6 on clean water and sanitation, lowering NRW is also of utmost importance for conducting good business. NRW thus has an impact on water utilities' ability to raise money for needed service additions, carry out necessary maintenance, and make technology investments. Therefore, NRW decreased water availability, lowered the water quality, decreased the financial viability of water utilities, and also shot up the cost as the charges were passed to the taxpayers.

2.5.3 Agricultural Sector

2.5.3.1 Water Demand for Irrigation and Water Pollution

If this trend of consumption of freshwater resources continues then two-thirds of the world's population will have to live under water stress by 2025. The current number predicts that there will be more than 50% rise in food demand as a result of the rising population (Malekian et al. 2017). So, increasing demand for food is directly linked with water demand, and the process is related to water pollution. One of the main worldwide water consumers is agriculture. Water withdrawals from rivers, lakes, and groundwater sources are mostly used for irrigation (Panday et al. 2023). Water scarcity and diminished water security for other industries and communities can result from the high-water demand, especially in areas with restricted water availability (Gupta 2019). On the one hand, there is the issue of quantity, and on the other, the degradation of water quality brought on by irrigation practices is another blunder.

2.5.3.2 Deforestation and Soil Erosion

Deforestation is a common part of expanding agricultural land, which upsets the natural water cycle. In order to restore groundwater reserves and regulate stream flows, forests are essential for controlling water flow because they absorb rainwater and release it gradually (Gupta 2019). Deforestation can have a detrimental effect on water availability and quality by causing increased runoff, soil erosion, and lower water infiltration. Soil erosion and deforestation activities have a significant negative influence on water supplies. When it comes to soil erosion, it is known that a percentage of fertile soil is taken away from the area; as a result, production falls and food security concerns also develop (Kasaro et al. 2019).

2.5.4 Industrialization

To understand the impact of industrialization on water security, it's important to consider its historical context. The Industrial Revolution, which began in the 18th century, marked a transition from an agrarian to an industrial society, with the emergence of factories, machinery, and mass production (Ray and Shaw 2019). This shift had profound implications for water resources, as industries required large quantities of water for various purposes, such as processing, cooling, and transportation. The Industrial Revolution brought significant advancements in technology, economic growth, and improved standards of living (Nepal et al. 2021). However, it has also had profound effects on the environment, particularly on water resources and water security. As industries continue to grow and expand worldwide, the impact of industrialization on water security has become a pressing concern (Besada and Werner 2015).

Groundwater, a critical source of freshwater, has also been impacted by industrialization. Industries often rely on groundwater for their water supply (Zhang et al. 2022), leading to increased abstraction and depletion of groundwater resources. Overpumping of groundwater can lead to falling water tables, land subsidence, and reduced availability of water for other uses, including domestic and agricultural water supply. Groundwater depletion can have long-term implications for water security, particularly in regions where groundwater is a major source of freshwater.

2.5.4.1 Water Pollution from Industrial Activities

Water pollution caused by industrial operations is a significant environmental concern with far-reaching consequences. Various industries contribute to water pollution through their manufacturing processes and waste disposal practices. These industries often discharge pollutants directly into water bodies or contaminate groundwater sources, leading to adverse effects on aquatic ecosystems and human health (Damania et al. 2019). Industrialization has also resulted in water pollution as a consequence of various industrial activities like chemical manufacturing, mining operations, power generation, and wastewater discharges from factories and processing plants. Industrial processes can generate large volumes of wastewater containing pollutants such as heavy metals, toxic chemicals, and organic pollutants. Improper disposal of industrial wastewater or inadequate treatment can result in contamination of water sources, posing risks to human health, aquatic ecosystems, and agriculture. This pollution can also affect downstream water users, leading to water quality and availability conflicts (Bartarya and Bahukhandi 2012).

Major industries which are the main contributors to water contamination are mining, chemical, power plants, manufacturing, refineries, electroplating, and paints. The mining industry is another major culprit. Mining operations generate a substantial amount of waste material known as tailings, which often contain toxic substances (Simpson et al. 2019), such as heavy metals and sulfides. The agricultural sector is also responsible for water pollution. The use of fertilizers, pesticides, and herbicides (Zarei et al. 2021) in large-scale farming operations leads to runoff, carrying these chemicals into rivers and streams. Excessive nutrient levels in the water, known as eutrophication, can cause algal blooms, deplete oxygen levels, and harm aquatic organisms.

One of the main contributors to water pollution is the chemical industry. Chemical manufacturing processes involve the use and production of hazardous substances that can find their way into water sources (Chowdhary et al. 2020). Effluents containing toxic chemicals, heavy metals, and synthetic compounds are discharged into rivers and lakes, causing contamination and disrupting aquatic life. Typically, more than 160 micropollutants of various classes in conventional wastewater treatment plants (WWTPs) were investigated, which significantly affect aquatic life. Hydrophobic pollutants such as heavy metals, persistent organic pollutants (POPs), brominated flame retardants, and several personal care products (PCPs) are usually well removed (>70%) in WWTPs, either by sorption onto sewage sludge or by biodegradation (Margot et al. 2015). However, because some of these substances are poisonous even at extremely low concentrations, good removal efficiencies do not guarantee that the effluent concentrations won't possibly harm aquatic life.

2.5.4.2 Over Extraction Scenario

Certain industries, such as agriculture, energy production, and mining, are particularly water-intensive, and their expansion as a result of industrialization has significant implications for water security. Agriculture, for example, requires large amounts of water for irrigation, and expansion of agricultural lands to meet growing food demands can strain water resources, leading to overextraction and depletion of water sources (Turley and Caretta 2020). According to the AQUASTAT database, 3856 km^3 of freshwater were withdrawn for domestic, industrial, and agricultural purposes globally around the year 2003. Most of it was used in the agricultural sector (70%) followed by the industry (19%) and domestic (11%) sectors (Florke et al. 2013). Similarly, energy production, including power generation and extraction of fossil fuels, requires substantial water inputs, and increased energy demand can exacerbate water stress in regions with limited water resources. Mining activities, which involve the extraction of minerals and metals from the earth, can also have significant impacts on water quality and availability (Diaz-Caravantes et al. 2020). The worldwide textile industry consumes over 1 trillion kilowatt hours and 9 trillion liters of water annually to produce 60 billion kilograms of textiles (Asok Raja and Shaw 2019). In food processing industry of United States, the total water withdrawals are approximately 1344 million m^3 per day (Compton et al. 2018).

2.5.4.3 Impact on Aquatic Ecosystems

Industrialization has had significant impacts on aquatic ecosystems, which are intricately connected to water security. Discharge of industrial pollutants increased water abstraction, and habitat destruction associated with industrial activities can have adverse effects on aquatic biodiversity, disrupting ecological balance and threatening the survival of many species (Opoku and Aluko 2021). Metal-sensitive groups such as crustaceans and mayflies were found to be negatively impacted by metal pollution from mining (Clements and Kiffney 1995). Mercury (Hg) is taken by fish and transferred up the food chain to other fish-eating species. Through bioaccumulation, it affects not just aquatic environments but also communities (Morcillo et al. 2017). Long-term exposure to Cd pollution causes multiple tissues in fish to absorb and store Cd, which alters the gills, liver, and gonad's structure and function (Liu et al. 2022). The most poisonous element for microorganisms and invertebrates is silver, which is also a potential biocide. Microorganisms can be killed at concentrations lower than 1 g Ag/L because they can be inhibited by enzymes that perform crucial tasks (Chen et al. 2010). Organic dyes are one of the critical threats to aquatic life. Dyes are generally released by a number of industries, including pharmaceuticals, textiles, paper, pulp, and tanneries. Various industries such as pharmaceuticals, textiles, paper, pulp, and tannery release organic dyes. In textile industries, about 2–21% of dye is released directly in to the aquatic system. The color of dyes makes the water lesser transparent then it is difficult for the aquatic life to get enough sunlight. Also, the degradation products of the dyes are quite toxic, carcinogenic, and mutagenic (Kumar et al. 2018).

Aquatic ecosystems provide important services such as water purification, nutrient cycling, and habitat for wildlife, and their degradation can have far-reaching consequences for water security and ecosystem health. Whenever an industrial effluent enters any lotic ecosystem then rapid drawdown of dissolved oxygen (DO) takes place due to the microbial degradation of organic matter (Duarte et al. 2020). When the DO level goes below 4 ppm, it becomes problematic for aquatic life. Organic matter not only causes a dramatic change in DO concentration but also warm water discharged from thermal power plants can rapidly deplete DO concentrations. These unhealthy practices not only pollute aquatic ecosystems (Duarte et al. 2021) but are also a threat to water security, especially for those who are totally dependent on surface water bodies directly or indirectly.

2.6 Opportunities and Challenges in Achieving Water Security

Opportunities: The sections below go into detail on many engineering opportunities to improve long-term water security in order to meet these critical and developing challenges.

Improve urban water distribution: Numerous techniques and instruments have been developed for managing losses in water distribution systems (Mutikanga et al. 2013). Active control methods, quick and high-quality repairs, and efficient pressure management are all necessary for managing physical losses (leakage) in distribution systems. Prerequisite metering and pricing of water is one of the technologies that may improve the usage of water. Water usage by urban residences and in rural areas, groundwater pumping for irrigation may be metered and billed directly and also indirectly using electricity (Sathre et al. 2022).

Pressure management is another way to reduce urban water losses and achieve water security. The rate of gradual leaking and the frequency of rupture pipes are directly physically related to water pressure. Automatic pressure-reducing valves (PRV) are the most widely used as it is a cost-efficient solution. PRVs are placed strategically throughout the network to lower or keep network pressure at a predetermined level. Regardless of changes in flow rate or upstream pressure, the valve maintains the preset downstream pressure (Prescott and Ulanicki 2008).

Improve competency of industrial water: Asian industries currently use water in a very inefficient manner, and adopting international best practices will greatly cut this usage. For example, in Asia, the steel sector typically uses more water to produce a tonne of the steel compared to global industry average. In steel and other industries, there are many chances to reduce the use of industrial water, such as dry coke quenching, seawater cooling, dry cooling, and zero liquid discharge (Sathre et al. 2022). The amount of wastewater produced in Asia and recycling this water for other uses is a key strategy for improving water security.

Utilization of grey water: Grey water is the word for domestic wastewater, such as that from sinks, showers, bathtubs, and washing machines, that does not include a large number of feces. This water can be used directly again for many purposes like agriculture, gardening, construction, and cooling application in industries (Vuppaladadiyam et al. 2019). It is estimated that if 80% of household uses grey water to flush their toilets then the total water withdrawal reduction would be 2.4 km^3/year (Sathre et al. 2022).

Challenges: The world is currently dealing with one of the most troublesome issues which is water security. In many countries, there is a change in rainfall patterns, alternating between drought and flood, a steady decline in the groundwater table, and pollution in water bodies is getting worse (Shan et al. 2020). Finding the causes and remedies to these issues is extremely challenging because of the impact of climate change, population increase, economic development, urbanization, and insufficient regulation. The severity of these issues is greatest in developing countries, particularly in Asia, where there are practically all of the major water-related issues (Marcoux 1996; Shan et al. 2020).

- The primary cause of water scarcity in emerging countries is the exponential growth rate of the human population, which is extensively and unsustainably uses renewable freshwater resources. According to projections, the urban population in Asia will nearly quadruple by 2050 as a result of both total population growth and rising urbanization. Therefore, urban water use for both residential and industrial purposes is quickly expanding in response to this population growth. In underdeveloped nations, untreated sewage is released into surface water bodies, which impacts the water's ecosystems and people's health (International Decade for Action 'Water for Life' (2005–2015). According to the River Health Index, approximately 80% of the rivers in the Asian region are in very poor health. The issue of the security of the water environment is more complex than just city pollution (Das et al. 2016; Shan et al. 2020).
- The geologic history of the area has a significant impact on South Asia's ability to secure its supply of water. The Himalayan Mountain range is being created by the steady collision of the Indian and Eurasian tectonic plates. As a result, hard rock and alluvial groundwater regimes have developed, creating two separate groundwater systems in South Asia. Hard rock groundwater is characterized by restricted well production, unpredictable well productivity fluctuations over relatively short distances, and in some places, poor water quality. Therefore, in hard rock regions, groundwater supply is limited and unreliable. To address the issue of rapidly declining water tables and seasonal depletion caused by excessive extraction, restrictions on groundwater extraction rates are imposed. This is also a main challenge faced in South Asian countries that lead to the condition of water scarcity and stress (Amarasinghe et al. 2016).
- By 2050, it is anticipated that the global water demand for agriculture – the single largest consumer of water – will have increased by 19%. By 2040, it is anticipated that water stress will be extremely high in 40% of all irrigated agriculture. Globally, about half of the water withdrawn for irrigation either falls to reach the desired crops or is applied excessively through methods like flood irrigation, resulting in minimal crop output. Consequently, irrigation and planting methods are often ineffective and unsustainable (Ungureanu et al. 2020). Poor soil, planting, or other agricultural practices like tilling also restrict water infiltration into the ground and increase evaporation. On average, local plant transpiration and evaporation account for around 40% of

all rainfall that falls on land; however, in some areas, this proportion is significantly higher. The local climate and water availability may be significantly impacted by degraded landscapes that alter plant mix or density.

- Nations in development are susceptible, but are not immune: Due to corruption, unenforced environmental laws, lax, high pollution levels, environmentally damaging development, transboundary or internal water disputes, and rising water demand from agriculture, textiles, and fishing, many nations in South and Central Asia as well as portions of Southeast Asia are also likely to experience declining water security. Geopolitical conflicts may put a strain on already-existing water-sharing arrangements or thwart future collaboration.

2.7 Conclusion

The overuse of freshwater in both developed and developing countries is posing increasing problems for its supply worldwide. The complexity of this issue is growing as a result of the unsustainable usage and poor management of water in developing countries. Urban household water demand is increasing rapidly, surface water supply in restricted river basins is being fully used, groundwater extraction is typically outpacing natural recharge rates, and even in places with plenty of water resources, the poorest citizens frequently lack the means to access it. These issues are the outcome of a symbiotic interaction between demographic, climatic, behavioral, economic, and political problems. Many other regions are dealing with comparable difficulties, so South Asia is not the only one. To tackle this issue and create a workable solution, proper cooperation and coordination between the public sector, the corporate sector, and the community are necessary. To address this issue, new policies should be implemented and there is a clear need for global governance. International negotiations and conventions will be useful in reducing the demand for various natural resources. The thoughtful and nuanced policy must go hand in hand with technological innovation to effectively handle some issues related to water security. A balance between social, environmental, and economic factors is necessary to achieve water security worldwide. Grey water systems and zero discharge practices may both make a big difference in sustaining water security.

Acknowledgment

This work is funded by the Department of Science and Technology (DST) and UPES under Science for Equity Empowerment and Development (SEED) Project.

References

Aboelnga, H., Khalifa, M., McNamara, I. et al. (2018). *The Water-Energy-Food Security Nexus: A Review of Nexus Literature and Ongoing Nexus Initiative for Policymakers*, 25–30. Bonn: Nexus Regional Dialogue Programme (NRD).

Aboelnga, H.T., Ribbe, L., Frechen, F.B., and Saghir, J. (2019). Urban water security: definition and assessment framework. *Resources* 8 (4): 178.

Akinyemi, P.A., Afolabi, O.T., and Aluko, O.O. (2022). The effects of seasonal variations on household water security and burden of diarrheal diseases among under 5 children in an urban community, Southwest Nigeria. *BMC Public Health* 22 (1): 1–10.

Ali, A.(2013). Indus basin floods: mechanisms, impacts, and management.

Amarasinghe, U.A., Muthuwatta, L., Smakhtin, V. et al. (2016). *Reviving the Ganges Water Machine: Potential and Challenges to Meet Increasing Water Demand in the Ganges River Basin; Research Report 167*. Colombo: International Water Management Institute.

Artioli, F., Acuto, M., and McArthur, J. (2017). The water-energy-food nexus: an integration agenda and implications for urban governance. *Political Geography* 61: 215–223.

Asok Raja, S.K., Panday, D.P., and Kumar, M. (2023). Decoding the enigma of 100-year record-breaking rainfall over Tamil Nadu using wavelet analysis. *Groundwater for Sustainable Development 23*: 101007. https://doi.org/10.1016/j.gsd.2023.101007.

Bartarya, S.K. and Bahukhandi, D.K. (2012). Impact assessment of urbanization and industrialization on surface and groundwater quality. *Global Journal of Engineering Design & Technology* 1 (1): 11–22.

Bengtsson, M. and Shivakoti, B.R. (2015). The role of water security in achieving the sustainable development goals: realising synergies, balancing trade-offs, chapter 6. *Achieving the Sustainable Development Goals: From Agenda to Action* 8 (1): 121 2021. https://www.iges.or.jp/en/pub/role-water-security-achieving-sdgs-realising/en (accessed 3 April 2021).

Benneyworth, L. M. (2016). *Factors Affecting Drinking Water Security in South-Western Bangladesh (Doctoral dissertation)*.

Besada, H. and Werner, K. (2015). An assessment of the effects of Africa's water crisis on food security and management. *International Journal of Water Resources Development* 31 (1): 120–133.

Burek, P., Satoh, Y., Fischer, G. et al. (2016). *Water Futures and Solution: Fast Track Initiative (Final Report)*. Laxenburg: International Institute for Applied Systems Analysis.

Carrard, N., Foster, T., and Willetts, J. (2019). Groundwater as a source of drinking water in Southeast Asia and the Pacific: a multi-country review of current reliance and resource concerns. *Water* 11: 1605.

Cashman, A. (2014). Water security and services in the Caribbean. *Water* 6 (5): 1187–1203.

Chen, W.H., Erker, B.T., Kanematsu, M., and Darby, J.L. (2010). Disposal of arsenic-laden adsorptive media: economic analysis for California. *Journal of Environmental Engineering* 136 (10): 1082–1088.

Chowdhary, P., Bharagava, R.N., Mishra, S., and Khan, N. (2020). Role of industries in water scarcity and its adverse effects on environment and human health. *Environmental Concerns and Sustainable Development: Air, Water and Energy Resources* 1: 235–256.

Clements, W.H. and Kiffney, P.M. (1995). The influence of elevation on benthic community responses to heavy metals in Rocky Mountain streams. *Canadian Journal of Fisheries and Aquatic Sciences* 52 (9): 1966–1977.

Compton, M., Willis, S., Rezaie, B., and Humes, K. (2018). Food processing industry energy and water consumption in the Pacific northwest. *Innovative Food Science&Emerging Technologies* 47: 371–383.

Damania, R., Desbureaux, S., Rodella, A.S. et al. (2019). *Quality Unknown: The Invisible Water Crisis*. World Bank Publications.

Das, N., Deka, J.P., Shim, J. et al. (2016). Effect of river proximity on the arsenic and fluoride distribution in the aquifers of Brahmaputra floodplains, Assam, Northeast India. *Journal of Groundwater for Sustainable Development* 2–3: 130–142.

Das, A., Kumar, M., Jha, P.K. et al. (2022). Isotopic and hydrogeochemical tracking of dissolved nutrient dynamics in the Brahmaputra River System: a source delineation perspective. *Chemosphere* 307: 135757. https://doi.org/10.1016/j.chemosphere.2022.135757.

Díaz-Caravantes, R.E., Zuniga-Teran, A., Martín, F. et al. (2020). Urban water security: a comparative study of cities in the arid Americas. *Environment and Urbanization* 32 (1): 275–294.

Duarte, C.M., Agusti, S., Barbier, E. et al. (2020). Rebuilding marine life. *Nature* 580 (7801): 39–51.

Duarte, C.M., Chapuis, L., Collin, S.P. et al. (2021). The soundscape of the Anthropocene ocean. *Science* 371 (6529): eaba4658.

Ewa, E.E., Iwara, A.I., Adeyemi, J.A. et al. (2011). Impact of industrial activities on water quality of Omoku creek. *Sacha Journal of Environmental Studies* 1 (2): 8–16.

Flörke, M., Kynast, E., Bärlund, I. et al. (2013). Domestic and industrial water uses of the past 60 years as a mirror of socio-economic development: a global simulation study. *Global Environmental Change* 23 (1): 144–156.

Food and Agriculture Organization of the United Nations (2016). AQUASTAT Website. http://www.fao.org/nr/water/aquastat/water_use/index.stm (accessed 30 January 2019).

Frauendorfer, R. and Liemberger, R.(2010). The issues and challenges of reducing non-revenue water.

Frederick, W.H. andLeinbach, T.R. (2020).Southeast Asia. Encycl. Br. https://www.britannica.com/place/SoutheastAsia (accessed 7 June 2021).

Gun, J. (2012). *Groundwater and Global Change: Trends, Opportunities and Challenges. United Nations World Water Assessment Programme*. Paris: United Nations Educational, Scientific and Cultural Organization.

Gupta, G.S. (2019). Land degradation and challenges of food security. *Review of European Studies* 11: 63.

International Decade for Action 'Water for Life'(2005–2015). https://www.un.org/water forlifedecade/.

IPCC (2007). IPCC Fourth Assessment Report: Climate Change 2007.

Kanu, I. and Achi, O.K. (2011). Industrial effluents and their impact on water quality of receiving rivers in Nigeria. *Journal of Applied Technology in Environmental Sanitation* 1 (1): 75–86.

Kasaro, D., Phiri, E., and Nyambe, I. (2019). Deforestation impact on ecosystem services in Kamfinsa sub-catchment of Kafue River Basin in Zambia. *Journal of Ecology and the Natural Environment* 11 (4): 33–45.

Kumar, A., Rana, A., Sharma, G. et al. (2018). Aerogels and metal–organic frameworks for environmental remediation and energy production. *Environmental Chemistry Letters* 16: 797–820.

Manish Kumar, Francisco Munoz-Arriola, Hiroaki Furumai, Tushara Chaminda G.G (2020). *Resilience, Response, and Risk in Water Systems: Shifting Management and Natural Forcings Paradigms*. https://link.springer.com/book/10.1007/978-981-15-4668-6 (Book by Manish sir).

Kumar, M., Panday, D.P., Bhagat, C. et al. (2023a). Demystifying the decadal shift in the extent of groundwater in the coastal aquifers of Gujarat, India: a case of reduced extent but increased magnitude of seawater intrusion. *Science of the Total Environment* 165451. https://doi.org/10.1016/j.scitotenv.2023.165451.

Kumar, M., Mazumder, P., Silori, R. et al. (2023b). Prevalence of PPCPs, microplastics and co-infecting microbes in the post-COVID-19 era and its implications on antimicrobial resistance and potential endocrine disruptive effects. *Science of the Total Environment* 166419. https://doi.org/10.1016/j.scitotenv.2023.166419.

Liu, Y., Chen, Q., Li, Y. et al. (2022). Toxic effects of cadmium on fish. *Toxics* 10 (10): 622.

Lohani, B.N. and Ait-Kadi, M. (2013). *Asian Water Development Outlook 2013: Measuring Water Security in Asia and the Pacific*. Asia-Pacific Water Forum.

Maheshwari, B., Singh, V., and Thoradeniya, B. (2016). *Balanced Urban Development: Options and Strategies for Liveable Cities*. Berlin: Springer.

Malekian, A., Hayati, D., and Aarts, N. (2017). Conceptualizations of water security in the agricultural sector: perceptions, practices, and paradigms. *Journal of Hydrology* 544: 224–232.

Maniam, G., Poh, P.E., Htar, T.T. et al. (2021). Water literacy in the Southeast Asian context: are we there yet? *Water* 13: 2311. https://doi.org/10.3390/w13162311.

Marcoux,A. (1996).Population change-natural resources-environment linkages in Central and South Asia.

Margot, J., Rossi, L., Barry, D.A., and Holliger, C. (2015). A review of the fate of micropollutants in wastewater treatment plants. *Wiley Interdisciplinary Reviews: Water* 2 (5): 457–487.

Morcillo, P., Esteban, M.A., and Cuesta, A. (2017). Mercury and its toxic effects on fish. *AIMS Environmental Science* 4 (3): 386–402.

Moumen, Z., ElIdrissi, N.E., Tvaronavičienė, M., and Lahrach, A. (2019). Water security and sustainable development. *Insights into Regional Development* 1 (4): 301–317.

Mutikanga, H.E., Sharma, S.K., and Vairavamoorthy, K. (2013). Methods and tools for managing losses in water distribution systems. *Journal of Water Resources Planning and Management* 139: 166–174.

Nelson, G.C., Rosegrant, M.W., Palazzo, A. et al. (2010). Food security, farming, and climate change to 2050: scenarios, results, policy options. *Intl Food Policy Res Inst.* 172.

Neog, N., Goswami, R., Panday, D.P. et al. (2024). Arsenic contamination in the groundwater of Northeastern India: critical understandings on geotectonic controls and the need for intervention. *Current Opinion in Environmental Science & Health* 100539. https://doi.org/10.1016/j.coesh.2024.100539.

Nepal, S., Neupane, N., Belbase, D. et al. (2021). Achieving water security in Nepal through unravelling the water-energy-agriculture nexus. *International Journal of Water Resources Development* 37 (1): 67–93.

Opoku, E.E.O. and Aluko, O.A. (2021). Heterogeneous effects of industrialization on the environment: evidence from panel quantile regression. *Structural Change and Economic Dynamics* 59: 174–184.

Panday, D.P. and Kumar, M. (2022). Climate indices and hydrological extremes: deciphering the best-fit model. *Environmental Research* 215: 114301. https://doi.org/10.1016/j.envres.2022.114301.

Panday, D.P. and Kumar, M. (2023). Application of remote sensing techniques to deal with scale aspects of GRACE data to quantify groundwater levels. *MethodsX* 102108. https://doi.org/10.1016/j.mex.2023.102108.

Panday, D.P., Khosa, R., Maheswaran, R. et al. (2021a). Game-theoretic-based modelling of Krishna waters dispute: equilibrium solutions by metagame analysis. *The European Physical Journal B* 94: 101. https://doi.org/10.1140/epjb/s10051-021-00107-w.

Panday, D.P., Khosa, R., Maheswaran, R. et al. (2021b). Game theoretic-based modelling of Krishna waters dispute: equilibrium solutions by hypergame analysis. *The European Physical Journal B* 94: 131. https://doi.org/10.1140/epjb/s10051-021-00135-6.

Panday, D.P., Singh, A., Siddhant, A.K. et al. (2023). Probable maximum precipitation analysis of high rainfall regions in India. *Groundwater for Sustainable Development* 100893. https://doi.org/10.1016/j.gsd.2022.100893.

Panella, T., Fernandez Illescas, C., Cardascia, S. et al. (2020). *Asian Water Development Outlook 2020: Advancing Water Security Across Asia and the Pacific*. Asian Development Bank.

Prescott, S.L. and Ulanicki, B. (2008). Improved control of pressure reducing valves in water distribution networks. *Journal of Hydraulic Engineering* 134: 56–65.

Ray, B. and Shaw, R. (2019). Developing water security index for urban areas. In: *Urban Drought: Emerging Water Challenges in Asia*, 53–68. Singapore: Springer Nature.

Ritchie, H. and Roser, M. (2017).Water use and stress. https://ourworldindata.org/water-use-stress (accessed 8 August 2021).

Sathre, R., Antharam, S.M., and Catena, M. (2022). Water security in South Asian cities: a review of challenges and opportunities. *CivilEng* 3: 873–894.

Scanlon, B., Ruddell, B., Reed, P. et al. (2017). The food-energy-water nexus: transforming science for society. *Water Resources Research* 53: 3550–3556.

Shan, V., Singh, S.K. and Haritash, A.K., 2020. Water crisis in the Asian countries: status and future trends.In: Manish Kumar, Francisco Munoz-Arriola, Hiroaki Furuma, Tushara Chaminda. *Resilience, Response, and Risk in Water Systems: Shifting Management and Natural Forcings Paradigms* 173–194. Springer

Simpson, G.B., Badenhorst, J., Jewitt, G.P. et al. (2019). Competition for land: the water-energy-food nexus and coal mining in Mpumalanga Province, South Africa. *Frontiers in Environmental Science* 7: 86.

Turley, B. and Caretta, M.A. (2020). Household water security: an analysis of water affect in the context of hydraulic fracturing in West Virginia, Appalachia. *Water* 12 (1): 147.

UN (2021). *The Sustainable Development Goals Report 2021*. New York: United Nations.

UN (United Nations, Department of Economic and Social Affairs, Population Division) (2019). *World Population Prospects: The 2019 Revision*. New York: UN (United Nations, Department of Economic and Social Affairs, Population Division).

UNESCO and UNESCO I-WSSM (2019). Water security and the sustainable development goals. The role of water security in attaining sustainable development.

Ungureanu, N., Vlăduţ, V., and Voicu, G. (2020). Water scarcity and wastewater reuse in crop irrigation. *Sustainability* 12 (21): 9055.

Vörösmarty, C.J., McIntyre, P.B., Gessner, M.O. et al. (2010). Global threats to human water security and river biodiversity. *Nature* 467 (7315): 555–561.

Vuppaladadiyam, A.K., Merayo, N., Prinsen, P. et al. (2019). A review on greywater reuse: quality, risks, barriers and global scenarios. *Reviews in Environmental Science and Bio/Technology* 18: 77–99.

Zarei, S., Bozorg-Haddad, O., Kheirinejad, S., and Loáiciga, H.A. (2021). Environmental sustainability: A review of the water–energy–food nexus. *Journal of Water Supply: Research and Technology-AQUA* 70 (2): 138–154.

Zhang, J., Wang, J., Zhang, L., and Zhao, L. (2022). Impact of industrialization on China's regional energy security in the new era. *Environment, Development and Sustainability* 24 (6): 8418–8440.

3

Water Security in Africa

Khem Chand[1], Ruchi Mehrotra[2], Urvashi Arya[2], Durgesh Bahuguna[3], and Rajesh Tiwari[2]

[1] *MM Institute of Management, Maharishi Markandeshwar (Deemed to be University), Ambala, Bharat, India*
[2] *Department of Management Studies, Graphic Era (Deemed to be University), Dehradun, Uttarakhand, India*
[3] *Graduate School of Business, Tulas Institute, Dehradun, Uttarakhand, India*

3.1 Introduction

Sufficient availability and accessibility to water of adequate quality and quantity, capable of meeting the year-round livelihood needs without compromising the needs of other users is referred to as water security (Calow et al. 2002). Water is crucial for the survival and sustainable prosperity of societies and ecosystems. United Nations sustainable development goals also incorporate the significance of water in the form of goal 6 which focuses on clean water and sanitation for all. Coleridge 1798 quoted "water every-where not any drop to drink" to highlight the issue of the essentiality of water but the looming danger of scarcity (Coleridge 1798). Not just uneven availability but the quality of water has been a concern (Oki and Kanae 2006). Africa has very low water security. However, a better global water security index can be achieved by better management practices (Gain et al. 2016). Unfortunately, water scarcity studies have been isolated from the policymaking leading to a disconnect (Bakker 2012). Human and organizational capabilities have been ignored (Bakker and Morinville 2013). Grey and Sadoff (2007) referred to water security as access to acceptable quality and quantity of water along with acceptable level of risk for biological, economic existence, and for environment. Comprehensive assessment frameworks are missing (Das et al. 2022; Kumar et al. 2023a, b). Now the human–water interaction has started getting attention comprehensively as water security (Gunda et al. 2015). Scarcity of water is emerging as a global threat (Vorosmarty et al. 2000). Community water security was evaluated by Dickson et al. (2016) using comprehensive indicators. Water security is referred to as the availability and accessibility of adequate quantity and quality of water resources (Gain et al. 2016).

The water shortage is going to become a critical issue for countries with very poor water security (Meissner et al. 2018). A score of 5 indicates extremely high-water stress, a score of

Water Scarcity Management: Enabling Technologies, First Edition. Edited by Kanchan D. Bahukhandi, Manish Kumar, Durga P. Panday, and Tushara G.G. Chaminda.

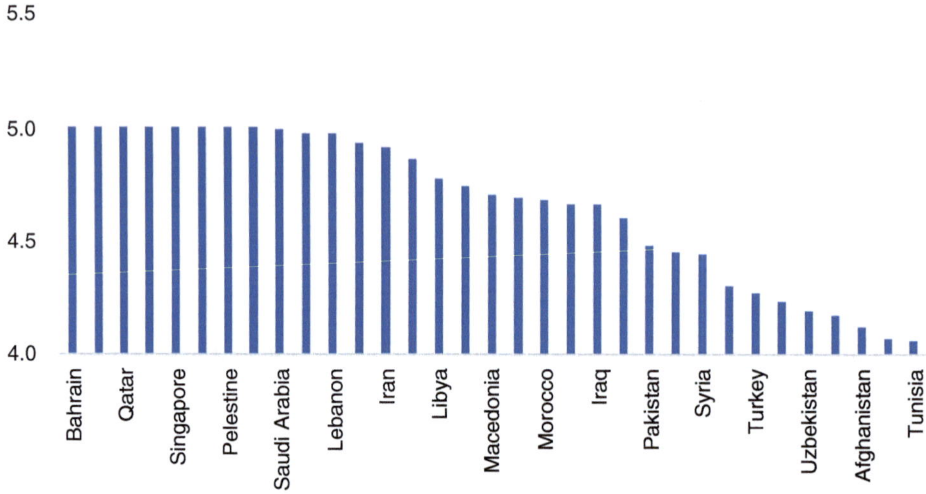

Figure 3.1 Prominent water stressed countries by 2040. *Source:* Adapted from Luo et al. (2015).

3–4 indicates high stress, a score of 2–3 indicates medium to high stress, a score of 1–2 indicates low to medium stress, and a score of 0–1 indicates low water stress. Figure 3.1 depicts the top 33 countries with the highest concern about water security by 2040.

High levels of fluoride have become a concern at the global level (Shakir et al. 2016). Although high fluoride levels are prevalent in many African countries, with the Great Rift Valley belt being recognized as the world's most severe fluoride belt (Sajidu et al. 2008), the African Union's Agenda 2063 aims to achieve water-secure Africa (DeGhetto et al. 2016). Surprisingly, the Africa Water Vision for 2025 does not explicitly acknowledge fluorosis as one of the natural threats to the sustainability of African water resources (La Banque Africaine and Bankgroup 2000; Lotter and Le Maitre 2021). Africa has 9% of the world's water resources and 18% of the population (FAO 2003). The population of Africa will double by 2050, and it will comprise more than 25% of the world population (Economist 2020). According to a report by Pew Research, by 2100 5 countries in the top 10 most populous nations will be from Africa (Chapman 2019).

Water quality is upheld by the physical and biological processes of capturing, filtering, and retaining pollutants and nutrients present in industrial effluents, agricultural runoff, and other discharges (Zalewski 2002; Malan et al. 2018; Chua et al. 2019). Proper management of riparian plants is crucial for preserving water quality, streamflow, and ensuring sufficient freshwater provision (Neog et al. 2024). The inadequacy of traditional management approaches highlights the need for a new paradigm and nature-based approach that enhances the carrying capacity for sustainable management in a cost-effective and integrated manner (Richard-Ferroudji et al. 2018). This is essential to maintain river stability and ensure the continued availability of freshwater resources (Raphael and Lalika 2022). The need for safe water is essential for life. The waterborne diseases are a common cause of mortality. Table 3.1 shows the countries in South Africa with high water, sanitation, and hygiene (WASH)-related mortality rates per 100,000 individuals.

Table 3.1 African countries with high WASH-related mortalities.

Country	Death Per 100,000
Lesotho	108.1
Chad	99.2
Somalia	99.2
Central African Republic	97
Nigeria	71.7
Niger	70.3
Sierra Leone	69.5
South Sudan	68.1
Eritrea	66.5
Mali	66.1
Burkina Faso	60.9
Benin	60.2

Source: Adapted from Global Water Security Assessment (2023).

Table 3.2 Per capita water availability in Africa.

Region	Per Capita Availability of Water (Cubic meters per capita)
Central Africa	More than 310,000
North Africa	Less than 500

Source: Adapted from United Nations University (2021).

The unequal distribution of water is also a concern for Africa. The central part and Northern part vary significantly in terms of per capita availability of water as shown in Table 3.2.

The unequal distribution of water is leading to four levels of water security in Africa. Some countries are experiencing low withdrawal to supply of water ratio (less than 10%); some are experiencing low-to-medium withdrawal to supply ratio (10–20%), medium to high withdrawal (21–40%), high withdrawal (41–80%) while there are some countries experiencing extremely high withdrawals (more than 80%). The levels of water stress by the year 2040 for Africa are depicted in Figure 3.2. Morocco, Western Sahara, Algeria, Tunisia, and Libya are in the extreme withdrawal category of 80–100%. South Africa, Botswana, Naminia, Eritrea are in the high withdrawal ratio of 40–80%. Egypt, Sudan, Mauritania, Somalia, Tanzania, and Zimbabwe are in the low-to-medium withdrawal category (10–20%) while the rest of Africa is in the low withdrawal category (less than 10%). Figure 3.2 shows the upper band of withdrawal for each category.

The climatic changes need urgent attention. In Africa, climate changes have affected a large number of people. The regions that are prone to damage due to climate change should be considered on priority for development aid. Eastern Africa is the most severely affected

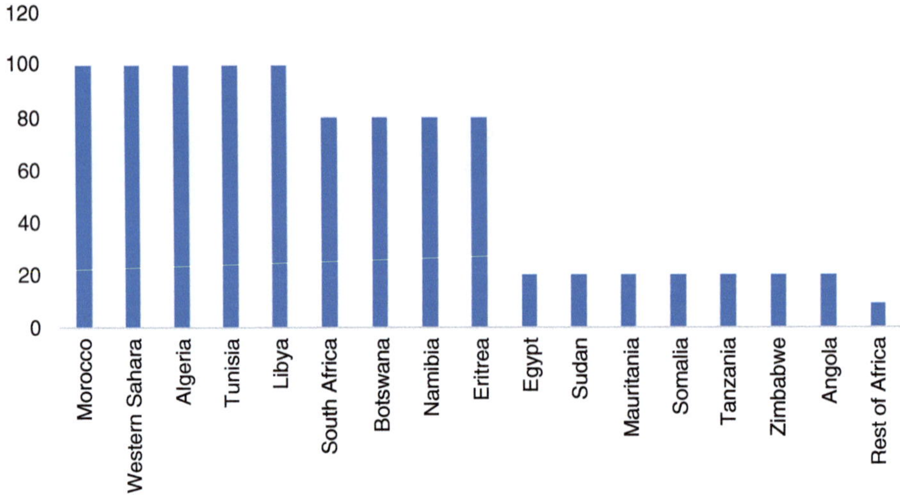

Figure 3.2 Withdrawal to supply ratio of African countries. *Source:* Adapted from Maddocks et al. (2015).

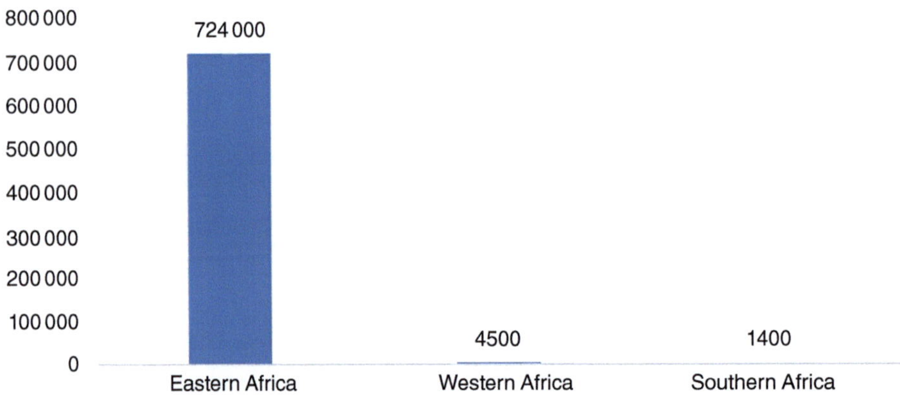

Figure 3.3 People affected in different regions of Africa due to storms (per year between 2011 and 2020). *Source:* Adapted from Climate Risks in Africa (2022).

region due to storms. In western Africa, 724,000 people were affected due to storms per year in the time period between 2011 and 2020 as shown in Figure 3.3.

Eastern and Western Africa are worst affected due to drought. In Eastern Africa 6,560,000 people were affected annually between 2011 and 2020, whereas in Western Africa, 3,896,000 were affected as shown in Figure 3.4.

Eastern and Western Africa had been worst affected due to floods. In Western Africa, 13,452,000 people were affected annually during 2011–2020, whereas in Eastern Africa 1,272,000 people were affected. Figure 3.5 illustrates the region-wise people affected by flood.

The dependence of the majority of the population on agriculture makes it imperative to have a well-developed irrigation system for agriculture. The unit cost of irrigation projects is high as shown in Table 3.3.

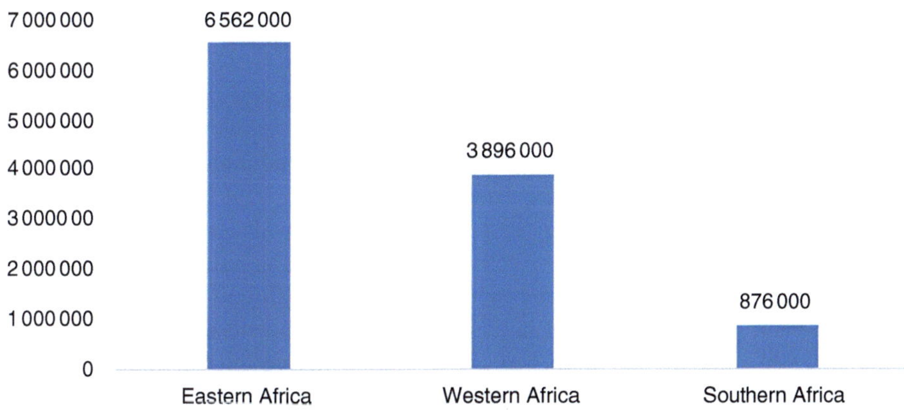

Figure 3.4 People affected in different regions of Africa due to drought (per year between 2011 and 2020). *Source:* Adapted from Climate Risks in Africa (2022).

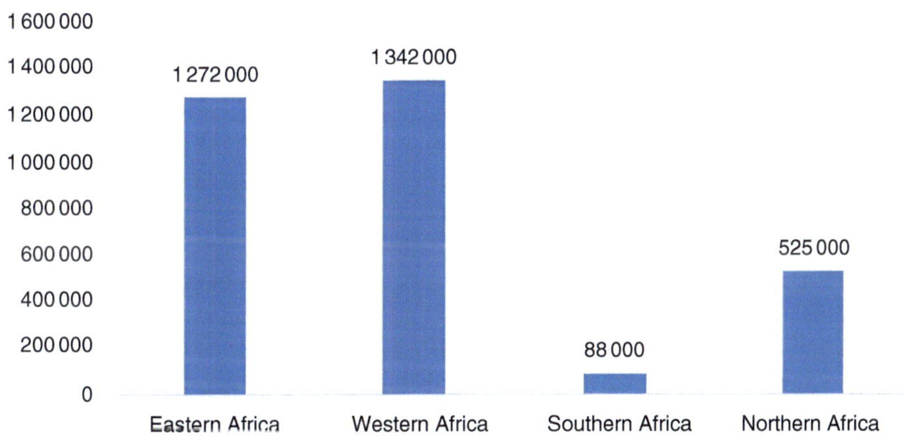

Figure 3.5 People affected in different regions of Africa due to flood (per year between 2011 and 2020). *Source:* Adapted from Climate Risks in Africa (2022).

Table 3.3 Cost of irrigation projects.

	Projects		Newly built		Modernization	
Region	**Unit total cost (US$ per ha)**	**Hardware unit cost (US$ per ha)**	**Unit total cost (US$ per ha)**	**Hardware unit cost (US$ per ha)**	**Unit total cost (US$ per ha)**	**Hardware unit cost (US$ per ha)**
Sub-Saharan Africa	11,828	8,188	14,455	10,473	8,233	5,059
Non-Sub-Saharan Africa	3,882	3,183	6,590	5,481	2,280	1,824
North Africa	6,311	5,251	8,780	7,542	4,582	3,648

Source: You (2008) and Inocencio (2007).

3.2 Factors Affecting the Water Security and Availability

The water security of any region is influenced by various factors, many of which hold particular relevance to Africa. These factors include population growth rates, population density, industrialization, expansion of irrigation, water demand management and efficiency, and ecological sustainability requirements. Water security is influenced by perceptions associated with contexts of perceptions held by individuals regarding threats linked with water and their impact on both individuals and the natural environment. These perceptions are shaped by various factors. First, the condition of the local natural environment plays a role, as different individuals highlighted concerns such as pathogens being washed into storm drains and the sea during summer rainfall (Panday et al. 2021a, b; Panday and Kumar 2023). Second, socioeconomic status is important, as some individuals have the capability to invest in tanks and rainwater harvesting to enhance water security. Third, experiences with local authorities and community interactions can impact perceptions, with dissatisfaction expressed regarding the response to reported water leakages and inconsistent answers regarding water supply issues (Karthick et al. 2023). It is evident that changing natural conditions alone, such as flooding and drought, do not solely shape water security perceptions (Panday and Kumar 2022). Instead, interpersonal relationships and practices related to water security enhancement, such as interactions with authorities and rainwater collection, also influence these perceptions to varying degrees. Good policies fail to deliver results due to poor participation of stakeholders (Tiwari et al. 2020; Panday et al. 2023; Asok Raja et al. 2023).

Nearly everywhere below the surface of the soil, there is groundwater. According to the geology and causes of contamination, the quality of groundwater is characterized by variability in different locations. The crystalline rocks create challenges for the extraction of significant amounts of groundwater in South Africa (Abiye 2023). In the 1960s, Elsie's River was a slum of 1 grade having 110,000 residents residing in an area intended for 60,000 individuals. An alarming population was placing pressure on the availability of clean drinking water which eventually led to a black spot in Elsie River (Erlank 1985). One of the deadliest droughts occurred in 2016–2017 in the Horn of Africa, which brought millions of people to the brink of famine. It not only worsened the water security but also put pressure on an already fragile economy (Chand et al. 2022). It also impacted the availability of water in urban areas (Fisher-Jeffes et al. 2017). Consequently, the 2017 Western Cape drought was exceptional in the region and was the worst since the drought of 1904 (Botai et al. 2017). It has been understood that the decrease in rainfall over those three years was the cause of the water supply shortage (Pascale et al. 2020). Due to the localized droughts, which resulted in significantly reduced water supply and much higher per-unit pricing, which together resulted in the under-recovery of money for maintenance and other operating costs, a tariff structure has been implemented.

3.3 Causes of Lack of Water Availability in Africa

Scarcity of water in Africa is primarily driven by a combination of natural and human-induced factors. One significant factor is climate variability and change. Africa is particularly vulnerable to droughts and erratic rainfall patterns, which are exacerbated by climate change. These conditions reduce the freshwater availability. Population growth and rapid urbanization are also key drivers. The continent's population is rapidly increasing, leading to a higher demand for water for farming, household, and industrial activities. In addition, unsustainable agricultural practices, such as extensive irrigation and inefficient water use, contribute to water scarcity. Land degradation, deforestation, and soil erosion also impact water availability by reducing the capacity of ecosystems to store and regulate water. Moreover, transboundary water management challenges and conflicts over shared water resources add to the complexities of water scarcity. Inadequate governance, weak institutions, and limited investment in water infrastructure and management exacerbate the problem. Addressing these causes requires a multifaceted approach, including sustainable water resource management, climate change adaptation strategies, improved infrastructure, efficient water use practices, and enhanced cooperation among countries and stakeholders.

3.4 Challenges in Water Security and Availability

1) **Climate change impacts**: Changing weather patterns, including droughts and floods, can disrupt water availability, exacerbating water scarcity and leading to increased competition for limited resources.
2) **Inefficient water management**: Inefficient water use practices, such as overirrigation in agriculture or high rates of nonrevenue water in urban areas, contribute to wastage and reduce the overall availability of water. Optimal utilization aimed to reduce wastage of resources enhances sustainability (Padmalosan et al. 2023).
3) **Water scarcity and stress**: Many regions face limited water availability due to factors like growth of population, changes in climate, and poorly managed water use, leading to scarcity of water.
4) **Inadequate infrastructure**: Insufficient or outdated water infrastructure, including dams, pipelines, and distribution networks, can impede the efficient delivery of water to communities and industries.
5) **Water pollution and contamination**: Pollution from industrial discharges, agricultural runoff, and inadequate sanitation systems can compromise water quality, making it unsafe for human consumption and ecological health.
6) **Transboundary water management**: Conflicts and disagreements over shared water resources between neighboring countries can impede effective management and cooperation, leading to tensions and inequitable water distribution.
7) **Population growth and urbanization**: Rapidly growing populations and urban expansion put pressure on existing water resources, leading to increased demand and strain on water availability.
8) **Socioeconomic disparities**: Marginalized communities and vulnerable populations often face limited water accessibility and sanitation due to socioeconomic inequalities, exacerbating water security challenges.
9) **Weak governance and institutions**: Inadequate policy frameworks, poor enforcement, and weak institutions can hinder effective water resource management, planning, and implementation of sustainable solutions.

10) **Lack of funding and investment**: Insufficient financial resources and investment in water infrastructure, research, and conservation efforts can impede progress in achieving water security and availability.

Addressing these challenges requires comprehensive approaches that prioritize sustainable water management, technological innovation, equitable access, stakeholder engagement, and effective governance at local, regional, and global levels.

3.5 Use Cases of Better Water Management

3.5.1 Case Study of South Africa

The mean annual rainfall in South Africa is 495 mm. This is 50% of the average of world (Hedden and Cilliers 2014). The varied rainfall has been a cause of concern. Many of the most significant water resources in the nation have already reached their full capacity (Hedden and Cilliers 2014). Accessibility to water is poor, with water shortages affecting the economically weaker populations. South Africa has implemented an integrated water resource management framework to address water security issues. The strategy focuses on sustainable development, social equity, and environmental protection. Key elements include:

a) **Management of catchment**: South Africa has a dedicated Catchment Management Agency for planning, allocation, and protection of water resources.
b) **Water allocation and licensing**: The strategy ensures fair and equitable water allocation through the Water Allocation Reform Program. Water users are required to obtain licenses, promoting responsible water use.
c) **Demand management**: Emphasis is placed on water demand management, promoting water-efficient practices and reducing water wastage through public awareness campaigns, education, and incentives.
d) **Development of infrastructure**: Enhanced focus on diverting investments for developing the required infrastructure.
e) **Water conservation and reuse**: Efforts are made to conserve water resources through efficient irrigation methods, wastewater treatment, and water reuse for non-potable purposes.
f) **Climate change adaptation**: The strategy incorporates climate change adaptation measures, such as storage of water and techniques of conservation, to mitigate the impact of changing rainfall patterns.
g) **Stakeholder engagement**: Collaboration and participation of various stakeholders, including communities, industries, and government agencies, ensure inclusive decision-making processes and effective implementation.
h) **Monitoring and data management**: Robust monitoring systems are established to assess water quality, quantity, and ecosystem health. Data collected is used for evidence-based decision-making and adaptive management.

The integrated management strategy in South Africa serves as a model for sustainable management of water, balancing socioeconomic development with environmental preservation. It addresses the challenges of water security and availability through a comprehensive and integrated approach, ensuring equitable access, efficient use of water resources, and long-term sustainability.

3.5.2 Case Study of Ethiopia

Ethiopia, a country prone to water scarcity and vulnerable to climate change impacts, launched the Water for Growth Initiative in 2013. The initiative aims to improve availability through integrated management of water resources, infrastructure development, and community participation. It focuses on enhancing water storage, increasing irrigation efficiency, and promoting sustainable water use practices.

As part of the initiative, Ethiopia has invested in the construction of large-scale dams and reservoirs to enhance water storage capacity, particularly in drought-prone regions. For instance, the Grand Ethiopian Renaissance Dam (GERD) on the River (Blue Nile) is one such flagship project aimed at bolstering water security and hydropower generation. In addition to infrastructure development, the initiative emphasizes the adoption of efficient techniques of irrigation, such as irrigation through drip, water-saving technologies, to enhance agricultural water use efficiency. This leads to better crop yield even in water-stressed areas.

Furthermore, community participation and engagement are central to the Water for Growth Initiative. Local communities are involved in decision-making processes of water-related activities. This participatory approach ensures that the strategies and interventions align with the specific needs and realities of the communities, fostering ownership and sustainability.

The policy initiative in Ethiopia highlights the integrated approach and community engagement. It demonstrates the potential for comprehensive strategies to address water challenges and promote sustainable water management practices.

3.6 Increasing Water Security and Availability in Africa

A comprehensive framework is required to enhance water security in Africa. Distribution, awareness, policymaking, capacity building, better management, community engagement, pricing, and public–private collaboration are the need of the hour. Integrating technology with human touch can deliver better outcomes (Tiwari et al. 2021, 2023). Stormwater harvesting (SWH) can prove to be beneficial, which provides additional benefits to collect and save the urban water runoff (Fisher-Jeffes et al. 2017). The Atlantis Water Resource Management Scheme (AWRM) has been a successful provision of water to both Atlantis residential and commercial districts for almost 30 years. It belongs to South Africa. It is economical. It is focused on artificial recharge. Industrial effluent with a greater salinity that is about 4000 m^3/day is treated before being released into coastal basins that are down-gradient from the wellfield near the ocean (Tredoux et al. 2010). Groundwater has become the lifeline of South Africa over time due to its increased usage for irrigation and agriculture (South African Department of Water Affairs and Forestry [DWAF] 2010).

However, due to the formation of crystalline rocks in the country, there is a need for the deployment of qualified skilled hydrogeologists. Hydrogeologists study the groundwater and identify the rocks and ways of water extraction (Abiye 2023). There is plenty of evidence to support the idea that proper catchment management would have considerably lessened the effects of the drought on those who were afflicted in terms of water supply and cost, and that it may be the best long-term solution to guarantee future water security. Proactive methods for EI management will be essential in guaranteeing water security as a result of more erratic climatic trends. Through situations like drought mitigation, flood protection, and water quality maintenance, intact ecological infrastructure can promote economic and social well-being (Juba et al. n.d.). Policymakers need to incorporate measures in Africa to overcome the water security issues. Develop and maintain water infrastructure, including dams, reservoirs, and pipelines. Promote water conservation and efficient water use in agriculture, industry, and households. Implement integrated water resource management practices. Enhance water storage capacity through the construction of additional reservoirs. Invest in modern irrigation techniques to improve agricultural water efficiency. Implement rainwater harvesting systems to capture and store rainwater. Promote the use of water-efficient technologies and appliances. Develop and enforce regulations to prevent water pollution and contamination. Improve wastewater treatment and promote water reuse.

Enhance groundwater management through monitoring and regulation. Strengthen transboundary water management and cooperation among neighboring countries. Integrate climate change adaptation strategies into water management plans. Invest in water infrastructure maintenance and repair. Enhance water governance and strengthen institutional capacities. Empower local communities in water management decision-making processes. Promote sustainable land and watershed management practices to protect water sources. Establish water pricing mechanisms that encourage responsible water use. Strengthen early warning systems for droughts and floods. Promote public awareness and education on water conservation and hygiene practices. Encourage research and innovation in water-related technologies and practices. Increase financial investment in water projects and infrastructure development. Foster public–private partnerships to enhance water management efficiency. Improve data collection and monitoring systems for better water resource management. Support capacity-building programs and knowledge exchange initiatives on water management. Implementing these measures can augment the existing efforts in improving water sustainability and availability in Africa, ensuring access to water for communities and fostering economic and social development.

3.7 Conclusion

The peaceful coexistence is dependent on water availability in Africa. Development by ignoring sustainability and natural resources will create imbalances and conflicts. The wastage of water resources needs the attention of all stakeholders. Optimal utilization of water resources for agricultural, industrial, and household activities needs to be achieved by the adoption of technology for the identification of the right quantity, right time, and right quality of water. Community engagement is crucial for making judicious use of water.

Rainwater harvesting, good governance, optimal storage and usage of rainwater, connecting and conserving rivers and natural resources will go a long way in achieving water security. Recycling of water should become a mission for every citizen and administrator of Africa. Penal action is needed for the misuse of water. High usage of water should be taxable to discourage wastage and encourage efficient usage. Every drop counts. Right from catchment areas, rivers, other natural resources, distribution systems, and end-use areas need to be properly monitored and controlled for proactive action. The efficiency of irrigation projects needs to be improved to reduce wastage and improve revenue collection. The development agencies should support investments in irrigation projects and the use of artificial intelligence to provide water based on specific needs for agricultural, personal, and industrial purposes. Drought-resistant seeds should be explored to augment the agriculture sector and ensure food security, besides managing water security. Life process reengineering is required to question the status quo for the manner in which water is used by individuals. Water management should be the top priority of leaders and citizens to ensure sustainable happiness and growth for the current and future generations.

References

Abiye, T.A. (2023). Contribution of hydrogeology to solving community water supply problems in South Africa. *South African Journal of Science* 119 (1–2): 1–3.

Asok Raja, S.K., Panday, D.P., and Kumar, M. (2023). Decoding the enigma of 100-year record-breaking rainfall over Tamil Nadu using wavelet analysis. *Groundwater for Sustainable Development* 101007. https://doi.org/10.1016/j.gsd.2023.101007.

Bakker, K. (2012). Water security: research challenges and opportunities. *Science* 337 (6097): 914–915.

Bakker, K. and Morinville, C. (2013). The governance dimensions of water security: a review. *Philosophical Transactions of the Royal Society A: Mathematical, Physical and Engineering Sciences* 371 (2002): 20130116.

Botai, C.M., Botai, J.O., De Wit, J.P. et al. (2017). Drought characteristics over the western cape province, South Africa. *Water* 9 (11): 876.

Calow, R., MacDonald, A., Nicol, A. et al. (2002). The struggle for water. *Drought, Water Security and Rural Livelihoods* CR/02/226N: 7–25.

Chand, K., Tiwari, R., and Sapna (2022). Effect of perception and satisfaction on preference for mobile wallet. *FIIB Business Review.* https://doi.org/10.1177/23197145221077365.

Chapman, W. (2019). Africa to lead world in population growth, US News (10 July 2019). https://www.usnews.com/news/best-countries/articles/2019-07-10/africa-to-lead-world-in-population-growth (accessed 2 August 2023).

Chua, E.M., Wilson, S.P., Vink, S., and Flint, N. (2019). The influence of riparian vegetation on water quality in a mixed land use river basin. *River Research and Applications* 35 (3): 259–267.

Climate Risks in Africa (2022). https://gca.org/wp-content/uploads/2023/01/GCA_State-and-Trends-in-Adaptation-2022_Climate-Risks-in-Africa.pdf (accessed 6 August 2023).

Coleridge, S.T. (1798). *The Rime of the Ancient Mariner Lyrical Ballads.* London: Lyrical Ballads.

Das, A., Kumar, M., Jha, P.K. et al. (2022). Isotopic and hydrogeochemical tracking of dissolved nutrient dynamics in the Brahmaputra River System: a source delineation perspective. *Chemosphere* 307: 135757. https://doi.org/10.1016/j.chemosphere.2022.135757.

DeGhetto, K., Gray, J.R., and Kiggundu, M.N. (2016). The African Union's agenda 2063: aspirations, challenges, and opportunities for management research. *Africa Journal of Management* 2 (1): 93–116.

Dickson, S.E., Schuster-Wallace, C.J., and Newton, J.J. (2016). Water security assessment indicators: the rural context. *Water Resources Management* 30: 1567–1604.

Economist (2020). Africa's population will double by 2050. https://www.economist.com/special-report/2020/03/26/africas-population-will-double-by-2050 (accessed 15 July 2023).

Erlank, D. (1985). A comparative quality of life survey in Elsies River and Basuto QwaQwa. Master's thesis, University of Cape Town.

FAO (2003). Review of world water resources by country, food and agriculture organization of the United Nations. https://www.fao.org/3/Y4473E/y4473e00.htm (accessed 3 July 2023).

Fisher-Jeffes, L., Carden, K., Armitage, N.P., and Winter, K. (2017). Stormwater harvesting: improving water security in South Africa's urban areas. *South African Journal of Science* 113 (1–2): 1–4.

Gain, A.K., Giupponi, C., and Wada, Y. (2016). Measuring global water security towards sustainable development goals. *Environmental Research Letters* 11 (12): 124015.

Global Water Security Assessment (2023). https://inweh.unu.edu/wp-content/uploads/2023/03/Global-Water-Security-Assessement-2023_F.pdf (accessed 30 March 2023).

Grey, D. and Sadoff, C.W. (2007). Sink or swim? Water security for growth and development. *Water Policy* 9 (6): 545–571.

Gunda, T., Benneyworth, L., and Burchfield, E. (2015). Exploring water indices and associated parameters: a case study approach. *Water Policy* 17 (1): 98–111.

Hedden, S. and Cilliers, J. (2014). Parched prospects-the emerging water crisis in South Africa. *Institute for Security Studies Papers* 2014 (11): 16.

Inocencio, A.B. (2007). *Costs and Performance of Irrigation Projects: A Comparison of Sub-Saharan Africa and Other Developing Regions*, vol. 109. IWMI.

Juba, R., Hiestermann, M., Dini, J., & Madikizela, B. (n.d.) Investing in Ecological Infrastructure in South Africa's Significant Catchments.

Karthick, M., Somesh, V., Gudadhe, N., et al. (2023). Structural analysis of motorcycle spokes design using finite element analysis with alloy materials. *Materials Today: Proceedings*.

Kumar, M., Panday, D.P., Bhagat, C. et al. (2023a). Demystifying the decadal shift in the extent of groundwater in the coastal aquifers of Gujarat, India: a case of reduced extent but increased magnitude of seawater intrusion. *Science of the Total Environment* 165451. https://doi.org/10.1016/j.scitotenv.2023.165451.

Kumar, M., Mazumder, P., Silori, R. et al. (2023b). Prevalence of PPCPs, microplastics and co-infecting microbes in the post-COVID-19 era and its implications on antimicrobial resistance and potential endocrine disruptive effects. *Science of the Total Environment* 166419. https://doi.org/10.1016/j.scitotenv.2023.166419.

La Banque Africaine, D.D.P. and Bankgroup, A. (2000). *The Africa Water Vision for 2025: Equitable and Sustainable Use of Water for Socioeconomic Development*. Economic Commission for Africa: Addis Ababa, Ethiopia.

Lotter, M. and Le Maitre, D. (2021). Fine-scale Delineation of Strategic Water Source Areas for Surface Water in South Africa Using Empirical Bayesian Kriging Regression Prediction: Technical Report.

Luo, T., Young, R., and Reig, P.2015). Aqueduct projected water stress rankings. Technical Note. Washington, DC: World Resources Institute (August 215). http://www.wri.org/publication/aqueduct-projected-water-stress-country-rankings (accessed 5 June 2023).

Maddocks, A., Young, R.S., and Reig, P. (2015). *Ranking the World's Most Water-Stressed Countries in 2040*. World Resources Institute. https://www.wri.org/insights/ranking-worlds-most-water-stressed-countries-2040 (accessed 6 April 2023).

Malan, J.A.C., Flint, N., Jackson, E.L. et al. (2018). Off stream watering points for cattle: protecting riparian ecosystems and improving water quality. *Agriculture, Ecosystems and Environment* 256 (1): 144–152.

Meissner, R., Funke, N., Nortje, K. et al. (2018). Water security at local government level in South Africa: a qualitative interview-based analysis. *The Lancet Planetary Health* 2: S17.

Neog, N., Goswami, R., Panday, D.P. et al. (2024). Arsenic contamination in the groundwater of Northeastern India: critical understandings on geotectonic controls and the need for intervention. *Current Opinion in Environmental Science & Health* 100539. https://doi.org/10.1016/j.coesh.2024.100539.

Oki, T. and Kanae, S. (2006). Global hydrological cycles and world water re-sources. *Science* 313: 1068–1072. https://doi.org/10.1126/science.1128845.

Padmalosan, P., Vanitha, S., Kumar, et al. (2023). An investigation on the use of waste materials from industrial processes in clay brick production. *Materials Today: Proceedings*.

Panday, D.P. and Kumar, M. (2022). Climate indices and hydrological extremes: Deciphering the best-fit model. *Environmental Research* 215: 114301. https://doi.org/10.1016/j.envres.2022.114301.

Panday, D.P. and Kumar, M. (2023). Application of remote sensing techniques to deal with scale aspects of GRACE data to quantify groundwater levels. *MethodsX* 102108. https://doi.org/10.1016/j.mex.2023.102108.

Panday, D.P., Khosa, R., Maheswaran, R. et al. (2021a). Game-theoretic-based modelling of Krishna waters dispute: equilibrium solutions by Metagame Analysis. *The European Physical Journal B* 94: 101. https://doi.org/10.1140/epjb/s10051-021-00107-w.

Panday, D.P., Khosa, R., Maheswaran, R. et al. (2021b). Game theoretic-based modelling of Krishna waters dispute: equilibrium solutions by hypergame analysis. *The European Physical Journal B* 94: 131. https://doi.org/10.1140/epjb/s10051-021-00135-6.

Panday, D.P., Singh, A., Gaur, S. et al. (2023). Probable maximum precipitation analysis of high rainfall regions in India. *Groundwater for Sustainable Development* 100893. https://doi.org/10.1016/j.gsd.2022.100893.

Pascale, S., Kapnick, S.B., Delworth, T.L., and Cooke, W.F. (2020). Increasing risk of another Cape Town "Day Zero" drought in the 21st century. *Proceedings of the National Academy of Sciences* 117 (47): 29495–29503.

Raphael, A. and Lalika, M.C.S. (2022). Ecohydrological phytoregulation of riverine ecosystems using riparian vegetation. *Youth and Water Security in Africa* 300.

Richard-Ferroudji, A., Raghunath, T.P., and Venkatasubramanian, G. (2018). Managed aquifer recharge in India: consensual policy but controversial implementation. *Water Alternatives* 11 (3): 749–769.

Sajidu, S.M.I., Masamba, W.R.L., Thole, B., and Mwatseteza, J.F. (2008). Groundwater fluoride levels in villages of Southern Malawi and removal studies using bauxite. *International Journal of Physical Sciences* 3 (1): 1–11.

Shakir, A.K., Thakur, S., Sarkar, A., and Shekhar, S. (2016). Worldwide contamination of water by fluoride. *Environmental Chemistry Letters.* https://doi.org/10.1007/s10311-016-05635.

South African Department of Water Affairs and Forestry (DWAF) (2010). *The Atlantis Water Resource Management Scheme: 30 Years of Artificial Groundwater Recharge*, 76. Pretoria: DWAF.

Tiwari, R., Chand, K., and Anjum, B. (2020). Crop insurance in India: a review of Pradhan Mantri Fasal Bima Yojana (PMFBY). *FIIB Business Review* 9 (4): 249–255. https://doi.org/10.1177/2319714520966084.

Tiwari, R., Chand, K., Bhatt, A. et al. (2021). Agriculture 5.0 in India: opportunities and challenges of technology adoption. *A Step Towards Society* 5: 179–198.

Tiwari, R., Agrawal, P., Singh, P. et al. (2023). Technology enabled integrated fusion teaching for enhancing learning outcomes in higher education. *International Journal of Emerging Technologies in Learning* 18 (7): 243–249. https://doi.org/10.3991/ijet.v18i07.36799.

United Nations University (2021). UN: 500 million live in 19 African nations deemed water insecure. https://inweh.unu.edu/un-500-million-live-in-19-african-nations-deemed-water-insecure/#:~:text=Per%20capita%20water%20availability%20is,water%20per%20capita%-20per%20annum (accessed 12 June 2023).

Vorosmarty, C.J., Green, P., Salisbury, J., and Lammers, R.B. (2000). Global water resources: vulnerability from climate change and population growth. *Science* 289 (5477): 284–288.

You, L.Z. (2008). *Africa: Irrigation investment Needs in Sub-Saharan Africa*, Africa Infrastructure Country Diagnostic Background Paper; no. 9. World Bank: Washington, DC.

Zalewski, M. (2002). Ecohydrology – the use of ecological and hydrological processes for sustainable management of water resources. *Hydrological Sciences Journal* 47 (5): 823–832.

4

Water Security in South America

Navjot Hothi[1] *and Saumyaranjan Sahoo*[2]

[1] *Department of Physics, School of Engineering, UPES, Dehradun, Uttarakhand, India*
[2] *Department of Civil Engineering, School of Engineering, UPES, Dehradun, Uttarakhand, India*

4.1 Introduction

South America is a continent surrounded by the Pacific Ocean to the west and the Atlantic Ocean to the east and ranks as the fourth-largest continent in terms of area and fifth in terms of population. The continent spans diverse latitudes and longitudes, showcasing a variety of ecosystems and landscapes, from the Amazon rainforest to the Andes Mountains to the Patagonian steppe (Bolius et al. 2006). South America consists of 12 independent countries, and 3 territories (Figure 4.1a). The independent countries include Argentina, Bolivia, Brazil, Chile, Colombia, Ecuador, Guyana, Paraguay, Peru, Suriname, Uruguay, and Venezuela. The territories encompass French Guiana (an overseas department of France), the Falkland Islands (a British Overseas Territory), and South Georgia and the South Sandwich Islands (a British Overseas Territory).

Water security and availability across South America exhibit significant variations, with some regions facing substantial challenges while others possess more abundant water resources (Mejia 2014). The continent is home to major rivers such as the Amazon, the Paraná, and the Orinoco, alongside numerous other rivers, streams, lakes, wetlands, and groundwater reserves. Figure 4.1b shows the major river basin in the country. Generally, the northern and northeastern parts of South America encounter water scarcity, while the southern and southeastern regions enjoy greater water resources (Lee 1995). Due to the diverse climates and ecosystems within South America, different areas face distinct water-related challenges (Klein Goldewijk et al. 2010). Some regions are prone to droughts, such as the arid Atacama Desert in Chile, the semiarid Brazilian Northeast, and the Andean region, which can experience water scarcity during dry seasons (Arias et al. 2021; Ferreira and Reboita 2022). Hydropower reliance among many countries in the region can impact downstream water availability and quality. Conflicts frequently arise among various water

Water Scarcity Management: Enabling Technologies, First Edition. Edited by Kanchan D. Bahukhandi, Manish Kumar, Durga P. Panday, and Tushara G.G. Chaminda.

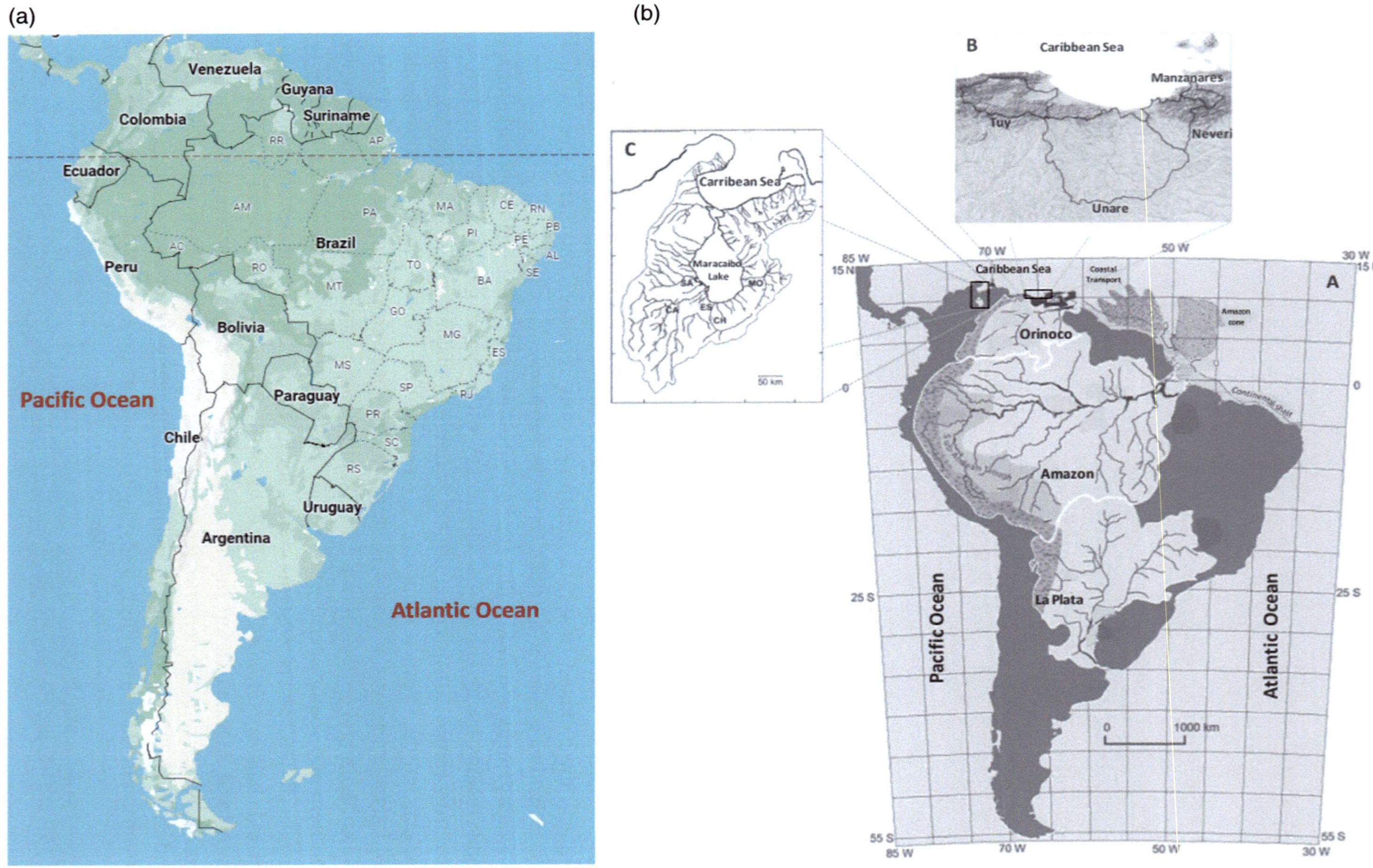

Figure 4.1 (a) Map of South America showing its countries. *Source:* Google LLC/https://www.google.com/maps//last accessed on 5 June 2024. (b) Major River basins in the South America. *Source:* Bustamante et al. (2015)/IOP Publishing/CC BY 3.0.

users, including farmers, industries, and urban areas, regarding access to and utilization of water resources (Das et al. 2022; Kumar et al. 2023a, b).

Climate change is significantly impacting water conditions in South America (Qin 2021). Alterations in precipitation patterns, increased evaporation rates, and rising temperatures are leading to changes in water availability, quality, and distribution (Slater 2019). The reduction in glacial coverage in the Andes due to melting glaciers has notably affected freshwater sources for South American countries (Neog et al. 2024). This phenomenon has resulted in changes in river flows, water temperature, and sediment levels, with significant implications for agriculture, hydropower generation, and ecosystems. Moreover, shifting precipitation patterns are causing more frequent and severe droughts in some areas while triggering increased flooding and erosion in others (Hothi and Hothi 2022). These variations have considerable impacts on water availability and quality, resulting in loss of biodiversity within many communities, reduced crop yields, and water shortages (Panday et al. 2021a, b; Panday and Kumar 2023).

To tackle these challenges, several nations in South America have instigated water management practices such as the construction of new water infrastructure, water recycling, and conservation measures. For example, Brazil has sanctioned programs to increase access to sanitation and clean water (Slater 2019), while Chile has established a national water strategy designed to refine water conservation and management. Active water management is important for guaranteeing the sustainability of South America's water resources and extenuating the effects of droughts on societies, economies, and ecosystems. Thus, water availability and security in South America present a complicated problem influenced by various factors such as uneven distribution of resources, water management policies, and climate change (Scherer and Pfister 2016). Continuing efforts to enhance water conservation, management, and access to sanitation and clean water are important for safeguarding a secure water and sustainable future for the continent (Jones and Carvalho 2013).

4.2 Natural Water Resources in South America

South America relies heavily on its natural water resources, which are vital for sustaining the ecosystems, economies, and societies of the continent (van Vliet et al. 2021). To ensure the longevity of these invaluable resources, effective water management practices are crucial (Al-Mashaqbeh et al. 2012). South America boasts abundant natural water resources. The Amazon River flows through several South American countries such as Brazil, Peru, and Colombia; the Amazon River holds the distinction of being the world's largest river in terms of discharge volume (Panday and Kumar 2022). It serves as a lifeline for the Amazon rainforest (Barlow et al. 2020). Found in Venezuela and Colombia, the Orinoco River ranks as the third-largest river in South America. The Paraná River runs through Brazil, Paraguay, and Argentina, and it holds the title of being the second-longest river in South America. Acting as a vast estuary, the Rio de la Plata forms the border between Argentina and Uruguay. It is renowned as the widest river in the world, stretching up to 220 km (140 miles) at its mouth. Furthermore, the Andes Mountain range, spanning various South American countries, boasts numerous glaciers that contribute freshwater to rivers and streams. Located in Chile, the Atacama Desert is known as one of the driest places on Earth.

Nonetheless, it houses oases and groundwater reserves that sustain human settlements and agriculture (Graham et al. 2020). Situated in northern South America, the Guiana Shield is a vast plateau brimming with water resources. It is home to numerous rivers and wetlands that support diverse flora and fauna. However, similar to many regions around the world, South America faces water management challenges that encompass issues like pollution, overuse, and the variability and changes brought about by climate (Hinkel 2011).

4.3 Water-Deficit Nations in South America

Several countries in South America confront water deficits, indicating limited access to freshwater resources relative to their population and water requirements (Damkjaer and Taylor 2017; Rodríguez et al. 2022). Chile ranks among the driest nations globally, with numerous regions facing chronic water scarcity. The country heavily relies on groundwater and surface water resources while implementing various policies and regulations to manage them effectively. Despite being home to the Amazon River, one of the largest rivers globally, Peru experiences water scarcity in several regions, especially along the arid coastal areas (Baraer et al. 2012). While the country has implemented policies and regulations for water resource management, overexploitation and contamination persist as significant obstacles. Ecuador is relatively rich in water resources; however, many highland regions encounter water scarcity. To address this, the country has established policies and programs, including the formation of water user associations and the promotion of integrated water resources management. Bolivia is one of the poorest countries in South America, and it faces acute water scarcity in various regions. Figure 4.2a shows that Argentina, Chile, and Uruguay are the three most water-stressed countries in South America. As per the predictions of the World Resources Institute (WRI), Chile and Peru would suffer extremely high water stress by the year 2040, Figure 4.2b. The country has employed programs and policies such as the formation of water user associations and the campaign of rainwater harvesting to cope with its water resources effectively. Paraguay is comparatively abundant in water resources, but experiences water scarcity in certain regions, chiefly in the semiarid Chaco area. The country has executed programs and policies focused on the establishment of water user associations and integrated water resources management (Panday et al. 2023; Asok Raja et al. 2023). These countries confront considerable challenges in achieving sustainable water resource management, including issues like overexploitation, pollution, and the impact of climate change. Tackling these challenges necessitates innovative policies, technologies, and enhanced cooperation among regional nations.

4.4 Water-Related Challenges in South America

South America encounters a range of water-related difficulties, encompassing issues such as water scarcity, pollution, and the consequences of climate change (Jones and Carvalho 2013). Consequently, numerous areas within the region have experienced significant water crises (Oliveira et al. 2017). The following examples shed light on these challenges.

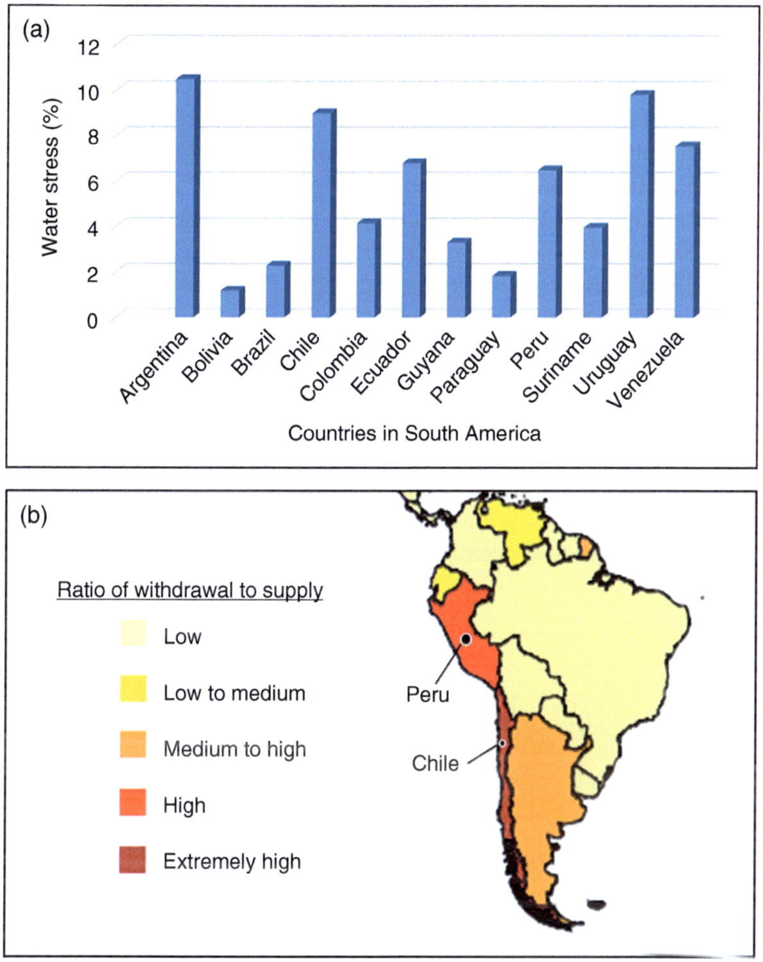

Figure 4.2 (a) Water stress in the South American countries by 2020. *Source:* Adapted from FAO Aquastat. (b) Predicted water stress in the South American countries by 2040. *Source:* Adapted from World Resources Institute.

4.4.1 Climate Change

One main challenge to water security in South America is climate change and variability (Liu et al. 2021). This can cause extreme weather events such as floods and droughts that can influence water availability and quality. Unpredictability and changes in climate are pertinent due to global warming. El Niño (Cai et al. 2014) and La Niña (Capotondi 2015) weather patterns also impact the water availability in South America. During El Niño events some areas face prolonged dry spells, while others receive excessive flooding and rainfall (Hansen 2005).

4.4.1.1 Major Droughts in South American Nations

Many South American nations have dealt with harsh droughts in recent years (Phillips et al. 2009). This section provides a brief account of some of these events. Between 2010 and 2015, Chile encountered a severe drought affecting vast regions, including the capital city of Santiago. The drought was caused by a combination of factors such as overuse of water resources, climate change, and reduced rainfall. It led to water shortages in several areas and forced farmers to abandon their crops. From 2014 to 2015, Brazil encountered a severe drought that affected millions of people across the country (Lewis et al. 2011). The drought was led by various factors, including excessive exploitation of water resources, deforestation, and climate change. It caused water shortages in several cities, including São Paulo (Coutinho et al. 2015), and brought Rio de Janeiro to the edge of water depletion (Filho et al. 2016). In 2016, Bolivia too faced a severe drought that affected over 125,000 families throughout the nation. During 2016 and 2017, Peru faced a severe drought that affected significant parts of the country, including the capital city of Lima. The drought was caused by factors like deforestation, excessive water resource usage, and climate change. It led to water shortages in many regions, forcing the government to declare a state of emergency in multiple areas. Satellite observations have also depicted droughts in South America starting from the year 2016 to 2020. Figure 4.3 represents a satellite image of the same in terms of the groundwater wetness profile. The east coast nations have a low water percentile, whereas Peru has a high water percentile.

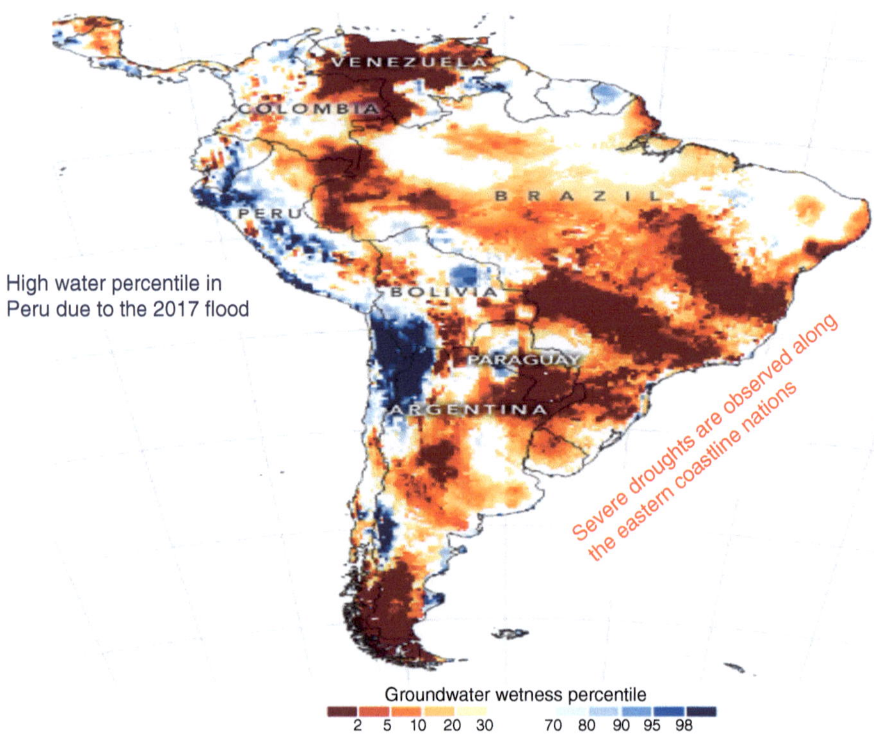

Figure 4.3 Satellite image of severe droughts in South America from the year 2016 to 2020. *Source:* NASA/https://landsat.visibleearth.nasa.gov/view.php?id=147480, last accessed 4 June 2024/ public domain.

These severe drought events have profoundly affected South American countries, influencing water availability, food production, and public health. They emphasize the urgency of implementing effective water management strategies and adopting measures for climate change adaptation to tackle the challenges arising from water scarcity and drought. Overall, understanding and addressing the diverse and intricate impacts of climate change on water resources in South America necessitate meticulous monitoring and prudent management to mitigate their effects on both human societies and the natural environment.

4.4.1.2 The 2017 Floods in Peru

The devastating floods that struck Peru in 2017 stand as one of the most severe natural calamities witnessed in the country's recent past (Son et al. 2020). These floods had a widespread impact, affecting 24 regions of Peru (Figure 4.1) and causing extensive damage to crucial infrastructure, including water supply systems and the natural water resources they rely upon (Ramírez and Briones 2017). The floodwaters resulted in the overflow of rivers and streams, triggering destructive landslides and mudslides that contaminated water sources. This contamination introduced debris, sediment, and pollutants into ecosystems, compromising water quality. As a consequence, both human populations and wildlife faced health issues, while agricultural production and food security were also adversely affected. The floods wreaked havoc on water supply systems, leaving many communities without access to clean and safe water. In certain regions, the destruction of water treatment plants further exacerbated the challenge of providing potable water.

In response to the devastating floods, the Peruvian government and international organizations swiftly took action to provide emergency assistance and implement measures aimed at restoring water supply systems and natural water resources. These measures encompassed water treatment and purification efforts, distribution of water filters and chlorine tablets, and the repair and reconstruction of critical infrastructure. Nonetheless, the profound impact of the 2017 floods on Peru's natural water resources underlines the vulnerability of such resources to natural disasters. It emphasizes the urgent need to construct resilient water systems capable of withstanding extreme weather events. Furthermore, it underscores the significance of implementing effective water management strategies that prioritize the safeguarding and conservation of natural water resources.

4.4.2 Uneven Distribution of Water Resources

An additional obstacle in South America is the inequitable distribution of water resources. The vast Amazon Basin, spanning a significant portion of northern South America, harbors the world's largest freshwater system, accounting for over 20% of the global freshwater supply. However, several countries in South America, including Peru, Bolivia, Ecuador, and Chile, contend with arid and semiarid regions that are susceptible to water scarcity. For instance, the Atacama Desert in Chile stands as one of the driest places on the planet, leaving numerous communities in the area grappling with insufficient water access for their fundamental needs. This uneven distribution of water resources often sparks competition among different sectors striving to secure their share of limited water resources. Consequently, conflicts can arise between agricultural, industrial, and municipal water

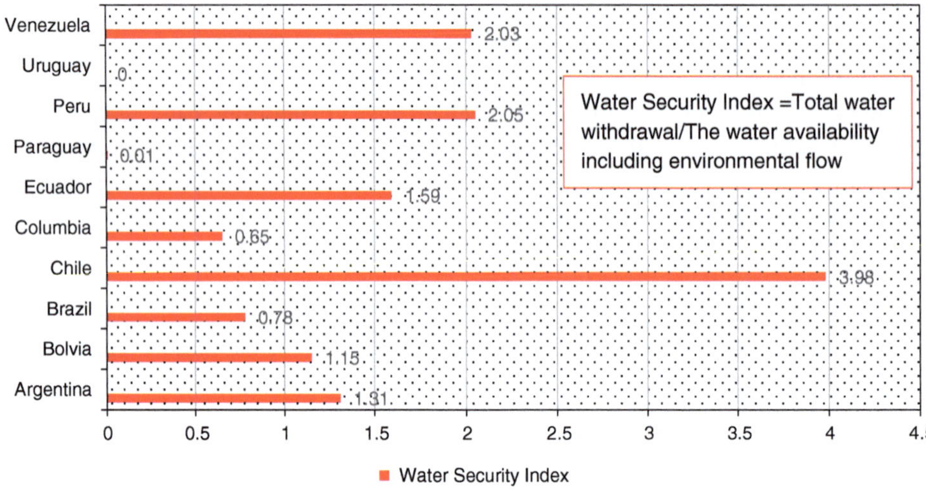

Figure 4.4 Water security index of 10 biggest nations of South America.

users, each vying to safeguard their access to these finite resources. The water security index chart of the 10 biggest nations of South America is displayed in Figure 4.4.

4.4.3 Water Pollution

Water pollution poses a significant challenge in South America, stemming from industrial activities, mining operations, and urban development (Cox et al. 2008). This contamination of water sources has resulted in adverse health effects for both humans and wildlife, as well as ecological and biodiversity disruptions. The pollution of lakes and rivers has had a deep effect on water resources throughout the region, consisting of both the quality and quantity of available water. This section provides some examples exemplifying the consequences of pollution on water resources in various regions of South America. The Amazon River Basin is one of the biggest global freshwater ecosystems. It harbors a rich variety of species found only in this particular region of the planet (Eichler et al. 2015). However, the basin copes with abundant pollution from urbanization, agriculture, and mining. Contamination resulting from mining activities releases toxic substances like cyanide and mercury into the water, thus posing severe health risks to wildlife and humans (Wilderer and Schreff 2000).

The Rio de la Plata, an important inlet bordering Uruguay and Argentina, plays a vital role in providing water for industry and agriculture in the area. However, this inlet confronts important pollution from agricultural runoff, industrial waste, and untreated sewage. Consequently, water quality has deteriorated, and the health of fish and other aquatic species has been affected. Located on the border between Peru and Bolivia, Lake Titicaca operated as an important source of irrigation and drinking water for neighboring communities. Regrettably, the lake is severely polluted by pollutants caused by agricultural runoff and untreated sewage. This is leading to a decline in water quality, algal blooms, and eutrophication. Pollution plagues the Pacific coast of South America, resulting from diverse sources such as sewage, agricultural runoff, and oil spills. For example, the 2010 Deepwater Horizon oil spill in the Gulf of Mexico had consequences along the Pacific coast of several South American countries. These countries include Ecuador, Peru, and Columbia (Zavala-Araiza et al. 2021). This spill had a substantial consequence on the local fishing industry, which assists as a crucial livelihood for many communities. The destructive effects of water pollution on South America's water resources spread beyond the environment and impact human well-being. Attempts are being made to cater to this tenacious issue, including regulation of mining activities, promotion of sustainable agricultural practices, and the improvement of wastewater treatment facilities.

4.4.4 Inadequate Infrastructure

In some regions of South America, access to clean and safe water is limited due to inadequate infrastructure. Rural areas and marginalized communities are often the most affected by this problem. Figure 4.5 demonstrates the inequitable access to healthy and safe drinking water in some of the South American countries. This figure shows that Peru and Chile having the highest Gross Domestic Product (GDP) per capita have about 100% access to healthy drinking water whereas Ecuador and Peru have low GDP and have as low as about 85% access to healthy drinking water. Thus, inadequate infrastructure is a major issue affecting water resources in South America. Here are some reasons why inadequate infrastructure is impacting water resources in the region. Many communities in South America (Mukherjee and Jensen 2020) do not have access to safe drinking water due to inadequate infrastructure. In some cases, water sources are contaminated by pollutants such as heavy metals or bacteria, while in other cases, there is simply a lack of infrastructure to provide clean water to these communities. In many parts of South America, water scarcity is becoming an increasingly pressing issue (Scherer and Pfister 2016). Inadequate infrastructure for water storage and distribution means that water is often wasted or lost before it can reach those who need it. This is predominantly problematic in regions with high rates of agricultural production, where water is necessary for irrigation. Insufficient infrastructure for handling stormwater and inhibiting erosion can lead to landslides and flooding (Marengo

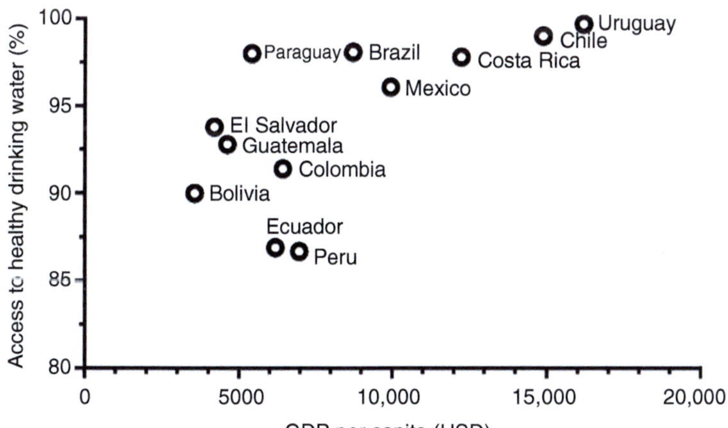

Figure 4.5 Relation between per capita GDP and access to healthy drinking water. *Source:* Rodriguez et al. (2022)/MDPI/CC BY 4.0.

and Espinoza 2016), which can cause noteworthy damage to ecosystems and communities. This is particularly valid in areas with heavy rainfall or steep terrain, where water management infrastructure is especially important. Insufficient infrastructure for handling industrial wastewater can steer the pollution of water resources. Untreated wastewater in some cases is discharged into other water bodies and rivers, indicating a range of damaging impacts on ecosystem health and water quality. Largely, insufficient infrastructure is a substantial challenge for water resources in South America. Attempts are being made to tackle these issues, such as refining wastewater treatment facilities, financing in new infrastructure for water storage, and promoting more sustainable water management practices.

4.5 Water Treatment Practices in South America

4.5.1 Indigenous Technology Adapted for Water Treatment in South America

To tackle these water-related challenges, many South American countries are implementing diverse water management strategies. These efforts include promoting water conservation, investing in water infrastructure, and enhancing water governance. Nonetheless, further actions are required to ensure sustainable and equitable access to water resources throughout the region. South America boasts a wealth of traditional knowledge and practices associated with water treatment, which has been developed and transmitted across generations by Indigenous communities (Imoro et al. 2021). Indigenous communities in South America harness the power of plants and herbs to treat water. Crushed leaves from the chagual plant are used for water purification in Chile. In the Andean highlands, people use clay pots to store and filter water. Clay pots are porous in nature and water seeps through them slowly and thereby removes impurities. In regions like the Amazon rainforest, charcoal is used as a natural filtering agent which can filter and also eliminate contaminants and impurities from the water. Several Indigenous communities in the Amazon region use natural materials such as rocks, gravel, and sand gravel to improve water quality. In water scare regions, native communities have established diverse water storage systems. These include reservoirs, underground tanks, and cisterns, to store and capture rainwater. These old-styled practices are often exemplified by their sustainability, affordability, and efficacy in refining water availability and security (Altieri 2004).

4.5.2 Modern Water Treatment Practices in South America

The modern water treatment practices in South America differ depending on the availability of resources and also the country (Jouravlev 2014). These practices continually evolve and adapt to address the region's water-related challenges, including pollution, scarcity, and variability (Bell et al. 2011). They play a crucial role in ensuring access to safe and dependable water resources for communities, industries, and the environment. There are several modern water treatment practices in South America. Numerous cities across South America have established water treatment plants that purify raw water sourced from rivers and other bodies of water to produce safe drinking water.

These plants employ advanced technologies such as filtration, coagulation, and disinfection to eliminate impurities and contaminants from the water. In coastal regions facing freshwater scarcity, certain South American countries have constructed desalination plants to convert seawater into freshwater (Deshpande et al. 2020). These plants utilize methods like reverse osmosis and distillation to remove salt and other impurities from seawater.

Various industries in South America, including mining and agriculture, necessitate substantial water usage. To minimize their environmental impact and conserve water resources, some industries have implemented water reuse systems. These systems treat wastewater generated from industrial processes and repurpose it for other applications like irrigation or cooling. In areas where water scarcity prevails, certain communities and households have adopted rainwater harvesting systems (Domínguez et al. 2017). These systems collect rainwater from roofs and store it for future use, such as drinking or irrigation purposes. South America is increasingly implementing water conservation measures to reduce water consumption and safeguard water resources. These measures encompass the use of water-efficient appliances and fixtures, water reuse systems, and public awareness campaigns to promote responsible water usage. Figure 4.6 shows a comparison of traditional and modern methods of water treatment. By embracing the modern water treatment practices, South America endeavors to address its water-related challenges effectively and ensure the sustainable management of its precious water resources.

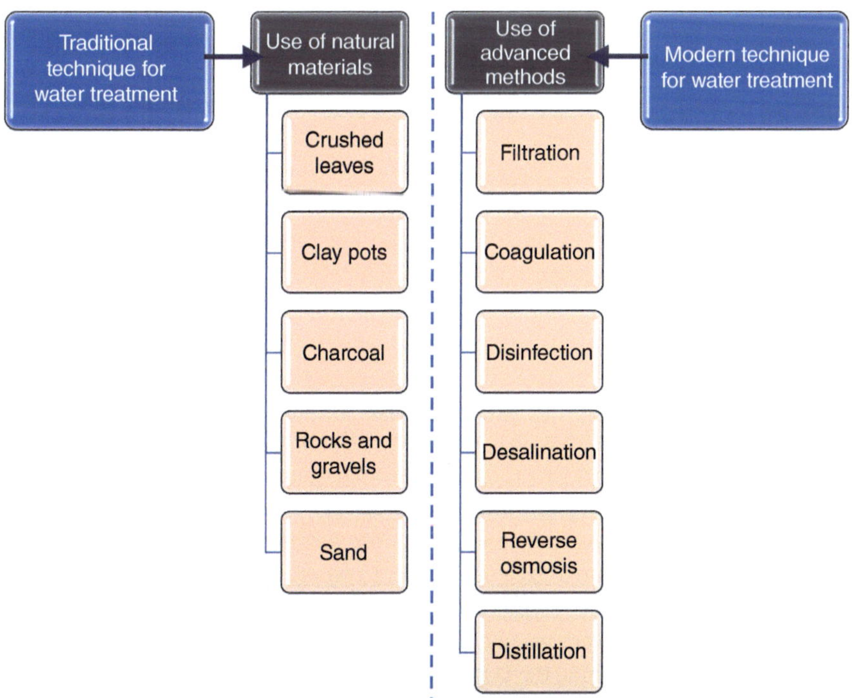

Figure 4.6 Comparison of traditional and modern techniques for water treatment in South America.

4.6 Water Treatment Treaties and Agreements in South America

South America has worked on several water treatment treaties and agreements for promoting sustainable and rightful management of water resources in the region (Water Governance in Latin America and the Caribbean 2012). Some of the details of prominent treaties and agreements are presented in this section (Sub-National Finances and Fiscal Consolidation 2013). The Amazon Cooperation Treaty was signed by eight countries in the Amazon Basin (Brazil, Bolivia, Colombia, Ecuador, Guyana, Peru, Suriname, and Venezuela) and targets to endorse sustainable development of the Amazon region. The La Plata Basin Treaty was signed by Argentina, Bolivia, Brazil, Paraguay, and Uruguay and aims to endorse the unified management of the La Plata Basin, which is one of the biggest river basins in South America. The Andean Community Treaty was signed by Bolivia, Colombia, Ecuador, and Peru and aims to promote social and economic integration among the Andean countries. The treaty embraces provisions for the sustainable management of natural resources. This includes the protection of the environment and protection of water resources. The Santiago Declaration was taken up by the countries of the Southern Cone (Argentina, Brazil, Chile, Paraguay, and Uruguay) to promote integration and cooperation in the management of water resources in the region. These treaties and agreements demonstrate the commitment of South American countries to work together to address water-related challenges and promote sustainable and equitable management of water resources in the region.

4.7 Wastewater Treatment Projects in South American Countries

In South American nations, various wastewater treatment plants have been employed to cater to water pollution and augment access to clean drinking water (Wade Miller 2006). In Santiago, the capital city of Chile, the Mapocho River suffered from high pollution due to industrial waste and untreated sewage. However, in the year 2000, the city commenced the Aguas Andinas wastewater treatment project (Wilderer and Schreff 2000). This enterprise included constructing a new treatment plant and laying a network of new pipelines to treat and collect sewage. The project has made incredible improvements in water quality and has assisted in the refurbishment of aquatic ecosystems. In Bogotá, the capital city of Colombia, the Salitre wastewater treatment plant was developed in the late 1980s to cater to the sewage generated by the growing population of the city (Matos et al. 2014). The plant deployed advanced treatment technologies such as activated carbon filtration and biological treatment to eradicate pollutants from the wastewater. The endeavor has greatly improved water quality in the Bogotá River and has abridged public health risks that are linked to water pollution. São Paulo, the largest city in Brazil (Davidson et al. 2012), once fought the pollution of the Tietê River caused by industrial waste and untreated sewage. However, since the late 20th century, the city has commenced various projects, embracing the construction of large-scale wastewater treatment plants. As a result, the river's water quality has improved immensely, with the treatment of over 90% of wastewater. Lima, the largest city in Peru, earlier met several challenges due to industrial waste and untreated sewage (Blackhurst et al. 2010), which were directly discharged into the Pacific Ocean. This led to prominent public health risks and water pollution. However, since the 1990s, the city has executed several wastewater treatment projects. These initiatives include the establishment of a sewerage system and the construction of large-scale treatment plants. These wastewater treatment projects have efficiently enhanced water quality and condensed pollution in South American countries.

4.8 Conclusions

In conclusion, water security and availability pose significant challenges in South America, demanding immediate attention. Despite the region's abundant natural water resources, issues such as uneven distribution, contamination, and rising water demands jeopardize the water security. Countries such as Bolivia and Peru experience severe water scarcity, making access to clean and safe water an everyday struggle. South America faces major water concerns such as pollution, inadequate infrastructure, and droughts. Industrial activities, mining, and inadequate sanitation systems contaminate water sources, posing severe risks to human health and the environment. Limited infrastructure, particularly in rural areas, hinders the provision of clean water and sanitation services. Moreover, more frequent and severe droughts resulting from climate change exacerbate water scarcity, impacting agriculture, livestock, and human settlements. Resolving these challenges necessitates coordinated efforts involving governments, civil society, and the private sector to ensure universal access to clean and safe water.

In water-deficit regions of South America, Indigenous technologies and practices have played a vital role in securing water availability. Traditional techniques like rainwater harvesting, terracing, and irrigation systems have permitted Indigenous communities to acclimate sustainably to water scarcity in arid regions. Modern water treatment exercises like desalination have gained applause, especially in areas facing acute water scarcity. International agreements and treaties have also helped promote water availability and security in South America. The United Nations Declaration on the Rights of Indigenous Peoples acknowledges the Indigenous people's right to manage and access water resources. The Escazú Agreement aims to improve public participation and transparency and public participation in decision-making concerning water resources. South American governments must give precedence to water availability and security by employing domestic initiatives and policies that commendably address water-related challenges. These policies should safeguard sustainable management practices, invest in advanced water treatment technologies, and lead to equitable distribution of water resources.

References

Akhmouch, A. (2012). Water Governance in Latin America and the Caribbean: A Multi-Level Approach (OECD Regional Development Working Papers No. 2012/04; OECD Regional Development Working Papers, Vol. 2012/04). https://doi.org/10.1787/5k9crzqk3ttj-en.

Al-Mashaqbeh, O.A., Ghrair, A.M., and Megdal, S.B. (2012). Grey water reuse for agricultural purposes in the Jordan Valley: household survey results in Deir Alla. *Water* 4 (3): 580–596. https://doi.org/10.3390/w4030580.

Altieri, M.A. (2004). Linking ecologists and traditional farmers in the search for sustainable agriculture. *Frontiers in Ecology and the Environment* 2 (1): 35–42. https://doi.org/10.1890/1540-9295(2004)002 [0035: LEATFI]2.0.CO2.

Arias, P.A., Garreaud, R., Poveda, G. et al. (2021). Hydroclimate of the Andes part II: hydroclimate variability and sub-continental patterns. *Frontiers in Earth Science* 8: 505467. https://doi.org/10.3389/feart.2020.505467.

Asok Raja, S.K., Panday, D.P., and Kumar, M. (2023). Decoding the enigma of 100-year record-breaking rainfall over Tamil Nadu using wavelet analysis. *Groundwater for Sustainable Development* 101007. https://doi.org/10.1016/j.gsd.2023.101007.

Baraer, M., Mark, B.G., McKenzie, J.M. et al. (2012). Glacier recession and water resources in Peru's Cordillera Blanca. *Journal of Glaciology* 58 (207): 134–150. https://doi.org/10.3189/2012JoG11J186.

Barlow, J., Berenguer, E., Carmenta, R., and França, F. (2020). Clarifying Amazonia's burning crisis. *Global Change Biology* 26 (2): 319–321. https://doi.org/10.1111/gcb.14872.

Bell, M.L., Cifuentes, L.A., Davis, D.L. et al. (2011). Environmental health indicators and a case study of air pollution in Latin American cities. *Environmental Research* 111 (1): 57–66. https://doi.org/10.1016/j.envres.2010.10.005.

Blackhurst, B.M., Hendrickson, C., and Vidal, J.S. (2010). Direct and indirect water withdrawals for U.S. industrial sectors. *Environmental Science & Technology* 44 (6): 2126–2130. https://doi.org/10.1021/es903147k.

Bolius, D., Schwikowski, M., Jenk, T. et al. (2006). A first shallow firn-core record from Glaciar La Ollada, Cerro Mercedario, central Argentine Andes. *Annals of Glaciology* 43: 14–22. https://doi.org/10.3189/172756406781812474.

Bustamante, M.M., Martinelli, L.A., Pérez, T. et al. (2015). Nitrogen management challenges in major watersheds of South America. *Environmental Research Letters* 10 (6): 065007.

Cai, W., Borlace, S., Lengaigne, M. et al. (2014). Increasing frequency of extreme El Niño events due to greenhouse warming. *Nature Climate Change* 4 (2): 111–116. https://doi.org/10.1038/nclimate2100.

Capotondi, A. (2015). Extreme La Niña events to increase. *Nature Climate Change* 5 (2): 100–101. https://doi.org/10.1038/nclimate2509.

Coutinho, R.M., Kraenkel, R.A., and Prado, P.I. (2015). Catastrophic regime shift in water reservoirs and São Paulo water supply crisis. *PLoS ONE* 10 (9): e0138278. https://doi.org/10.1371/journal.pone.0138278.

Cox, P.M., Harris, P.P., Huntingford, C. et al. (2008). Increasing risk of Amazonian drought due to decreasing aerosol pollution. *Nature* 453 (7192): 212–215. https://doi.org/10.1038/nature06960.

Damkjaer, S. and Taylor, R. (2017). The measurement of water scarcity: defining a meaningful indicator. *Ambio* 46 (5): 513–531. https://doi.org/10.1007/s13280-017-0912-z.

Das, A., Kumar, M., Jha, P.K. et al. (2022). Isotopic and hydrogeochemical tracking of dissolved nutrient dynamics in the Brahmaputra River System: a source delineation perspective. *Chemosphere* 307: 135757. https://doi.org/10.1016/j.chemosphere.2022.135757.

Davidson, E.A., de Araújo, A.C., Artaxo, P. et al. (2012). The Amazon basin in transition. *Nature* 481 (7381): 321–328. https://doi.org/10.1038/nature10717.

Deshpande, A., Miller-Petrie, M.K., Lindstedt, P.A. et al. (2020). Mapping geographical inequalities in access to drinking water and sanitation facilities in low-income and middle-income countries, 2000–17. *The Lancet Global Health* 8 (9): e1162–e1185. https://doi.org/10.1016/S2214-109X(20)30278-3.

Domínguez, I., Ward, S., Mendoza, J. et al. (2017). End-user cost-benefit prioritization for selecting rainwater harvesting and greywater reuse in social housing. *Water* 9 (7): 516. https://doi.org/10.3390/w9070516.

Eichler, A., Gramlich, G., Kellerhals, T. et al. (2015). Pb pollution from leaded gasoline in South America in the context of a 2000-year metallurgical history. *Science Advances* 1 (2): e1400196. https://doi.org/10.1126/sciadv.1400196.

Ferreira, G.W.S. and Reboita, M.S. (2022). A new look into the South America precipitation regimes: observation and forecast. *Atmosphere* 13 (6): 873. https://doi.org/10.3390/atmos13060873.

Filho, J.P.D., Lapola, D.M., Torres, R.R., and Lemos, M.C. (2016). Socio-climatic hotspots in Brazil: how do changes driven by the new set of IPCC climatic projections affect their relevance for policy? *Climatic Change* 136 (3–4): 413–425. https://doi.org/10.1007/s10584-016-1635-z.

Graham, N.T., Hejazi, M.I., Chen, M. et al. (2020). Humans drive future water scarcity changes across all Shared Socioeconomic Pathways. *Environmental Research Letters* 15 (1): 014007. https://doi.org/10.1088/1748-9326/ab639b.

Hansen, J. (2005). Efficacy of climate forcings. *Journal of Geophysical Research* 110 (D18): D18104. https://doi.org/10.1029/2005JD005776.

Hinkel, J. (2011). Indicators of vulnerability and adaptive capacity: towards a clarification of the science–policy interface. *Global Environmental Change* 21 (1): 198–208. https://doi.org/10.1016/j.gloenvcha.2010.08.002.

Hothi, N. and Hothi, G. (2022). Water crisis of Shimla Town: past and present scenario. *Materials Today Proceedings* 64: 1250–1254. https://doi.org/10.1016/j.matpr.2022.03.715.

Imoro, Z.A., Imoro, A.Z., Duwiejuah, A.B., and Abukari, A. (2021). Harnessing indigenous technologies for sustainable management of land, water, and food resources amidst climate change. *Frontiers in Sustainable Food Systems* 5: 691603. https://doi.org/10.3389/fsufs.2021.691603.

Jones, C. and Carvalho, L.M.V. (2013). Climate change in the South American monsoon system: present climate and CMIP5 projections. *Journal of Climate* 26 (17): 6660–6678. https://doi.org/10.1175/JCLI-D-12-00412.1.

Jouravlev, A. (2014). Drinking water supply and sanitation services on the threshold of the XXI century. *Serie Recursos Naturales e Infraestructure*.

Klein Goldewijk, K., Beusen, A., and Janssen, P. (2010). Long-term dynamic modeling of global population and built-up area in a spatially explicit way: HYDE 3.1. *The Holocene* 20 (4): 565–573. https://doi.org/10.1177/0959683609356587.

Kumar, M., Panday, D.P., Bhagat, C. et al. (2023a). Demystifying the decadal shift in the extent of groundwater in the coastal aquifers of Gujarat, India: a case of reduced extent but increased magnitude of seawater intrusion. *Science of the Total Environment* 165451. https://doi.org/10.1016/j.scitotenv.2023.165451.

Kumar, M., Mazumder, P., Silori, R. et al. (2023b). Prevalence of PPCPs, microplastics and co-infecting microbes in the post-COVID-19 era and its implications on antimicrobial resistance and potential endocrine disruptive effects. *Science of the Total Environment* 166419. https://doi.org/10.1016/j.scitotenv.2023.166419.

Lee, T.R. (1995). The management of shared water resources in Latin America. *Natural Resources Journal* 35: 541–553.

Lewis, S.L., Brando, P.M., Phillips, O.L. et al. (2011). The 2010 Amazon drought. *Science* 331 (6017): 554–554. https://doi.org/10.1126/science.1200807.

Liu, X., Liu, W., Liu, L. et al. (2021). Environmental flow requirements largely reshape global surface water scarcity assessment. *Environmental Research Letters* 16 (10): 104029. https://doi.org/10.1088/1748-9326/ac27cb.

Marengo, J.A. and Espinoza, J.C. (2016). Extreme seasonal droughts and floods in Amazonia: causes, trends and impacts: extremes in Amazonia. *International Journal of Climatology* 36 (3): 1033–1050. https://doi.org/10.1002/joc.4420.

Matos, C., Pereira, S., Amorim, E.V. et al. (2014). Wastewater and greywater reuse on irrigation in centralized and decentralized systems – an integrated approach on water quality, energy consumption and CO_2 emissions. *Science of the Total Environment* 493: 463–471. https://doi.org/10.1016/j.scitotenv.2014.05.129.

Mejia, A. (2014). Water scarcity in Latin America and the Caribbean: myths and reality. In: *Water for the Americas*, 40–60. Routledge.

Mukherjee, M. and Jensen, O. (2020). Making water reuse safe: a comparative analysis of the development of regulation and technology uptake in the US and Australia. *Safety Science* 121: 5–14. https://doi.org/10.1016/j.ssci.2019.08.039.

Neog, N., Goswami, R., Panday, D.P. et al. (2024). Arsenic contamination in the groundwater of Northeastern India: critical understandings on geotectonic controls and the need for intervention. *Current Opinion in Environmental Science & Health* 100539. https://doi.org/10.1016/j.coesh.2024.100539.

Oliveira, P.T., Santos e Silva, C.M., and Lima, K.C. (2017). Climatology and trend analysis of extreme precipitation in subregions of Northeast Brazil. *Theoretical and Applied Climatology* 130 (1–2): 77–90. https://doi.org/10.1007/s00704-016-1865-z.

Panday, D.P. and Kumar, M. (2022). Climate indices and hydrological extremes: deciphering the best-fit model. *Environmental Research* 215: 114301. https://doi.org/10.1016/j.envres.2022.114301.

Panday, D.P. and Kumar, M. (2023). Application of remote sensing techniques to deal with scale aspects of GRACE data to quantify groundwater levels. *MethodsX* 102108. https://doi.org/10.1016/j.mex.2023.102108.

Panday, D.P., Khosa, R., Maheswaran, R. et al. (2021a). Game-theoretic-based modelling of Krishna waters dispute: equilibrium solutions by metagame analysis. *The European Physical Journal B* 94: 101. https://doi.org/10.1140/epjb/s10051-021-00107-w.

Panday, D.P., Khosa, R., Maheswaran, R. et al. (2021b). Game theoretic-based modelling of Krishna waters dispute: equilibrium solutions by hypergame analysis. *Eur. Phys. J. B* 94: 131. https://doi.org/10.1140/epjb/s10051-021-00135-6.

Panday, D.P., Singh, A., Gaur, S. et al. (2023). Probable maximum precipitation analysis of high rainfall regions in India. *Groundwater for Sustainable Development* 100893. https://doi.org/10.1016/j.gsd.2022.100893.

Phillips, O.L., Aragão, L.E.O.C., Lewis, S.L. et al. (2009). Drought sensitivity of the Amazon rainforest. *Science* 323 (5919): 1344–1347. https://doi.org/10.1126/science.1164033.

Qin, Y. (2021). Global competing water uses for food and energy. *Environmental Research Letters* 16 (6): 064091. https://doi.org/10.1088/1748-9326/ac06fa.

Ramírez, I.J. and Briones, F. (2017). Understanding the El Niño Costero of 2017: the definition problem and challenges of climate forecasting and disaster responses. *International Journal of Disaster Risk Science* 8 (4): 489–492. https://doi.org/10.1007/s13753-017-0151-8.

Rodríguez, C., García, B., Pinto, C. et al. (2022). Water context in Latin America and the Caribbean: distribution, regulations and prospects for water reuse and reclamation. *Water* 14 (21): 3589. https://doi.org/10.3390/w14213589.

Scherer, L. and Pfister, S. (2016). Dealing with uncertainty in water scarcity footprints. *Environmental Research Letters* 11 (5): 054008. https://doi.org/10.1088/1748-9326/11/5/054008.

Slater, D. (2019). Water scarcity in Brazil: a case study. *Expeditions with MCUP.* https://doi.org/10.36304/ExpwMCUP.2019.02

Son, R., Wang, S.-Y.S., Tseng, W.-L. et al. (2020). Climate diagnostics of the extreme floods in Peru during early 2017. *Climate Dynamics* 54 (1–2): 935–945. https://doi.org/10.1007/s00382-019-05038-y.

Vammalle, C. (2013). Sub-national Finances and Fiscal Consolidation: Walking on Thin Ice (OECD Regional Development Working Papers No. 2013/02; OECD Regional Development Working Papers, Vol. 2013/02). https://doi.org/10.1787/5k49m8cqkcf3-en.

van Vliet, M.T.H., Jones, E.R., Flörke, M. et al. (2021). Global water scarcity including surface water quality and expansions of clean water technologies. *Environmental Research Letters* 16 (2): 024020. https://doi.org/10.1088/1748-9326/abbfc3.

Wade Miller, G. (2006). Integrated concepts in water reuse: managing global water needs. *Desalination* 187 (1–3): 65–75. https://doi.org/10.1016/j.desal.2005.04.068.

Wilderer, P.A. and Schreff, D. (2000). Decentralized and centralized wastewater management: a challenge for technology developers. *Water Science and Technology* 41 (1): 1–8. https://doi.org/10.2166/wst.2000.0001.

Zavala-Araiza, D., Omara, M., Gautam, R. et al. (2021). A tale of two regions: methane emissions from oil and gas production in offshore/onshore Mexico. *Environmental Research Letters* 16 (2): 024019. https://doi.org/10.1088/1748-9326/abceeb.

5

Water Security in Australia

Madhu Sharma[1] *and Apoorva*[2]

[1] *Centre of Excellence – Energy and Eco-sustainability Research, Uttaranchal Institute of Technology, Uttaranchal University, Dehradun, Uttarakhand, India*
[2] *Sustainbaility Cluster, School of Advanced Engineering, UPES, Dehradun, Uttarakhand, India*

5.1 Water in Australia – Background

Water is essential to life and plays a significant part in practically every area of human existence. Not only is Australia the largest country in the world, but it is also the driest continent and faces a huge crisis in water availability. The country faces a major challenge with its water availability and security for the past few decades. Access to water is essential for human well-being, food security, ecological health, and economic growth (Taylor 2019).

The climate, demographics, and geography of the country put it in a state of dry conditions, intermittent and irregular rainfalls, and subsequent droughts. Due to the tremendous variability of our rainfall, Australia's freshwater supplies are extremely limited and subject to significant year-to-year variations. The ever-increasing population and economic growth of the country act as a barrier to water management. Together, these factors make water availability a severe problem and hence make water security as one of the top priorities for the people and policymakers of the country. The government and related agencies have been continuously working in devising (or implementing) plans, policies, and strategies to cope with the challenge and develop a sustainable water infrastructure.

Owing to its climatic conditions, the major sources of water in Australia are rainfall, streamflows, stored water, and groundwater. The growing population, however, increases overall water consumption and demand for agriculture, industries, and communities as well. Agriculture contributes to 70% of total water extraction in Australia. Around 21% of water consumption is urban water, which supplies our home drinking water sources (Australian Government Bureau of Meteorology 2023). Estimates and projections state that this demand for water will increase in the coming years with a very low probability of an increase in the supply or availability of water.

Multiple surveys conducted in the recent past indicate that water security is of utmost priority to government water agencies and professionals. The policy and strategies on water

Water Scarcity Management: Enabling Technologies, First Edition. Edited by Kanchan D. Bahukhandi, Manish Kumar, Durga P. Panday, and Tushara G.G. Chaminda.

management of the country refer to "water security" as one of the important policy goals. In light of the same, the government has come up with several water reforms since the 1990s till today, 2023. "Water Management" in Australia is a state responsibility and the central government's power as per the constitution does not mention water or the environment.

Australia, which comprises 5.6% of the world's landmass, has only 1% of the world's freshwater resources. Substantial variance in precipitation is observed year after year, and it is anticipated that the consequences of natural rainfall variability will be compounded by human-caused climate change. The densely populated regions of southwest and southeast Australia are undergoing a persistent dry spell, which is expected to result in a 15% decrease in cool season precipitation by 2030. On the other side, the northern regions of the country are anticipated to have more intense storms; however, it is questionable if this would result in an overall increase in precipitation or effect groundwater recharge. In light of these changes, it is anticipated that Australia's water consumption will continue to rise, despite the fact that water availability is becoming increasingly uncertain and scarce. Demographic expansion in Australia is unusually high for a developed nation, averaging 1.6% each year. By 2050, it is anticipated that the global demand for crops will increase by at least 25% and may even double (Taylor 2019).

The water resources in the country include surface water, groundwater, desalinated and recycled water, and stored water. However, the majority of the supply in the country comes from surface sources such as lakes and rivers. Except for Perth city, which relies significantly on other water sources to meet its water needs, the key water reservoirs that serve large metropolitan regions concluded 2022 at over 82% of capacity (Annual Australian Climate Statement 2022, 2023a). Water consumption in the country is mostly attributed to households, agriculture, and industries, with around three-quarters of total water used for agricultural and industrial activities. Statistics reveal that more than 70 million megaliters of water has been consumed by the industry in the last seven years with household consumption being less than 2 million megaliters.

Being a relatively dry country, Australian history depicts multiple conditions of drought and unreliable rainfall. Environmental events such as the recent forest fires are mainly responsible for such conditions (Bureau of Meteorology 2023a).

5.2 Water Availability and Resources

5.2.1 Overview of Australia's Water Resources

With an average annual rainfall of only 587.8 millimeters (mm) compared to a global average of roughly 860 mm per year, Australia is the world's driest populated continent (Bureau of Meteorology 2023b). Total water used for consumptive purposes in Australia was 14,270 GL in 2020. Australia's two-thirds are considered to be arid or semiarid, meaning they receive less than 500 mm of precipitation annually (DCCEEW 2023). It has one of the most fluctuating rainfall patterns in the world. Frequent and severe droughts are common in the nation, and they can have a big impact on how much water is available for agriculture, industry, and home use. In addition, climate change is expected to exacerbate Australia's already severe water shortages.

The area-averaged mean temperature for Australia in 2022 was 0.50 °C above the 1961–1990 average. 2022 was the 22nd hottest year on record for Australia, compared to all years since 1910, and the coldest year since 2012. Australia experienced the ninth-wettest year on record in 2022. With 587.8 mm, rainfall in Australia was 26% above the national average (compared to the 1961–1990 average of 466.0 mm). Throughout the majority of mainland Australia, annual precipitation was above average, and for the southeastern quarter of the continent, it was much above average. Western Tasmania, much of the central and eastern Top End in the Northern Territory, and the extreme southwest of Western Australia had below-average precipitation. Water storage levels were high in much of Australia in 2022 because of abundant rainfall, moist soils, and streamflows. Northern Australia and southeast Queensland, which started the year well below capacity, had considerable rises in February. Central coast Queensland, western Tasmania, southeast New South Wales, and western Victoria have low water storage (Annual Australian Climate Statement 2022, 2023b).

Flooding is Australia's second worst natural catastrophe after heatwaves. Substantial flooding occurred in southeastern Queensland in January, followed by record-breaking flooding in southeast Queensland and eastern New South Wales as a result of significant multiday rainfall in late February and March. From June to December, the Murray–Darling Basin was flooded, with substantial amounts of flood water flowing downstream into South Australia in December. Several southern Australian areas, especially in New South Wales and some in the Kimberley and west coast of Western Australia, saw high streamflows. Southwest Western Australia, the Top End of the Northern Territory, and the northeast coast of Queensland saw low streamflows (Annual Australian Climate Statement 2022, 2023c).

Australia's streamflows demonstrate a level of unpredictability that is among the highest in the world. Many locations in southern Australia had high streamflows, with New South Wales, the Kimberley, and the west coast of Western Australia being severely affected. Low streamflows were primarily noted in southwest Western Australia, the Northern Territory's Top End, and along the northeast coast of Queensland. Despite the floods, several water storages in Australia remained at or above their maximum supply capacity for most of 2022, especially in the Murray-Darling Basin (MDB). Yet, sections of central Queensland, western Victoria, western and central Tasmania, southwest Western Australia, and southeast New South Wales recorded low storage levels for the whole of the year.

Groundwater comprises around 17% of Australia's accessible water resources and accounts for more than 30% of the country's overall water usage. In 2022, groundwater storage levels varied significantly across Australia. In regions of heavy precipitation and floods, such as the Murray–Darling Basin and southeast Queensland, groundwater levels soared, with several bores above typical levels. In contrast, groundwater levels were below normal at the Victoria–South Australia boundary and in the Northern Territory's Top End (Annual Australian Climate Statement 2022, 2023b).

The country has also experienced alone some harsh droughts in the last two to three decades including the millennium drought during the period from 1997 to 2009 affecting the southernmost part of Australia which experienced extreme dryness for a long time. Recently, a drought occurred between 2017 and 2019 when rainfall in the basin of Murray-Darling was lower on average. The MDB's average rainfall was significantly less than it was during the second-driest time period, at more than 100 mm (January 1965–December 1967).

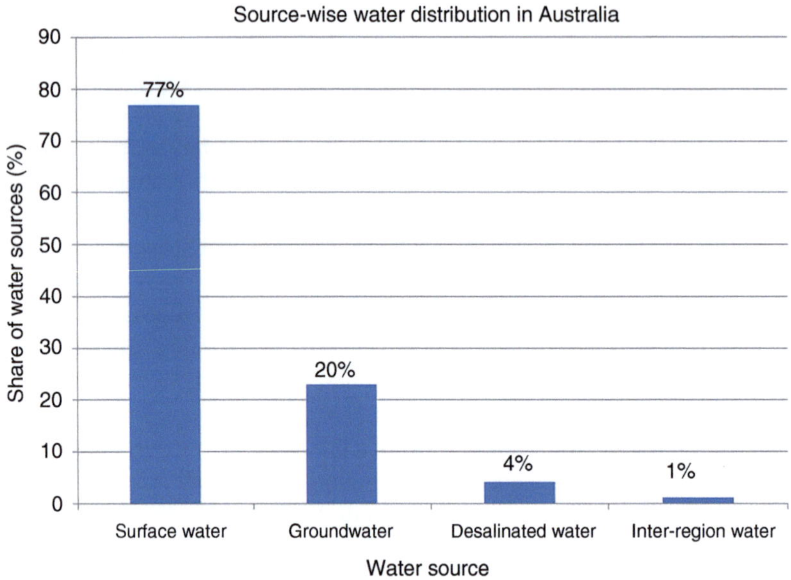

Figure 5.1 Distribution of Australia's water sources year 2020.

In comparison to the following driest season, New South Wales saw a decrease in rainfall of about 170 mm. The 36-month period from January 1900 to December 1902, known as the Federation drought, is when that was noted. This drought had a tremendous impact, reducing the amount of water available for household consumption, industry, and agriculture, as well as harming the environment and ecosystems. The drought highlighted the crucial role of water management and conservation in ensuring Australia's water resources are used sustainably(Bureau of Meteorology 2023c). Despite these difficulties in the country, it has a wide variety of water assets including surface water, groundwater, harvested rainwater, and desalinated water. Australia's greatest source of freshwater is surface water, which provides around 75% of the country's total water consumption while groundwater provides about 20%. Rainwater collection is a crucial supply of water for homes, especially in rural and remote places, while desalinated water and inter-region water encompasses about 4% of the total water distribution as shown in Figure 5.1 and is used in coastal regions with a shortage of freshwater (Bureau of Meteorology 2023d).

5.2.2 Major River System

River systems play a very important role in maintaining biodiversity and habitat; in addition to this, they also help in energy generation. Australia contains 439 rivers, according to the Geographical Names Board of New South Wales. Yet, many of these rivers are extremely small and run into bigger rivers as tributaries.

The Murray-Darling River System – The principal rivers in the Murray-Darling River Basin are the River Murray and its tributary, the Darling River. This drainage basin contains the majority of Australia's interior lowlands, including more than a million square kilometers, or around 14% of Australia's land area. Australia's longest river is the Murray with 2508 km, while the Darling is third at 1472 miles. There are over 20 other significant rivers in the Basin, including the Murrumbidgee, Goulburn, Lachlan, Macquarie, and Ovens. Before draining into the ocean, the river system travels through New South Wales, Victoria, and South Australia after rising in the Australian Mountains. Water for industry, town water supplies, and irrigation is provided by the Murray-Darling River System (Murray–Darling Basin Authority 2015a).

Southeastern Australia's second-longest river, **the Murrumbidgee**, flows into the Murray River near Peppercorn Hill. Its valley has animal meadows, vineyards, orchards, wheat farms, and cotton plantations. The river's rich floodplain produces world-class rice production. Canberra, Australia's capital city, relies on the Murrumbidgee River's economy (Murray–Darling Basin Authority 2015a).

The Lachlan River rises in the Great Dividing Mountains and flows through New South Wales before joining the Murray River at its mouth. The river has been dammed at Cowra to form the Wyangala Reservoir. Wheat growing and sheep grazing are the primary agricultural activities supported by the Lachlan River (Twinkl 2023).

The **Paroo River** begins near Augathella and empties into the Darling River. The Paroo River consists of several water holes that, during rainy weather, combine and flow together to form the river. From Queensland to New South Wales, it flows. The Paroo River Wetlands were designated as a wetland of national significance in 2007.

Cooper Creek River traverses South Australia from Queensland. The river originates from the Great Dividing Mountains and empties into Lake Eyre. Due to its unreliability as a result of unpredictable precipitation, the river is not utilized to cultivate crops (Twinkl 2023).

The Warrego River begins at the foot of Mount Ka Mundi and flows north to meet the Darling. Queensland's Warrego River flows into New South Wales, passing through the cities of Augathella, Charleville, Wyandra, and Cunnamulla (Twinkl 2023).

The Fitzroy River System, which spans an area of about 93,829 km^2, is Western Australia's biggest river system located in the west Kimberley area. At a length of 733 km, the Fitzroy River is the longest river in Western Australia. The Kimberley region's river system supplies both agriculture and industry with water. The Fitzroy River system is home to several aquatic species and environments and is an important source for irrigation, mining, and other industrial purposes (Department of Water and Environmental Regulation 2023).

The Ord River System is a network of rivers in Western Australia's Kimberley region that spans around 56,000 km^2. At a length of 650 km, the Ord River is the longest river in the Kimberley area of Western Australia. Water for agriculture is provided by the river system, especially for the Ord River Irrigation Areathe Ord River is a significant river system that supports hydropower generation. The river is home to several fish species, including barramundi (Ayre 2000).

5.2.3 Water Storage Facilities

A steady supply of water for diverse uses, including domestic, agricultural, and industrial use, depends on the presence of water storage structures like dams, reservoirs, and aquifers. Australian water storage facilities come in a variety of sizes, locations, and uses. Australia has more than two million farm dams, over 500 main storages, a few thousand small storages, and more. Extremely unpredictable precipitation and high heat levels that are common in parts of Australia require large storage. The national per capita surface water storage capacity is a comparatively high 3.25 ML, together with the overall available storage being around 81,000 GL (Australian Government Bureau of Meteorology 2023).

Tasmania has 54 major storages, mostly for the purpose of supplying water for the production of hydroelectric power. In the Australian mainland, storages are mostly concentrated in the southeast, which is also home to the majority of the country's major cities and irrigation areas. Here, direct water supply for agricultural, urban, and industrial needs is the primary usage of water (Bureau of Meteorology 2019–2020).

Over 48.8% of Australia's biggest dams' total storage capacity in water was present there at the conclusion of the fiscal year 2020 as depicted in Figure 5.2. Australia experienced a minor decrease in the overall storage capacity of major dams this year as compared to the previous year. The largest percentage of storage capacity, 84.20%, was recorded in 2012, indicating a rather plentiful supply of water. In contrast, the years with the lowest proportions were 2019 (47%) and 2020 (48.80%), indicating lesser water availability in those years.

Australia has the biggest per capita storage capacity of any country in the world and is the driest inhabited continent in the world. Due to Australia's aridity and highly variable rainfall, there are a great number and volume of water storage facilities as shown in Figure 5.2. The accessible volume is 61,813,683 ML and the accessible capacity is around 80,376,235 ML, a change of 4.7% occurred since last year 2021 as shown in Figure 5.3 (Bureau of Meteorology 2023e).

With the conclusion of the Millennium Drought in the middle of 2010, Australia's total accessible storage capacity climbed to roughly 80% by the end of 2011 and maintained at or above this level until the end of 2012 as demonstrated in Figure 5.4. Nevertheless, the

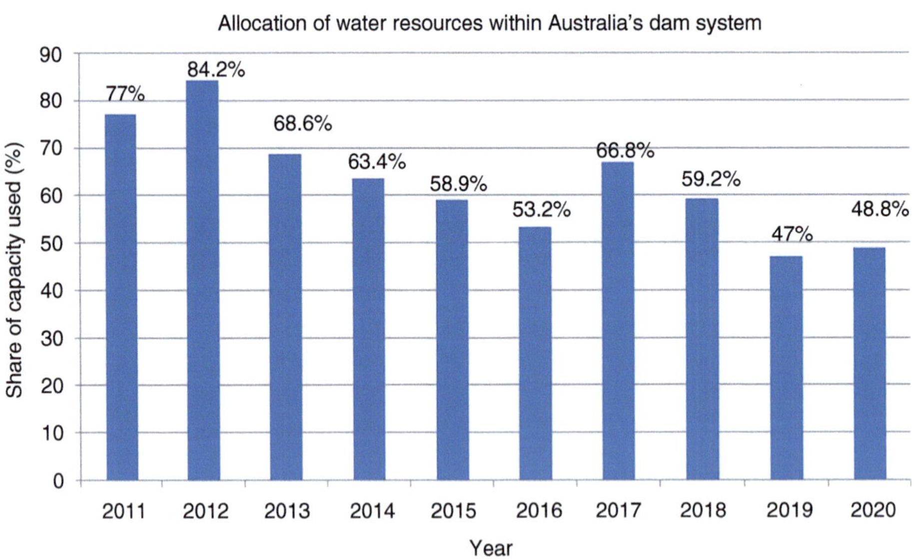

Figure 5.2 Year-wise share capacity of water stored in the dam.

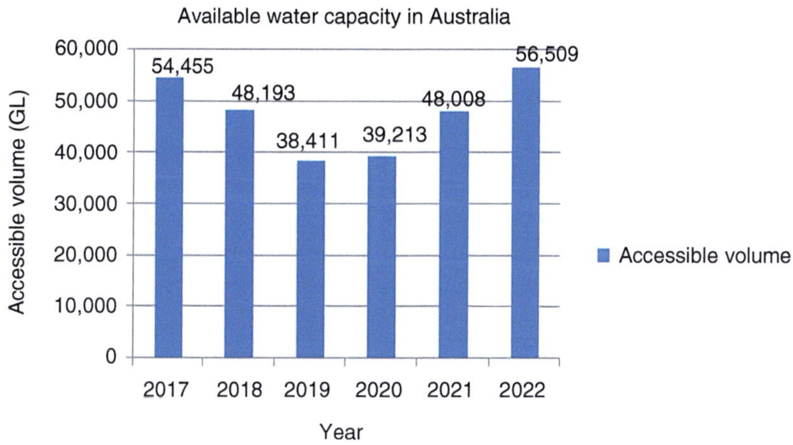

Figure 5.3 Year-wise accessible volume of water in Australia.

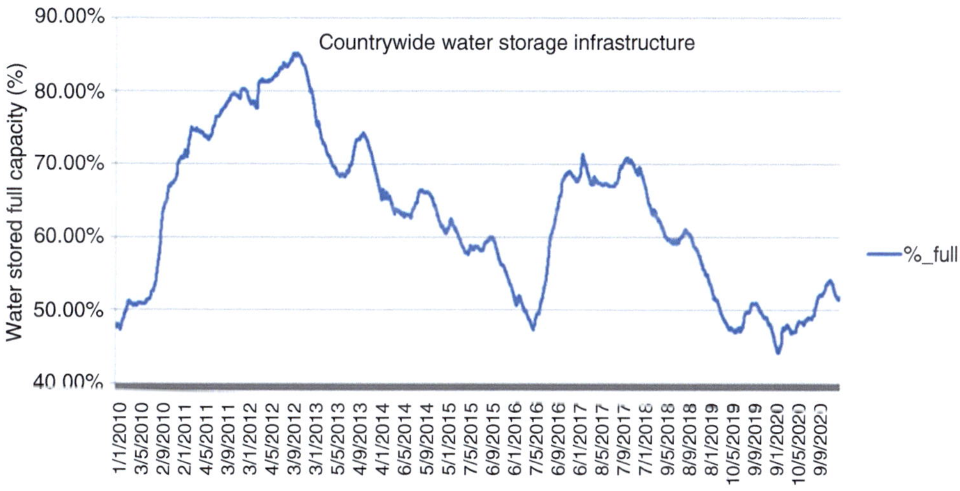

Figure 5.4 Year-wise national storage for water in Australia.

volume stored rapidly decreased, falling below 50% of capacity by March 2016, where it remained until June 2016. Early in 2017 and later in spring, rainy conditions throughout the majority of Australia led to a 70% increase in available storage capacity. Yet, as a result of continuing dry weather and ongoing diversions from storage, the available storage dropped, reaching its lowest level in more than a decade in January 2020, at around 45% of capacity. The accessible storage capacity improved considerably after February 2020, reaching 52% by the end of that year as demonstrated in Figure 5.4 due to increased rainfall in southeast Australia.

Due to the fluctuation of climate, ever-growing population, increase in building cost, environment challenges and very restricted and limited sites for water storage, it is difficult to estimate the availability of water, which can have an impact on water storage and management. As a result, the Australian government and other stakeholders are seeking new methods to manage and maximize the nation's water resources, such as water conservation, infrastructure upgrades, and more effective water management strategies.

One of the plans initiated by the government was launched in 2012, the MDB Plan intended to improve water management in the region through a variety of initiatives, such as water trading, infrastructure enhancements, and enhanced environmental management. The MDB Authority reports that over 2100 GL of water had been collected for the environment through the plan's water recovery program, and over 1500 GL of water have been returned to irrigators through the water trading program (Murray–Darling Basin Authority 2015b).

5.2.4 Groundwater

In Australia, groundwater is a significant supply of freshwater, especially in regions with insufficient or unstable surface water. Almost 23% of all water used in Australia is derived from groundwater, according to the Australian Bureau of Statistics (ABS). Groundwater is frequently the only dependable source of water in many areas of Australia, thus it requires careful planning and management to ensure its continued usage. Groundwater in Australia can be extracted by wells and bores from sedimentary rock formations such as sandstone. Climate conditions and extractions exert pressure on the sustainability of groundwater use. Recharge rates are often negligible relative to aquifer quantities. Rain and river water eventually penetrate into the ground and accumulate beneath rock fissures and between sediment particles. The groundwater system, which consists of aquifers and aquitards of varying structures and layering, is essential for managing the flow of water under the surface and maintaining the health of rivers, wetlands, and surface vegetation (Geoscience Australia 2014).

In the previous two years, there have been substantial changes in the groundwater level circumstances. In certain regions, heavy precipitation and floods refilled depleted supplies caused by extended drought conditions between 2017 and 2020. Since 2020, the proportion of studied bores with groundwater levels "above average" has climbed from 14% to 41% over the entirety of Australia, while the proportion of analyzed bores with groundwater levels "below average" has reduced from 57% to 27%. In 2022, the remaining 31% of bores recorded "average" groundwater levels (Annual Australian Climate Statement 2022, 2023d).

Australian Ground Explorer is a web-based application that allows access to groundwater data for various Australian locales. This contains thorough data including bore logs, groundwater levels, salinity levels, hydrochemistry data, 3D groundwater models, groundwater management regions, and landscape traits. Individuals can visualize and analyze groundwater-related data in their chosen area of interest by using the program. To access it, go to the Geoscience Australia website, pick "Maps & Data," then "Interactive Maps," and then click on the link labeled "Access the Australian Groundwater Explorer." On the Australian Groundwater Explorer website, you may use the search bar or the map to locate the desired place and explore the available groundwater data. In Figure 5.5, ground water level and their trend is focused which shows the measurement of groundwater in different areas of the state of Australia. The water level layer shown in the image below

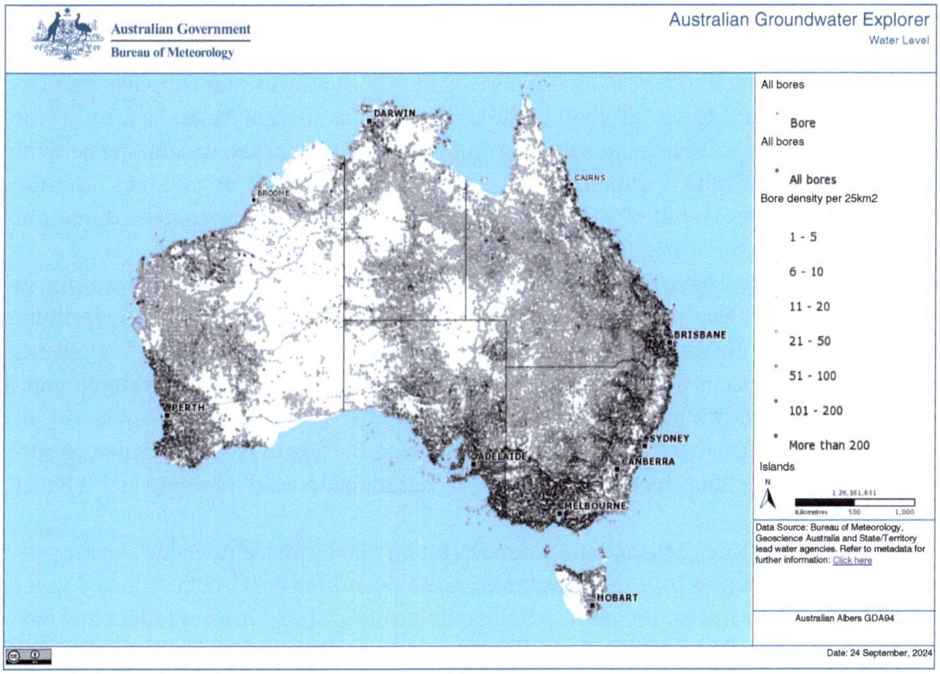

Figure 5.5 An overview of ground water level in Australia.

displays the bores containing groundwater level data, which are measured and reported as either depths or elevations. The coloration of the layer corresponds to the number of measurements recorded, ranging from dark blue (indicating more than 200 measurements) to light blue (representing fewer than 5 measurements). Most of the area in Australia lie in the water level range of more than 200 m. Some parts of Perth and Broome lie in a 5–200 m water level range where accessibility of groundwater comparatively becomes easy.

Australia's greatest groundwater system is the Great Artesian Basin. The area encompasses the majority of Queensland, the northwestern regions of New South Wales, a portion of the Northern Territory, and almost half of South Australia, making up around 20% of the country's total landmass. It is a significant source of water for springs, many of which sustain unique ecosystems, and is estimated to contain over 65,000 million megaliters of water or roughly 130,000 Sydney Harbours.

The Surat Basin in Queensland, the Perth Basin in Western Australia, and the MDB, which has both surface water and groundwater resources, are further significant groundwater resources in Australia. Moreover, the groundwater from the basin is a crucial resource for numerous mining operations, oil exploration, and in some cases, irrigation. Future water-dependent activities in the area can only move forward with the assurance of water access, knowledge of the implications on the Basin, and adherence to the management strategy (Geoscience Australia 2015).

Desalinated Water-Australia's urban areas have a huge problem because of the increase in population and deterioration in the dependability of current water resources. This is due to the deterioration of infrastructure, sedimentation of water storage facilities, climate pattern changes, and a decline in water quality. In response to this difficulty, climate-resilient options such as desalination of saltwater and recycling of water have been introduced thereby increasing the protection of municipal water supplies, particularly during dry years. Over 270 small-scale desalination facilities in Australia are used for a range of reasons to desalinate saltwater and brackish water. The annual capacity of desalination in Australia is around 880 GL. Australia's five largest urban areas have the ability to desalinate 534 GL of seawater every year (Australian Government Bureau of Meteorology 2023).

Perth's streamflows have decreased gradually across the previous 40 years, resulting in the construction of two major desalination facilities in 2006 and 2013. These facilities have a combined annual capacity of 145 GL (gigalitres per year). In 2019–2020, both plants operated at full capacity and produced approximately 140 GL of desalinated water, making desalination Perth's principal source of urban water supply. In 2019–2020, 47% of Perth's water supply came from desalination, allowing the city to reduce reliance on surface water storage and increase the use of desalination and groundwater (Water Technology 2023).

Local reservoirs, River Murray, and a desalinization plant provide Adelaide's metropolitan water supply. During 2019–2020, however, Adelaide utilized 40 GL of desalinated water under the Water for Fodder program, enabling the planting of 25,470 ha of fodder and pasture in the southern linked MDB. In Melbourne, 84 GL of desalinated water was provided to the distribution system at the end of 2018–2019, and 118 GL will be supplied in 2019–2020 as depicted in Figure 5.6.

In 2020–2021, the Victorian government has ordered 125 GL of desalination plant water. In 2019–2020, Sydney's desalination plant will provide 71 GL of desalinated water, which is equivalent to 13% of the city's urban water supply. South East Queensland produced 14 GL of desalinated water in 2019–2020, the highest amount since 2010–2011 and equivalent to 4% of the entire urban water supply (Statista 2023a).

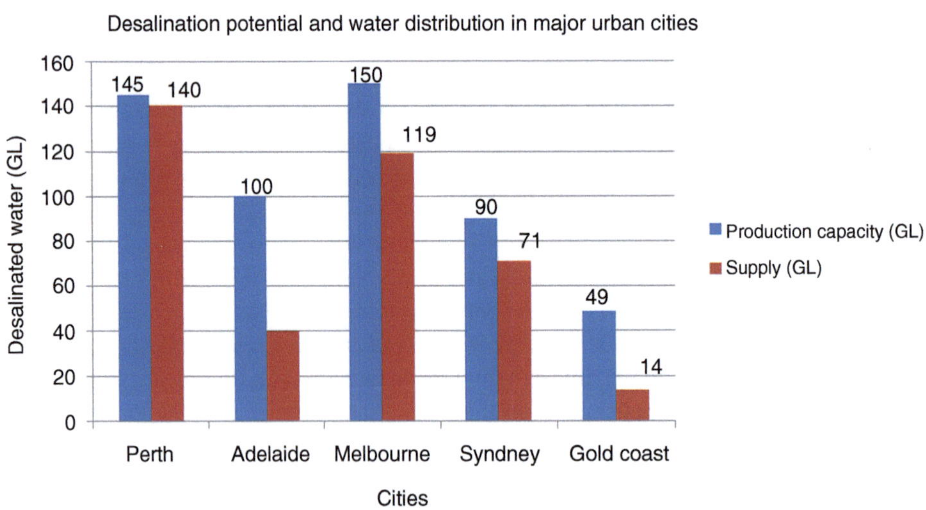

Figure 5.6 Capacity (GL) and supply (GL) of desalination in major urban cities in 2019–2020.

5.2.5 Water Recycling

Recycling of water involves processing wastewater to a quality that makes it suitable for a variety of applications, including industrial uses, irrigation of public spaces, agricultural irrigation, groundwater recharge, and non-potable domestic uses. Recycling water is an attractive solution in situations when existing sources cannot satisfy the growing demand since it assures a reliable supply throughout the year by harnessing the flow of municipal wastewater. In addition, water recycling contributes to environmental preservation by reducing the quantity of treated wastewater released into downstream rivers (Australian Government Bureau of Meteorology 2023).

The total amount of recycled water used in Australia's major urban areas was 145.91 GL, more than double the 70 GL used in 2010–2011. Recycled water accounts for 8% of the total water supply in urban areas. When new housing estates in metropolitan growth areas adopt dual pipe systems to supply recycled water, this usage will increase. When current water supplies cannot match the rising demand, water recycling is a desirable solution. It provides a steady water supply because it is continuously accessible all year long and is derived from the flow of sewage in urban areas. In addition, by reducing the amount of treated wastewater released into undeveloped bodies of water and relieving pressure on other water sources, water recycling has the potential to improve environmental results. Notably, the use of recycled water has significantly increased in Australia's largest cities, especially in Melbourne and Adelaide. Melbourne, Australia, retrieved roughly 45,500 ML of water from recycled water sources during the 2019 fiscal year. Sydney followed with around 44,000 ML of water derived from recycled water sources. Adelaide generated about 30,530 ML of water. The least amount of water was extracted from Canberra; the storage volume of Canberra has steadily declined since 2016 and is almost the lowest in these 10 years followed by Darwin as shown in Figure 5.7 (Statista 2023b).

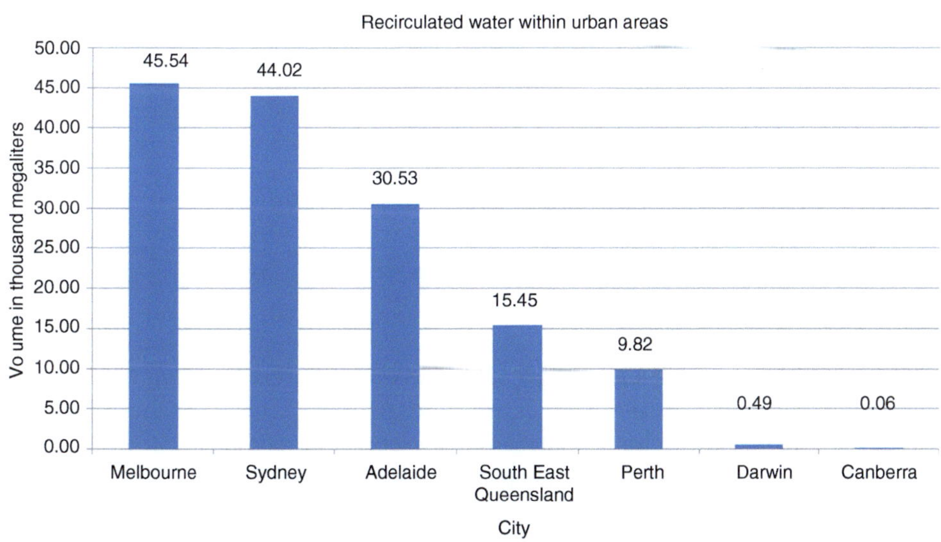

Figure 5.7 Recycled water capacity through urban areas, 2019.

Due to a significant decrease in winter rainfall in the southwest of Western Australia since the 1970s, the Shire of Augusta–Margaret River has developed safe and sustainable alternatives to meet the community's water requirements, and the Western Australian Government is investing $19 million to double the wastewater treatment plant's capacity to treat up to three million liters of wastewater per day. Treated wastewater will be utilized to irrigate a variety of ovals, parks, and other recreational facilities in the Shire, as well as an 87-ha pine plantation owned by the government (Australian Government Bureau of Meteorology 2023).

5.2.6 Water Quality

Due to the arid climate and heavily worn environment, soils, surface water, and groundwater in many parts of Australia have a high salt concentration. For instance, streams in Western Australia are by nature saltier in contrast to the increased rainfall in northern Australia and eastern Spiti which dilutes the salinity of streams in these places, resulting in lower salt concentrations. Australia's watershed water balance has been altered by dryland and irrigated agriculture, as well as by destroying native, perennial vegetation. These alterations to the topography promote the dissolution and leaching of highly soluble salts from saline aquifers or shallow water tables into streams. This unnatural rise in salinity poses a threat to the health of aquatic ecosystems and vegetation.

The control and monitoring of salt levels in the MDB and other locations are becoming more prominent in Australia. This is accomplished through the approval of Land and Water Management Plans by governing bodies, which prioritize water quality in their execution. These plans offer a structure for establishing salinity goals and targets as well as for creating practical salinity improvement strategies. These actions include ensuring enough water flows, making investments in revegetation, enhancing irrigation and dry land agricultural techniques, and implementing salt interception schemes to reduce salt transport (Annual Australian Climate Statement 2022, 2023d). In order to enhance water quality monitoring and planning, the Australian Government, in collaboration with State and Territory governments, has produced the Water Quality Australia website. The website provides resources to help in the formulation and execution of water quality plans and strategies by water managers, researchers, industry, state, and municipal governments (Annual Australian Climate Statement 2022, 2023c). The varied salinity classifications that govern the potability of water for various applications are presented in Table 5.1 (NHMRC 2011).

5.3 Emerging Technologies for Water Resource Management

Droughts have become more frequent and severe across the nation, necessitating new approaches to freshwater management. Australians have started investing in saltwater desalination technology because worries are so great. These actions are being taken to make sure there is adequate water available in the event that a severe drought depletes our reservoirs. As environmental worries about water supply increase, innovative water management strategies must be put into practice. These sophisticated systems not only operate more effectively, but they also aid in reducing water loss due to unrepaired, persistent leaks. Every year, millions of liters of water are lost due to these leaks.

Table 5.1 Water Salinity classifications and applications.

Type of salinity	Level of TDS (mg/L)	Applicability
Low salinity	0–500	Good-quality water suited for all irrigation and drinking
Slightly brackish	500–1,000	Fair to poor drinking water quality, the majority of irrigation, and potential harm to ecosystems
Moderately saline	1,000–3,000	Water of poor quality; good for most cattle; restricted irrigation to specific crops
Saline	3,000–35,000	Poor drinking water quality: usage for some cattle may be restricted
Highly saline	>35,000	Undrinkable seawater or more salinity, various mining and industrial purposes

- **Smart water metering**: Water usage can be tracked in real time with smart water meters. Based on IOT technology, they offer precise and timely data on water consumption that can aid in better understanding water usage patterns and the discovery of water-saving solutions for businesses and families. Smart water metering is being implemented in some areas of Australia to facilitate more effective management of water resources (IoT For All 2023). Precision irrigation is a technology that enables farmers to more accurately administer water to crops based on variables including crop growth stages and soil moisture levels. This can increase agricultural yields and decrease water waste. The application of temporally independent reactions appears to have been the focus of the majority of precision farming research, which was ostensibly encouraged by proponents of yield mapping and GPS/GIS technologies (Ayre 2000).
- **Desalination**: Desalination is the process of purifying freshwater by removing salt and other minerals from seawater. Desalination has emerged as a key technology in Australia, where access to water is frequently limited, to provide a consistent source of drinking water. One of the biggest desalination facilities in the world, the Perth Seawater Desalination Plant, for instance, supplies about 17% of Perth's drinking water (Wikipedia 2022).

5.4 Water Consumption

Australia's total water use for consumptive purposes is anticipated to be 13,874 GL in 2020–2021, which is 24.8% more than that in 2019–2020. 67% of the total was allocated to agricultural uses, 22% to urban uses, and 11% to industrial uses. The majority of the water was surface water, with groundwater making up the remaining 20%. The contribution from groundwater increased over the previous year as a result of the lack of surface water available for agricultural use. Due to the dry circumstances and limited surface water supply for urban usage, the use of desalination also grew from 1% to 4%. The drop in water use for agriculture in southeast Australia was ascribed to the region's persistent dry weather and limited surface water supply. The water consumption from different sources is shown clearly for better understanding of water uses in agriculture, urban, and industrial in Figure 5.8 (Bureau of Meteorology 2023d).

During the time of higher rainfall in Australia between 2020 and 2021, the agricultural sector boosted irrigation, resulting in an increase in overall water use. In 2019–2020, industrial water use climbed by 25%, from 11,120 GL to 13,874 GL, while agricultural water consumption increased by 37%, to 9978 GL as shown in Figure 5.9. In contrast, residential water use declined by less than 0.5% to 1808 GL. With a 23% increase in the main dam's storage capacity, the desalination process used 18% less seawater. According to the Bureau of Meteorology, the Australian industrial sector consumed around 72.6 million megaliters of water in fiscal year 2021, while the residential sector used just 1.8 million megaliters as discussed in Figure 5.10. Overall, compared to the prior year when the industrial and household sectors consumed about 75.46 million megaliters and 1.82 million megaliters, both industries had reduced their water uses (Bureau of Meteorology 2023e). It is crucial to remember that water consumption patterns are subject to fluctuations over time and can be

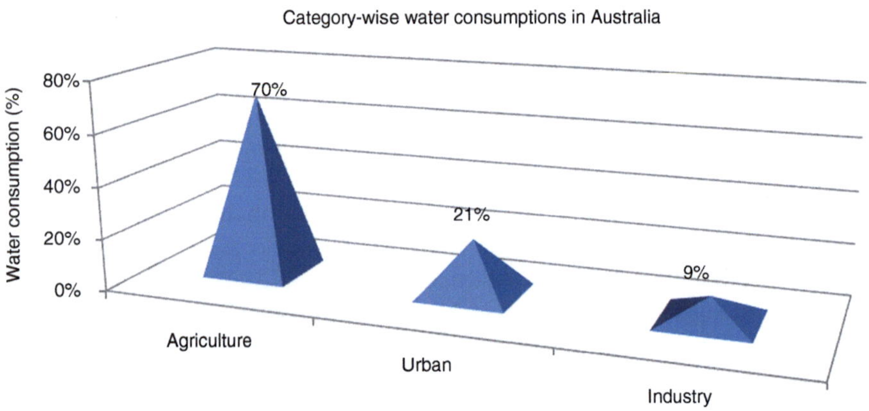

Figure 5.8 Sector-wise water consumption in 2021.

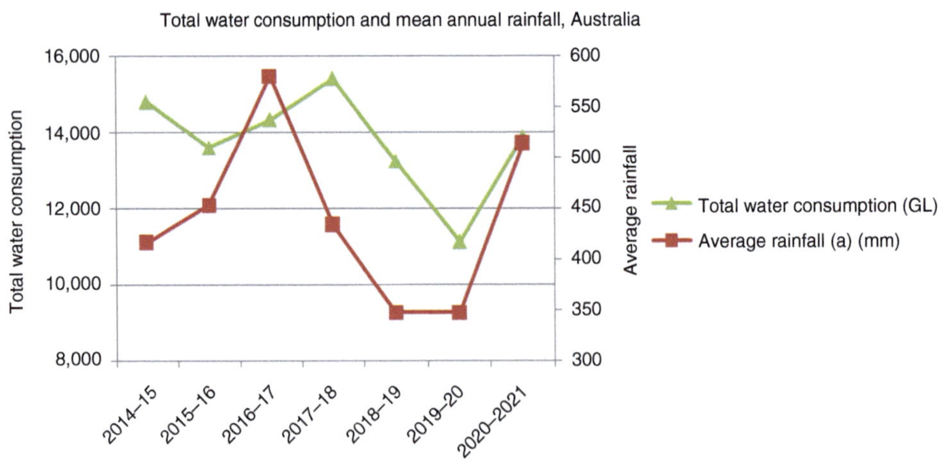

Figure 5.9 Year-wise water usage and average precipitation in Australia.

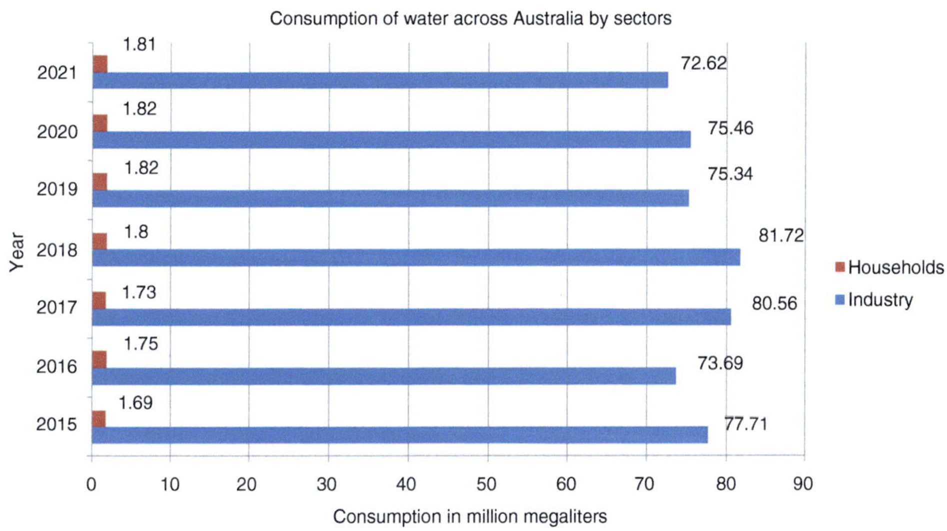

Figure 5.10 Water consumption by industry and households in Australia.

impacted by various factors, including climate conditions, governmental policies, and advancements in technology.

Due to the ongoing drought, Australia's water security has become a serious concern, as the majority of the country's water supply is derived from surface water sources like rivers and lakes. Agriculture is the main water-consuming industry, accounting for almost three-quarters of the nation's water use, along with industrial processes. Water use, however, varies greatly by area, with the Northern Territory having the greatest per-household consumption about 0.31 ML. The average home water use in other states and territories varied, with Victoria having the lowest average at 0.17 ML. In 2020–2021, 1.9 million hectares of agricultural land were irrigated, and 7.8 million megaliters of water were applied to pastures and crops. With 20,700 farms using water, the number of farms applying water was unchanged from the previous year.

There are many factors contributing to changes in water consumption as shown in the following:

1) **Climate**: Australia has a changeable climate, and variations in the weather can have a big impact on how much water is used. The total annual rainfall in Australia was around 513 mm in 2021 slightly higher than in 2020. For instance, prolonged droughts can reduce the availability of surface water, which may increase dependency on groundwater and desalinated water. Particularly in regions of New South Wales, annual flow in certain rivers in southern Australia was the lowest ever. Even in Murray-Darling basin, most of the rivers were on the lowest level of running. In contrast, wetter times can lead to more surface water being available and less groundwater and desalinated water being used. Large climate changes in Australia have already been noticed over the past 20 years, including decreases in the average winter rainfall in southern Australia and overall increases in temperature (Bureau of Meteorology 2023f).
2) **Growth in population**: Australia's population has been continuously growing throughout the years, from 18.9 million in 2003 to 25.7 million in 2021, as stated by the

ABS. The demand for water increases along with the population, which may cause variations in how much water is used. This could involve modifications to the common water-using activities in urban and industrial regions as well as adjustments to the amount of water used for these activities. The annual rainfall across Australia has decreased by roughly 16% since 1970, according to the Bureau of Meteorology of the Australian Government, which has increased pressure on water supplies. As a result, the Australian government has developed a number of laws and initiatives, such as water restrictions and water price schemes, to regulate the demand for water (Australian Bureau of Statistics 2023a).

3) **Agriculture practices**: Australia's water use is high; in 2019–2020, agriculture accounted for almost 67% of all water used in the nation. The manner in which agricultural procedures are carried out greatly affects how much water is used in this industry. The amount of water required for agriculture can be decreased, for instance, by switching to irrigation systems with greater water efficiency or altering crop selection. On the other hand, inefficient farming methods or growth in crops that require a lot of water may result in higher water usage(Department of Agriculture 2023).
4) **Investment in infrastructure**: Water resources may be more readily available and consumption patterns may alter as a result of investment in new infrastructure, such as desalination plants. For instance, multiple desalination plants were constructed in coastal communities to supplement surface water supplies during Australia's Millennium Drought, which lasted from the late 1990s to the early 2010s. Some localities now rely more on desalinated water and have altered their water usage habits as a result of this investment in new infrastructure(Australian Bureau of Statistics 2022).
5) **Policy and regulation**: Australia's water use may be significantly impacted by water policy and regulation. For instance, water trading programs enable the transfer of water entitlements between users, which may alter how water resources are distributed and how it is used. By placing restrictions on the amount of water that can be utilized for certain purposes, water allocation laws can also have an impact on water use. All sectors, including agriculture, urban, and industrial, might experience changes in water use patterns as a result of policy and regulatory changes (IoT For All 2023).

Water use has a tremendous impact on Australia's environment, economy, and society. With limited water resources and frequent droughts, managing water consumption in a sustainable and effective manner is critical. Agriculture consumes most water, and effective irrigation techniques can assist minimize water consumption. Water use in cities must also be properly monitored, and water conservation measures are required. Water consumption patterns can have an impact on the ecosystem and wildlife, and change in weather and climate patterns is projected to increase the occurrence and intensity of droughts. Hence, proper water resource management and the adoption of sustainable practices are essential for guaranteeing Australia's water supply's long-term sustainability. One of the schemes for promoting the sustainable use of water is Water Efficiency Labeling and Standards is a national program that aims to increase water efficiency in Australia by informing consumers about the water efficiency of various items. The program encompasses a variety of products, such as showerheads, toilets, and washing machines (One Planet Network 2023).

5.5 Dual Challenges

Australia is the driest inhabited continent, with an annual average rainfall of 470 mm that is unevenly distributed. The irregular and inadequate rainfall in Australia is being exacerbated by climate change, which is causing temperatures to rise by one degree Celsius compared to a century ago and precipitation to drop significantly. With the exception of Tasmania and Western Australia, every Australian state saw below-average rainfall in 2018, while Queensland recorded record-breaking evaporation rates due to high temperatures. The Millennium Drought, which lasted from 1996 to 2010, prompted a focus on water efficiency and new water sources in the southeastern and southwestern regions of the United States. Australia is making attempts to safeguard its water future notwithstanding declining supplies and increasing demand (National Geographic 2015a).

Despite Australia's vast terrain most people live in cities (urban zone), which are projected to grow by 20 million people over the next three decades, resulting in a 73% increase in water use in major cities. Australia is creating huge desalination facilities that can offer up to 674 Gt of additional freshwater to meet this need and safeguard city dwellers from population expansion and drought. However, desalination is an expensive, contentious, and energy-intensive process, earning it the nickname "bottled electricity." Even when idle, it costs A$500,000 a day to operate Sydney's plant. Yet, these plants are essential for preventing severe water shortages and securing Australia's water destiny.

In rural Australia, where rainfall and groundwater are highly relied upon, securing water supply may be a costly and complicated process, often including the shipment of emergency supplies or the construction of lengthy pipelines from remote reservoirs. Isolated rural communities are evaluating the viability of technological innovations such as hydro panels that catch moisture from the air to address this problem. Around 70% of Australia's water is consumed by agriculture; however, initiatives such as the MDB Plan have sought to promote more efficient and sustainable water management. This approach, which encompasses the region that produces the majority of Australia's food, involves acquiring water entitlements from landowners and implementing more efficient irrigation systems to avoid water loss. The strategy has successfully decreased water consumption by a third while balancing competing consumer demands across multiple states and benefiting the environment.

The universal recognition of environmental sustainability as a primary objective to administer the water is undoubtedly Australia's most significant water policy achievement. Australia has been lauded for enhancing its water security through improving public awareness and investing in infrastructure, innovation, and conservation. Nonetheless, many communities have suffered again this year. Water restrictions have been enforced in Brisbane, Sydney, and Melbourne due to inadequate inflows to Sydney's dams and the activation of the desalination plant in Queensland, New South Wales, and Victoria (National Geographic 2015b).

To ensure the long-term viability of urban water sources, Australia is actively recharging aquifers with treated wastewater. Most urban areas are increasing their use of recycled water, and houses are installing water-saving measures and collecting rainfall to relieve demand for municipal supplies. Residential wastewater systems filter and reuse greywater for non-potable purposes, while most cities limit outdoor water usage through Water Smart Measures. Melbourne has established a daily water use target of 155 liters per person, substantially less than the national average of 340 liters.

5.6 Cause of Inadequate "Water Security"

In recent decades, the term "water security" has become more prevalent (One Planet Network 2023). Although the concept of "water security" is growing in popularity, its definition and attainment remain disputed (National Geographic 2015a).

Water security is objective-driven and has the capacity to create a "vision" of what policy is attempting to accomplish. Integrated Water Resources Management (IWRM) is process-oriented in comparison. Many definitions of water security emphasize two primary objectives: maximizing water benefits and mitigating negative impacts. Various academic disciplines emphasize various facets of water security. Cook and Bakker identify different ways to water security, as detailed in Table 5.2 (Cook and Bakker 2012).

Water insecurity is the absence of water security. Water insecurity poses an ever-increasing hazard to societies. The main factors leading to inadequate water security in Australia is:

Table 5.2 Academic topic matter and emphasis on water security.

Area of interest	Focus	Key areas
Farming	Contribution to agriculture and food security	Agricultural resources and food availability
Engineering	Prevention from water-related risks	Water-related threats, such as flooding, drought, pollution, and terrorism, must be mitigated.
Sustainability science	Availability to water for human and environmental activities and services	Ensuring equitable access water services and functions for both human and environmental requirements, managing hydrological variability
Aquatic resource	Variation in groundwater hydrology	Variability of hydrologic (groundwater) conditions; safety of the entire hydrologic cycle
Community health	Water supply security and access	Supply security and availability to potable water; evaluation and prevention of water pollution in distribution networks
Anthropology, economics, geography, history, law, management, political science	Governance, inter-disciplinary connections, sustainable development	Contribution to food production and health/well-being of humans; armed conflict (driver of occupation or obstacle to collaboration and/or peace); reducing (household) susceptibility to hydrological variation; interdisciplinary (food, climate, economics, human security) sustainable development; protection against water-related dangers; protection of water systems and against floods and droughts; sustainable development of water resources to provide access to water functions and services
Water supplies	Water deficit, supply assurance	Limited water availability, water deficiency, drought circumstances, inadequate water supply, water scarcity "water security: management and governance"

Climate change: Australia is one of the nations most affected by climate change, as seen by rising temperatures, altering precipitation patterns, and a rise in the frequency of extreme weather events, such as droughts. The Bureau of Meteorology reports that 2019 was Australia's driest year on record, with the lowest precipitation levels in more than a century.

Research indicates that rising temperatures are contributing to a drop in rainfall in southern Australia due to climate change. The Intergovernmental Panel on Climate Change (IPCC) forecasts that the frequency and severity of droughts in Australia will rise in the future, placing further strain on the country's already scarce water supplies. Climate change has significantly reduced the quantity of water flow into the MDB, which feeds water to over 3 million people and irrigates over 40% of Australia's agricultural area, according to a 2018 study published in the journal *Nature*. This drop was connected to alterations in the atmospheric circulation patterns that influence precipitation in the region, according to the study. Overall, climate change is one of the primary causes of Australia's poor water security, and immediate action is needed to limit greenhouse gas emission (GHG) emission and adapt to the changing environment to ensure future generations' water supply (Bureau of Meteorology 2023g).

Population: Rapid population expansion and urbanization in Australia are putting pressure on the country's water supplies. According to the survey, it is projected that Australia's population will reach 30 million by 2031, with the majority of this growth occurring in metropolitan regions. Population growth is accompanied by an increase in water demand, which is exacerbated by climate change and drought (Australian Bureau of Statistics 2023b).

The MDB, which is Australia's largest river system and provides water to over 30% of the country's agricultural production, is one of the regions most impacted by the water shortage. The water levels in the basin's rivers and reservoirs have hit all-time lows as a result of the region's severe droughts. In addition to the overallocation of water licenses, unsustainable irrigation techniques have contributed to the basin's water crisis. The growing demand for water in metropolitan areas is also exerting pressure on water supplies. To preserve water in cities such as Melbourne and Sydney that are experiencing water shortages, water restrictions have been implemented. Population growth is also causing the extension of urban areas into water-sensitive zones, such as wetlands and watershed areas, further limiting water supply (DCCEEW 2011).

The government has adopted numerous steps to alleviate the water crisis, including investing in water infrastructure, enhancing water efficiency, and promoting water conservation. To manage water allocation in the MDB, the government has also established water trading schemes and water entitlements (Australian Government 2019). Nonetheless, these regulations have been criticized for favoring large-scale irrigators and failing to solve the water crisis adequately.

Australia may now have enough freshwater to satisfy its needs, but the inconsistency of its rainfall, the unpredictability of climate change, and the difficulties of feeding fast-increasing cities and outlying rural areas present obstacles. The expected 10–25% decline in river flows over the next decade would put a burden on Australia's water systems as the population expands. Australia must continue to adapt in order to preserve its water security.

5.7 Strategies to Strengthen Water Security (Government Acts/Policies and Regulation)

1) **National water initiative**: The ACT Government is committed to the 2004 National Water Initiative (NWI) of the Council of Australian Governments (COAG), which intends to address the overallocation of Australia's water resources and promote water usage efficiency through a unified national approach. As detailed in the 2014 National Water Reform Assessment, some NWI activities have already been implemented in the ACT under the water planning processes of the Water Resources Act 2007. The insertion of a formal water plan into the Act was a notable accomplishment, since it assisted in achieving a balance between environmental needs and human use. While these gains are acknowledged, additional work may be required on some of the remaining NWI reforms, such as acknowledging Indigenous values in water planning and enhancing research capacity (A. G. P. S. and R. C. and P. Directorate 2023).
2) **Urban water reform by Council of Australian Governments**: The ACT Government is committed to the 2004 COAG NWI, which aims to address the overallocation of Australia's water resources and promote water usage efficiency via a unified national strategy. Many of the NWI initiatives have already been implemented in the ACT through the water planning processes under the Water Resources Act 2007, as detailed in the 2014 National Water Reform Assessment. The introduction of a formal water plan into the Act has been a major accomplishment, as it helps to balance environmental needs with human consumption. Notwithstanding these gains, there may be a need for additional work on some of the remaining NWI reforms, such as incorporating Indigenous values in water planning and enhancing knowledge and research capacity (A. G. P. S. and R. C. and P. Directorate 2023).
3) **Water Act**: In order to support the NWI's implementation, the Australian government approved the Water Act in 2007. The Water Act gave the Australian government greater authority to manage water more effectively across the country and to deal with problems in the MDB. Significantly, the MDB Authority was founded by the Water Act to create a Basin Plan for the sustainable management of water throughout the MDB.
4) **Murray Darling Basin plan**: The MDB has obstacles such as excessive water allocation, extended drought, natural climate variability, and climate change, resulting in the deterioration of rivers, wetlands, forests, and floodplains. To address these concerns, the MDB Authority has produced the MDB plan, which establishes restrictions on water use that are environmentally sustainable, strives to promote effective water trading systems, and enhances water security for all users. On a Basin-wide scale, the strategy aims to strike a balance between environmental, economic, and social issues.
   The MDB's water resources must be managed in order to achieve healthy and productive rivers, ecosystems, and communities. Key to this effort is the Water Resource Plan, which outlines how water resources for a specific portion of the basin will be managed in accordance with the Basin Plan requirements. To maintain uniform and standardized management of water resources, all Basin jurisdictions, including the ACT, are required to produce Water Resource Plans (Murray–Darling Basin Authority 2015b).

5) **Water trading**: Essential to the NWI, water trading permits water entitlement holders to buy and sell water allocations. This market-based approach has contributed to more effective water allocation and the distribution of water resources to their highest-value uses. Many factors, including water demand, weather patterns, available allocations, storage quantities, legal arrangements, and commodities market circumstances, influence water trade prices. In general, dry climate conditions result in higher prices for water entitlement and allocation trades, whereas increased storage levels and carryover from the previous year result in lower prices (Australian Government Bureau of Meteorology 2023).
6) In some sections of the MDB, allocation prices reached record highs in 2019–2020 as a result of the dry circumstances that dominated the first half of the year. However, early 2020 rainfall relief resulted in a decline in allocation prices in the southern MDB. The entire amount of surface water allocation trading in Australia grew by 14% compared to the previous year, with the majority of the volume traded in the southern MDB. The volume of groundwater allocation trade grew by 11% in the MDB, while the volume of entitlement trade climbed by 13%, with entitlement trade growing in the MDB and allocation trade increasing in the remainder of Australia. The Australian Water Markets Report(BOM 2021) offers a more comprehensive evaluation of water commerce.
7) Off-farm Effective program.

The Australian Government's Off-farm Efficiency Initiative is a major program targeted at modernizing water infrastructure to reduce water losses and boost water availability for environmental, urban, and industrial uses. There are three financing streams for the program: $1.33 billion for state-led projects, $150 million for direct grants to applicants (already closed), and $60 million for on-farm initiatives (not being progressed by the NSW Government).

The program is anticipated to generate long-term advantages by enhancing water infrastructure, decreasing water losses, and increasing water reliability. In addition, the program is anticipated to generate job opportunities and prospects for infrastructural development in local areas. One of the key objectives of the initiative is to upgrade water delivery infrastructure and reduce water losses, thereby increasing the quantity of water available for environmental, urban, industrial, and irrigation uses. In addition, the program is consistent with the New South Wales government's commitment to the Murray Darling Basin Plan, which seeks to restore 450 GL of water to the environment by June 2024 (Viostream 2023).

5.8 Security vs Availability

Water availability and water security are closely linked terms, although they pertain to significantly distinct facets of Australia's water situation.

Water availability refers to the quantity of water available for human use without damaging the environment. Water resources are the natural resources of water. The quantity of water physically present in a country, whether in rivers, lakes, aquifers, or other sources. Australia is the driest inhabited continent in the world, with limited surface water supplies, a climate that is very changeable, and a history of droughts. The supply of water in Australia is crucial, especially for agricultural, urban water supplies, and environmental sustainability. On the other hand, Australia's water security refers to the ability of individuals, communities, and companies to access safe and dependable water supplies. This includes topics such as water quality, water infrastructure, and water management strategies, in addition to the physical availability of water. Protection from waterborne pollution and water-related calamities, as well as preservation of ecosystems, relies on an atmosphere of peace and political stability. Water security is a major concern in Australia, particularly in regions where water is scarce or where water resources are widely exploited.

The current status of water supply and its security is at stake. According to a report by the Australian Bureau of Meteorology, the nation is experiencing its worst severe drought in decades, resulting in widespread water shortages, especially in rural and regional areas. According to the analysis in 2018, the MDB, which generates one-third of the nation's food supply, has witnessed a 41% reduction in flow over the past two decades. Sydney and Melbourne have been the most affected areas in the decades due to reduced rainfall causing a diminished water supply. The water scarcity in the MDB has reduced agriculture production, increased water stress on ecosystems, and limited community access to water. This has emphasized Australia's need for improved water management techniques and policies. The government has implemented a number of measures, including investments in water-saving infrastructure and technologies, enhancements to water trading systems, and encouragement of water-use efficiency in agriculture. However, the implementation of these policies has been difficult due to political and economic concerns, as well as opposing stakeholder interests.

Australia is facing a major challenge in managing its water resources, and the blame for the current and future crises is often attributed to government policies. The reality is that the southern and eastern regions, as well as some coastal areas, do not have sufficient potable water to meet the needs of their growing populations. As a result, many cities and regional centers have turned to desalination plants, while inland areas have resorted to drilling deeper bores and demineralizing fossil waters. However, this has led to a range of environmental problems, including interference with river flows, salination of soils, and leaching of contaminants into waterways.

Although there are areas with plenty of water and arable land, such as the Ord in the northwest, these places are not desirable for settlement. In addition, the high evaporation rates and lack of significant mountain ranges in Australia make deep water dams an impractical storage solution. Piping water over long distances to major urban centers is also not economically feasible, and growing populations tend to be located in areas where water resources are. Australia's 13-year-long drought, which broke numerous records, forced a reevaluation of existing water management policies. In 2007, the Water Act was enacted, mandating the establishment of a Basin Plan to set environmentally acceptable levels of water extraction and eliminate the overallocation of water entitlements, which threatened water security. Due to the legislative context in which it was developed, the Act was exceptional in that it initially saw environmental considerations as an inflexible limitation on other water uses. This framework influenced future negotiations throughout the formation of the Basin Plan and provides useful lessons for drafting legislation that promotes sustainable water management in interjurisdictional river basins. In addition, it opens up new opportunities for continual improvement, such as the reciprocal benefits derived through joint management of environmental and irrigation water.

5.9 Worst Hit Areas

The worst affected areas of Australia were during the Millennium Drought which burnt parts of southern Australia from 1996 to 2010. The agricultural core of the MDB and the densely populated southeast and southwest felt its consequences strongly. Perth, Adelaide, Canberra, Sydney, Brisbane, and even Tasmania's infamously wet state capital Hobart had severe water shortages; Melbourne's water reserve fell to 33%. Since 2016, parts of northern and interior New South Wales and southern Queensland have been experiencing drought. The MDB is the largest river system in Australia and provides water to more than 30% of the country's population. Unfortunately, because of excessive water withdrawals for agriculture, industry, and domestic usage, the basin is experiencing acute water shortages. As a result of climate change, droughts are becoming increasingly frequent and severe.

Over the past two to three decades, Southeast Australia has suffered a 15% drop in late autumn and early winter precipitation and a 25% decline in average April and May precipitation. The population centers of Brisbane, Sydney, Canberra, Melbourne, and Adelaide are included in this region (Climate Council 2023). Perth is Australia's largest city situated in a semiarid environment. Perth – where major declines in water storage inflows are occurring and numerous water infrastructure projects are presently being implemented. The city's water supply relies mainly on groundwater, but aquifers are being depleted at an alarming rate. Climate change contributes to the problem by reducing precipitation and increasing evaporation rates. Delaide – where the majority of drinking water is derived from the River Murray and water restrictions are prevalent. Sydney, which has endured many droughts throughout the years; Brisbane – where a severe drought in the 2000s prompted the development of a potable recycled water plan to distribute recycled water to dams if water storage levels fall below 40% of capacity; Melbourne, where a large-scale desalination plant is being constructed despite strong public opposition, as a result of an extended drought in the 2000s. The Mallee – a region in the state of Victoria with relatively low average precipitation that experienced a significant drought in the 2000s; and Toowoomba – a rural–urban area in the state of Queensland that experienced a severe drought in the 2000s and whose residents voted against a drinkable recycled water system in a referendum.

The Northern Territory is a huge and sparsely populated territory in the center of Australia. The region undergoes extended periods of drought, and many isolated settlements rely on bore water as their only source of water. The region is also impacted by climate change, with rising temperatures resulting in increased evaporation rates and decreased precipitation. The northern Queensland region of Cape York Peninsula is a rural, sparsely populated area. The region has seasonal rainfall, and water is scarce during the dry season. Several settlements in the region obtain their water supply from rainwater tanks and bore water.

Western New South Wales is an agricultural region that has endured drought conditions for several years. The region depends significantly on the MDB for its water supply, but because the basin is experiencing severe water shortages, so is the region (Dolnicar et al. 2012). In many regions of Australia, water scarcity is a severe issue, and climate change is increasing the situation. To ensure that water is available for future generations, it is necessary that governments, communities, and individuals take action to reduce water usage and promote water efficiency.

5.10 Public's Role/Responsibility/Behavior to the Challenge

The Australian public has a key part in addressing the problem of water availability. It is essential for individuals to practice responsible behavior and take steps to save water in light of the increasing demand on water supplies. The Australian government has adopted a number of water conservation regulations and programs. In addition to boosting its water supply, Australia has made tremendous progress in decreasing its water use by introducing measures such as rating and labeling items for water efficiency, installing water-saving features in residences such as flow-regulating showerheads and dishwashers that use just 12 L of water each load, which is only 10% of the amount of water required for conventional rinsing and washing, and adopting integrated water management systems that utilize rainfall and treat non-potable greywater. In addition to the collecting and storage of rainwater by more than a fifth of households, which contributes 177 billion liters to residential water supply? Residential greywater treatment and reuse systems also lessen the demand on non-potable municipal water supplies. Most communities have optional Water Smart Measures in place to further reduce outdoor water usage, and in Melbourne, permanent legislation have resulted in a large drop in daily water use to a target of 155 L per person, which is far lower than the national average of 340 L (National Geographic 2015b).

Dolnicar and Hurlimann's "Australians' Water Conservation Behaviours and Attitudes" examines the public's role and duty in relation to water conservation in Australia. Many factors, including water shortage, media campaigns, and the perception of the efficacy of conservation measures, have influenced the water conservation behaviors and attitudes of Australians, according to the study. Here are some particular instances. In reaction to severe droughts and dwindling water resources, many cities and areas in Australia have enacted water restrictions. In Melbourne, for instance, the city's water authority implemented stage 4 restrictions, which limited outdoor watering, car washing, and swimming pool filling. In six months, the limits were successful in cutting water consumption by 36%.The study indicated that media programs advocating water conservation had a favorable effect on the attitudes and actions of Australians. In Victoria, for instance, the "Target 155" initiative encouraged inhabitants to limit their daily water consumption to 155 L per person. 78% of respondents indicated that they had taken action to conserve water as a result of the campaign's success in raising awareness and encouraging behavior modification. The study also discovered that Australians are more inclined to engage in water conservation practices if they believe their efforts would have a significant impact. For instance, households that received regular feedback on their water consumption were more willing to lower their consumption because they could see the immediate impact of their conservation efforts.

5.11 Conclusion

Australia's water security and availability remain a dual problem due to the effects of climate change and population expansion. The Australian government has developed a number of initiatives to address these issues, including investment in water infrastructure, improvement of water efficiency, and promotion of alternate water sources, such as desalination and recycling. Notwithstanding these efforts, water shortage remains a significant problem in many regions of the country, especially during droughts. During the Millennium Drought, for instance, some towns were forced to enforce stringent water restrictions, and the agriculture sector suffered enormous losses. Moreover, the recent bushfires and floods have emphasized the need for good water management in the face of extreme weather occurrences.

It is essential to continue investing in water infrastructure, promoting water efficiency, and investigating new sources of water, such as stormwater harvesting and groundwater recharge, to ensure water security and availability in the future. Also, it is essential to evaluate the effects of climate change on water resources and to adapt accordingly. To ensure water security and availability in Australia, all stakeholders, including governments, corporations, and individuals, must collaborate. By working together and taking action to protect and manage our water resources, we can ensure that all Australians will have access to sustainable water in the future.

5.12 Way Ahead

The general recognition that environmental sustainability is a key objective of water management is possibly the most significant development in Australia's water policy. Australia has been lauded for enhancing its water security due to a rise in public consciousness and investments in infrastructure, innovation, and conservation, as well as a rise in public awareness.

References

A. G. P. S. and R. C. and P. Directorate (21 February 2023). Water strategies and plans. https://www.environment.act.gov.au/water/water-strategies-and-plans (accessed 19 March 2023).

Annual Australian Climate Statement 2022 (8 February 2023a). Scheme=AGLSTERMS A. corporateName=Australian G.-B. of Meteorology. http://www.bom.gov.au/climate/current/annual/aus/#tabs=Overview (accessed 21 March 2023).

Annual Australian Climate Statement 2022 (8 February 2023b). Scheme=AGLSTERMS A. corporateName=Australian G.-B. of Meteorology. http://www.bom.gov.au/climate/current/annual/aus/#tabs=Rainfall (accessed 21 March 2023).

Annual Australian Climate Statement 2022 (8 February 2023c). Scheme=AGLSTERMS A. CorporateName=Australian G.-B. of Meteorology. http://www.bom.gov.au/climate/current/annual/aus/#tabs=Water&waterSection=Surface-water-storages (accessed 21 March 2023).

Annual Australian Climate Statement 2022 (8 February 2023d). Scheme=AGLSTERMS A. corporateName=Australian G.-B. of Meteorology. http://www.bom.gov.au/climate/current/annual/aus/#tabs=Water&waterSection=Groundwater (accessed 21 March 2023).

Australian Bureau of Statistics (20 October 2022). Water Account, Australia, 2020–21 Financial Year | Australian Bureau of Statistics. https://www.abs.gov.au/statistics/environment/environmental-management/water-account-australia/latest-release (accessed 19 March 2023).

Australian Bureau of Statistics (16 March 2023a). https://www.abs.gov.au/ (accessed 20 March 2023).

Australian Bureau of Statistics (2023b). Population Projections, Australia, 2017 (Base) – 2066 | Australian Bureau of Statistics. https://www.abs.gov.au/statistics/people/population/population-projections-australia/latest-release (accessed 19 March 2023).

Australian Government Bureau of Meteorology (2023). Water in Australia 2019–20.pdf. http://www.bom.gov.au/water/waterinaustralia/files/Water-in-Australia-2019-20.pdf (accessed 19 March 2023).

Ayre, M. (January 2000).Water Planning in the Ord River of Western Australia. https://www.academia.edu/24929353/Water_planning_in_the_Ord_River_of_Western_Australia (accessed 19 March 2023).

BOM (2021). Annual Statement 2021. http://www.bom.gov.au/climate/current/annual/aus/2021/ (accessed 19 March 2023).

Bureau of Meteorology (2019–2020). Water in Australia 2019–2020.

Bureau of Meteorology (18 March 2023a). Wikipedia. https://en.wikipedia.org/w/index.php?title=Bureau_of_Meteorology&oldid=1145321547 (accessed 20 March 2023).

Bureau of Meteorology (2023b). Australia's Official Weather Forecasts & Weather Radar – Bureau of Meteorology. http://www.bom.gov.au/ (accessed 19 March 2023).

Bureau of Meteorology (2023c). Previous Droughts. http://www.bom.gov.au/climate/drought/knowledge-centre/previous-droughts.shtml (accessed 20 March 2023).

Bureau of Meteorology (2023d). Water in Australia: Water Information: Bureau of Meteorology. http://www.bom.gov.au/water/waterinaustralia/ (accessed 19 March 2023).

Bureau of Meteorology (2023e). Water in Australia: Water Information: Bureau of Meteorology. http://www.bom.gov.au/water/waterinaustralia/#wateruse (accessed 20 March 2023).

Bureau of Meteorology (2023f). State of the Climate 2022: Bureau of Meteorology. http://www.bom.gov.au/state-of-the-climate/ (accessed 20 March 2023).

Bureau of Meteorology (2023g). State of the Climate 2022: Bureau of Meteorology. http://www.bom.gov.au/state-of-the-climate/ (accessed 19 March 2023).

Climate Council (2023). Deluge &Drought: Water Security in a Changing Climate. https://www.climatecouncil.org.au/resources/water-security-report/ (accessed 19 March 2023).

Cook, C. and Bakker, K. (2012). Water security: debating an emerging paradigm. *Glob. Environ. Change* 22 (1): 94–102. https://doi.org/10.1016/j.gloenvcha.2011.10.011.

DCCEEW (2011). The National Water Initiative – Securing Australia's.pdf. https://www.dcceew.gov.au/sites/default/files/sitecollectiondocuments/water/nwi-assessment-2011.pdf (accessed 19 March 2023).

DCCEEW (2023). Home – DCCEEW. https://www.dcceew.gov.au/ (accessed 19 March 2023).

Department of Agriculture (2023). Water and the Environme.pdf. https://www.agriculture.gov.au/sites/default/files/documents/annual-report-2020-21-awe-oct-2021.pdf (accessed 20 March 2023).

Department of Water and Environmental Regulation (2023). Home – Department of Water and Environmental Regulation. https://www.der.wa.gov.au/ (accessed 19 March 2023).

Dolnicar, S., Hurlimann, A., and Grün, B. (August 2012). Water Conservation Behavior in Australia. https://www.sciencedirect.com/science/article/pii/S0301479712001612 (accessed 19 March 2023).

Geoscience Australia (15 May 2014). c\=AU\;o\=Australia G. Australia. What is groundwater? https://www.ga.gov.au/scientific-topics/water/groundwater/basics/what-is-groundwater (accessed 20 March 2023).

Geoscience Australia (21 December 2015). c\=AU\;o\=Australia G. Australia. Navigating Australia's Largest Groundwater Resource. https://www.ga.gov.au/news-events/features/navigating-australias-largest-groundwater-resource (accessed 19 March 2023).

IoT For All (2023). Smart Meters Help Prevent Water Loss in Australia – and Everywhere Else. https://www.iotforall.com/smart-meters-help-prevent-water-loss-in-australia-and-everywhere-else (accessed 19 March 2023).

Murray–Darling Basin Authority (30 September 2015a). The Murray–Darling Basin and Why It's Important. https://www.mdba.gov.au/importance-murray-darling-basin (accessed 19 March 2023).

Murray–Darling Basin Authority (5 November 2015b). A Plan for the Murray–Darling Basin. https://www.mdba.gov.au/basin-plan/plan-murray-darling-basin (accessed 19 March 2023).

National Geographic (6 August 2015a). Failing Rains and Thirsty Cities: Australia's Growing Water Problem. Environment. https://www.nationalgeographic.com/environment/article/partner-content-australia-water-problem (accessed 20 March 2023).

National Geographic (4 August 2015b). Making Every Drop Count: How Australia is Securing Its Water Future. Environment. https://www.nationalgeographic.com/environment/article/partner-content-how-australia-is-securing-its-water-future (accessed 19 March 2023).

NHMRC (2011). Australian Drinking Water Guidelines.

One Planet Network (6 March 2023). Water Efficiency Labelling and Standards Scheme. https://www.oneplanetnetwork.org/knowledge-centre/policies/water-efficiency-labelling-and-standards-scheme (accessed 20 March 2023).

Statista (2023a). Australia –Desalinated Water Sourced by Urban Center 2019. https://www.statista.com/statistics/981966/australia-desalinated-water-sourced-by-urban-center/ (accessed 19 March 2023).

Statista (2023b). Australia – Recycled Water Sourced by Urban Center 2019. https://www.statista.com/statistics/981971/australia-recycled-water-sourced-by-urban-center/ (accessed 19 March 2023).

Taylor, K.S. (October 2019). What Does 'Water Security' Mean for Australia A r.pdf (accessed 7 February 2023).

Twinkl (2023). Australia Rivers | Seven Rivers of Australia. https://www.twinkl.co.in/teaching-wiki/the-rivers-of-australia (accessed 21 March 2023).

Viostream (2023). Off-Farm Efficiency –Video 02. https://publish.viostream.com/play/bgoo5gydmb3y94 (accessed 19 March 2023).

Water Technology (2023). Perth Seawater Desalination Plant – Water Technology. https://www.water-technology.net/projects/perth/ (accessed 20 March 2023).

Wikipedia (8 August 2022). Seawater Desalination in Australia. https://en.wikipedia.org/w/index.php?title=Seawater_desalination_in_Australia&oldid=1103013742 (accessed 19 March 2023).

6

Water Security in North America and Europe

Ganesh Datt Bhatt[1] and Deepali Rana[2]

[1] *College of Agriculture Sciences, An ICAR Accredited College, Teerthanker Mahaveer University, Moradabad, Uttar Pradesh, India*
[2] *Department of Zoology, Dolphin (PG) Institute of Biomedical & Natural Sciences, Dehradun, Uttarakhand, India*

6.1 Introduction

Managing too little, too much, or too contaminated water is the foundation of water security. Water security pertains to the growing significance of managing water resources sustainably, safeguarding human health and drinking water, and protecting property and lives from water-related calamities. Economic growth and ecosystem health are related to water security. Sustainable development requires both water security and the development of a "smart society," where water is valued for what it is and is managed to minimize pollution and maximize resource efficiency (Water Europe. WssTP 2017). The Sustainable Development Goal (SDG) of clean water and sanitation is essential to meeting Agenda 2030 (UN General Assembly 2015; Makarigakis and Jimenez-Cisneros 2019; Romano and Akhmouch 2019).

Water is fundamental to the health of our economy and its expansion, and it plays a critical role in the creation and preservation of jobs in industries that rely on it, such as construction, forestry, fishing, energy, industry, recycling, and transportation (Gleick 2018). In one way or another, water is necessary for more than three out of every four jobs (United Nations World Water Assessment Programme 2016). Raising living standards, fostering more social inclusion, and generating employment are all facilitated by sustainable water management and the provision of infrastructure for access to a clean, safe water supply and proper sanitation. Ignoring water-related problems will hurt the economy and eventually endanger lives. It is predicted that by 2050, 4 billion people will live in severely water-stressed areas, putting at risk 40% of the world's grain production and 45% of the world's gross domestic product (UN-Water n.d.) (Figure 6.1). It also significantly raises the possibility of exceedingly large expenses as a result of natural disasters and development setbacks. The cost of doing nothing in important environmental areas might be far higher than the

Water Scarcity Management: Enabling Technologies, First Edition. Edited by Kanchan D. Bahukhandi, Manish Kumar, Durga P. Panday, and Tushara G.G. Chaminda.

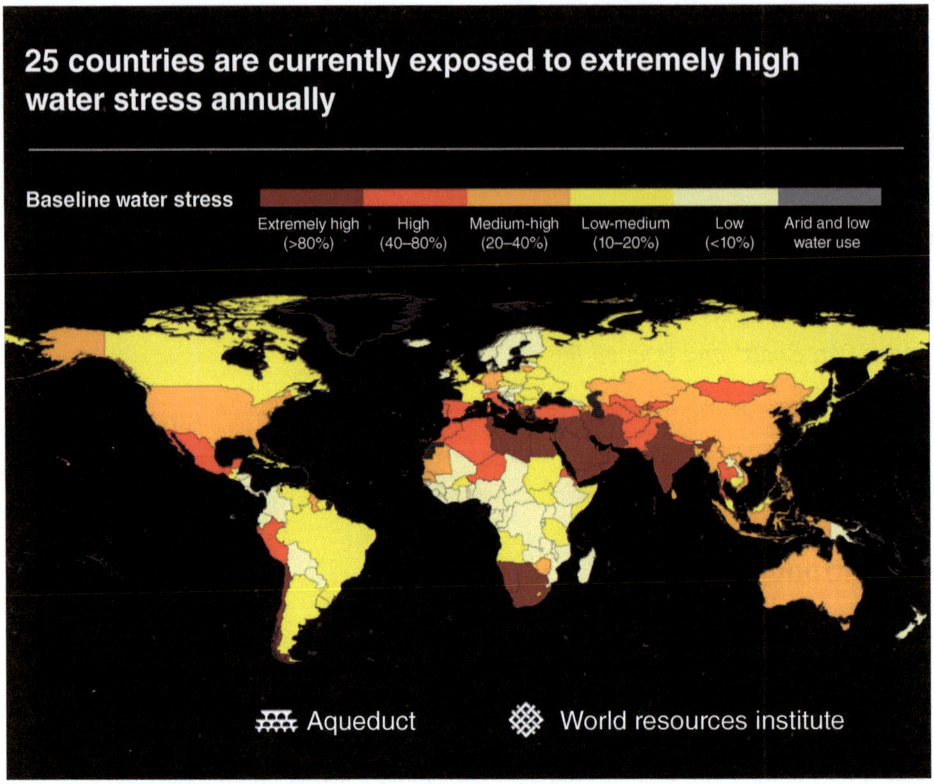

Figure 6.1 Countries exposed to extremely high water stress.

cost of investing in adaptation and mitigation strategies (OECD 2008, 2013); the consequences of water risk, for example, can be five times more than the cost of taking mitigation measures (CDP n.d.).

Many continents are experiencing significant population expansion and urbanization (UN-DESA 2019), which will put further strain on water security because these sectors depend on water for industry, agriculture, and water delivery. Currently, 17 nations account for 25% of the world's population and suffer from exceptionally high levels of water stress, using over 80% of their available resources to provide water for cities, industries, and agriculture (FAO n.d.; WRI – World Resources Institute 2021). The Middle East and North Africa are home to 12 of these nations, demonstrating the unequal distribution of these forces around the world (WRI – World Resources Institute 2021). In addition to increasing pollution and health concerns, urban expansion and industrial activity also degrade water supplies, rendering them unusable for other purposes Water-related disasters are frequently caused by droughts and altered precipitation patterns brought on by climate change (UNESCO World Water Assessment Programme 2019); in the past 20 years, water-related disasters have accounted for 90% of allnatural disasters (CRED – Centre for Research on the Epidemiology of Disasters; UNISDR – UN Office for Disaster Risk Reduction 2015).

6.2 North America

6.2.1 Droughts and Water Stress

- Certain regions, particularly the western United States and parts of Canada, face recurrent droughts, leading to water stress and scarcity.
- Overexploitation of water resources for agriculture, industry, and urbanization exacerbates these issues.

6.2.2 Groundwater Depletion

- Groundwater supplies, particularly in areas like the High Plains Aquifer (Ogallala Aquifer), are being depleted faster than they can be replenished, primarily due to agricultural irrigation practices.

6.2.3 Climate Change Impacts

- Climate change is altering precipitation patterns, leading to more extreme weather events such as floods and droughts.
- Glacier retreat in regions like the Rocky Mountains affects water availability downstream.

6.2.4 Water Quality Concerns

- Contamination of water sources from agricultural runoff, industrial discharge, and urban pollutants poses significant challenges to water quality and public health.

6.2.5 Infrastructure Challenges

- Aging water infrastructure, including dams, pipelines, and treatment plants, requires substantial investment for maintenance and upgrades to ensure reliable water supply and distribution.

6.2.6 Interstate and Transboundary Water Management

- Management of shared water resources between states and countries, such as the Colorado River Compact and the Great Lakes Compact, requires cooperation and effective governance mechanisms.

6.3 Europe

6.3.1 Droughts and Water Scarcity

- Southern Europe, including countries like Spain, Italy, and Greece, faces chronic water scarcity, exacerbated by population growth, tourism, and agricultural demands.
- Eastern European countries also experience water stress, particularly during dry summers.

6.3.2 Water Pollution

- Pollution from agricultural runoff, industrial discharges, and inadequate sewage treatment affects surface water quality and threatens ecosystems and human health.
- Nutrient pollution, including nitrates and phosphates, leads to the eutrophication of water bodies, such as the Baltic Sea.

6.3.3 Water Management and Governance

European Union directives, such as the Water Framework Directive (WFD), aim to improve water quality, promote sustainable water use, and enhance ecosystem health through integrated water management approaches.

6.3.4 River Basin Management

- Transboundary river basins, such as the Danube and Rhine, require coordinated management strategies among riparian countries to address water quantity and quality issues effectively.

6.3.5 Urban Water Challenges

- Urbanization and population growth place pressure on water resources and infrastructure, necessitating investments in water efficiency measures, leak detection, and stormwater management.

6.3.6 Climate Change Adaptation

European countries are adapting water management strategies to cope with the impacts of climate change, including more frequent and intense droughts, floods, and heatwaves.

Both North America and Europe are implementing a mix of regulatory, technological, and behavioral measures to enhance water security and ensure sustainable water management for current and future generations. Collaboration among stakeholders, including governments, communities, industries, and environmental organizations, is crucial to addressing these complex challenges effectively.

6.4 Importance of Addressing These Challenges for Sustainable Development

Addressing water security challenges is crucial for sustainable development in North America and Europe for several reasons:

6.4.1 Economic Stability

Water is fundamental to numerous economic sectors, including agriculture, manufacturing, energy production, and tourism. Ensuring water security supports stable economic growth by safeguarding these industries against disruptions caused by water scarcity, pollution, or inadequate infrastructure.

6.4.2 Public Health

Access to clean and safe water is essential for public health. Addressing water security challenges helps prevent waterborne diseases and ensures adequate sanitation, improving overall well-being and reducing healthcare costs associated with water-related illnesses.

6.4.3 Ecosystem Health

Healthy aquatic ecosystems provide essential services, such as water purification, flood control, and habitat for biodiversity. By maintaining water security, we protect these ecosystems, which in turn support resilient and productive landscapes, benefiting both humans and wildlife.

6.4.4 Climate Resilience

Sustainable water management practices, such as water conservation, groundwater recharge, and natural floodplain restoration, contribute to climate resilience by mitigating the impacts of droughts, floods, and extreme weather events associated with climate change.

6.4.5 Social Equity

Water security is closely linked to social equity and justice. Ensuring equitable access to clean water and sanitation services, particularly for marginalized communities, helps reduce disparities and promote social cohesion and inclusivity.

6.4.6 Food Security

Agriculture relies heavily on water for irrigation, livestock, and processing. Sustainable water management practices support food security by ensuring reliable water supplies for crop production and minimizing the risks of water scarcity-induced crop failures.

6.4.7 Resource Efficiency

Efficient water use and wastewater treatment technologies help conserve water resources and reduce energy consumption, thereby promoting resource efficiency and mitigating environmental impacts associated with water extraction and pollution.

6.4.8 Responsible Governance

Addressing water security challenges requires effective governance structures, policies, and regulations at local, national, and international levels. By promoting transparent and accountable water management practices, governments can foster trust and cooperation among stakeholders while promoting SDGs.

6.5 Key Factors Influencing Water Security and Availability in these Regions

6.5.1 Climate Change

Both North America and Europe are experiencing shifts in precipitation patterns, including changes in the frequency and intensity of rainfall, prolonged droughts, and increased temperatures. These changes affect water availability, quality, and distribution, impacting water security.

6.5.2 Water Management Policies

Government policies and regulations play a significant role in managing water resources. Effective water management strategies, such as sustainable water use practices, water conservation measures, and integrated water resource management (IWRM), can enhance water security and ensure equitable access to water.

6.5.3 Water Infrastructure

The condition and capacity of water infrastructure, including dams, reservoirs, pipelines, and treatment facilities, influence water availability and reliability of water supply systems. Aging infrastructure, inadequate maintenance, and insufficient investment in infrastructure upgrades can compromise water security.

6.5.4 Population Growth and Urbanization

Population growth and rapid urbanization exert pressure on water resources, leading to increased water demand for domestic, industrial, and agricultural purposes. Urbanization alters land use patterns, affects natural hydrological processes, and contributes to surface water runoff and pollution, impacting water availability and quality.

6.5.5 Agricultural Practices

Agriculture is a significant water user, accounting for a large portion of water withdrawals in both regions. Irrigation practices, crop choices, and agricultural water management techniques influence water demand and availability. Sustainable agricultural practices, such as precision irrigation and soil conservation, can mitigate water stress and enhance water security.

6.5.6 Industrial Water Use

Industrial activities, including manufacturing, energy production, and mining, require substantial amounts of water for cooling, processing, and manufacturing purposes. Efforts to improve water efficiency, recycle and reuse water, and adopt cleaner production technologies can reduce industrial water demand and minimize water-related risks.

6.5.7 Water Pollution and Contamination

Pollution from urban runoff, industrial discharges, agricultural runoff, and wastewater effluents can degrade water quality, rendering water sources unfit for human consumption and ecological health. Contamination of groundwater and surface water bodies poses challenges to water treatment and increases the cost of ensuring safe drinking water supplies.

6.5.8 Transboundary Water Issues

Many water resources in North America and Europe are shared across political boundaries, leading to complex transboundary water management challenges. Cooperation, negotiation, and conflict resolution mechanisms are essential for addressing issues related to water allocation, usage rights, and environmental protection across international borders.

6.5.9 Natural Disasters

Extreme weather events, such as floods, hurricanes, and droughts, can disrupt water supply systems, damage infrastructure, and exacerbate water scarcity and contamination issues. Preparedness, resilience, and adaptive management strategies are necessary to mitigate the impacts of natural disasters on water security.

6.5.10 Social and Economic Factors

Socioeconomic factors, including income levels, access to water services, institutional capacity, and governance effectiveness, influence water security outcomes. Socioeconomic disparities can exacerbate inequalities in water access and vulnerability to water-related risks, underscoring the importance of equitable water governance and stakeholder engagement.

6.6 Solutions to Water Security

North America and Europe both require a comprehensive and multifaceted approach to combat water security challenges involving various stakeholders and integrating innovative solutions. Here are some key solutions to enhance water security in these regions.

6.6.1 Investment in Infrastructure

Upgrade and modernize water infrastructure, including dams, reservoirs, pipelines, treatment plants, and distribution networks, to improve the efficiency, reliability, and resilience of water supply systems. Investing in green infrastructure, such as rainwater harvesting systems, permeable pavements, and constructed wetlands, can also help manage stormwater runoff and replenish groundwater resources.

6.6.2 Water Conservation and Efficiency

Promote water conservation measures and efficiency practices across sectors, including agriculture, industry, and households, to reduce water demand and minimize wastage. Implement water-saving technologies, such as drip irrigation, low-flow fixtures, and water-efficient appliances, and adopt water reuse and recycling strategies to maximize the use of available water resources.

6.6.3 Integrated Water Resource Management

Adopt IWRM approaches that consider the interconnectedness of water systems, ecosystems, and socioeconomic factors. Develop watershed-based management plans, stakeholder engagement processes, and adaptive governance mechanisms to balance competing water uses, optimize water allocation, and protect ecosystems while ensuring equitable access to water resources.

6.6.4 Climate Resilience and Adaptation

Enhance resilience to climate change impacts by implementing adaptation measures, such as floodplain restoration, drought preparedness plans, and nature-based solutions. Incorporate climate risk assessments into water management strategies, invest in climate-resilient infrastructure, and promote ecosystem-based adaptation strategies to mitigate the impacts of extreme weather events and changing hydrological patterns.

6.6.5 Water Quality Management

Implement pollution prevention measures, enforce environmental regulations, and invest in advanced water treatment technologies to safeguard water quality and protect human health and ecosystems. Monitor water quality indicators, conduct pollution source tracking, and promote best management practices to reduce pollutant loads from agriculture, industry, and urban runoff.

6.6.6 Transboundary Cooperation

Foster cooperation and dialogue among neighboring countries to collaboratively address transboundary water management challenges. Establish joint monitoring and data-sharing mechanisms, negotiate equitable water-sharing agreements, and promote cross-border cooperation on water-related issues to enhance regional stability, peace, and sustainable development.

6.6.7 Community Engagement and Capacity Building

Empower local communities, Indigenous peoples, and marginalized groups to participate in water governance processes, decision-making, and management initiatives. Provide education and training on water conservation, watershed stewardship, and sustainable water management practices to build local capacity, raise awareness, and foster community resilience.

6.6.8 Policy and Regulatory Reforms

Strengthen water governance frameworks, enhance regulatory enforcement, and align policies with SDGs, environmental objectives, and human rights principles. Promote integrated land and water planning, establish water pricing mechanisms that reflect the true value of water, and incentivize sustainable water use practices through economic instruments and incentives.

6.6.9 Research and Innovation

Support research, innovation, and technology development in water science, engineering, and management to advance knowledge, improve decision support tools, and develop innovative solutions to water security challenges. Foster collaboration between academia, industry, and government agencies to promote technology transfer, knowledge exchange, and capacity building (Das et al. 2022; Kumar et al. 2023a, b; Neog et al. 2024).

6.6.10 International Cooperation and Aid

Provide assistance, funding, and technical support to developing countries facing water security challenges, including access to safe drinking water, sanitation, and hygiene services. Collaborate with international organizations, donor agencies, and multilateral institutions to mobilize resources, share best practices, and implement sustainable water management projects globally.

By implementing these solutions in a coordinated manner, North America and Europe can enhance water security, build resilience to water-related risks, and ensure sustainable management of water resources for current and future generations.

6.7 Case Studies: Examples of Successful Water Management Initiatives and Projects in North America

6.7.1 Water Conservation and Efficiency Programs

Many municipalities across North America have implemented water conservation and efficiency programs to reduce water consumption. These initiatives include public education campaigns, rebates for water-saving appliances, and water-efficient landscaping ordinances. For example, the City of Los Angeles has successfully reduced per capita water use through programs like turf removal rebates and efficient fixture installations (Panday et al. 2021a, b; Panday and Kumar 2022, 2023).

6.7.2 Stormwater Management and Green Infrastructure

Cities like Portland, Oregon, have implemented innovative stormwater management strategies using green infrastructure. Green roofs, permeable pavements, rain gardens, and bioswales help capture and treat stormwater runoff, reducing pollution and alleviating pressure on sewer systems. Portland's "Grey to Green" initiative has transformed over 8000 acres of impervious surfaces into green infrastructure since its inception.

6.7.3 Water Recycling and Reuse

Some regions have invested in water recycling and reuse projects to augment water supplies. For instance, the Orange County Water District in California operates one of the largest water reuse facilities in the world, which purifies wastewater to produce high-quality drinking water. This project has helped reduce dependence on imported water and enhance local water resilience.

6.7.4 Desalination Plants

In response to water scarcity, several coastal communities have turned to desalination to supplement freshwater supplies. The Claude "Bud" Lewis Carlsbad Desalination Plant in California is one such example. It produces up to 50 million gallons of freshwater daily, providing a drought-proof water source for the region while reducing reliance on imported water from distant sources.

6.7.5 Integrated Water Resource Management

Some regions have adopted IWRM approaches that consider the entire water cycle and balance competing water demands. For example, the Colorado River Basin states have collaborated to develop water management plans that address the needs of agriculture, urban areas, and the environment while ensuring the sustainability of water resources.

6.7.6 Watershed Protection and Restoration

Organizations and government agencies work to protect and restore watersheds to maintain water quality and ecosystem health. The Chesapeake Bay Program, a partnership among federal, state, and local entities, has made significant progress in reducing nutrient pollution and restoring habitats in the Chesapeake Bay watershed through targeted conservation efforts and regulatory measures (Asok Raja et al. 2023; Panday et al. 2023).

6.7.7 Transboundary Cooperation

International agreements and collaborations have been successful in managing shared water resources along the U.S.-Canada border and between the U.S. and Mexico. For example, the Great Lakes Water Quality Agreement between the U.S. and Canada has led to coordinated efforts to address pollution and protect the Great Lakes ecosystem.

6.8 Examples of Successful Water Management Initiatives and Projects in Europe

6.8.1 European Water Framework Directive

The European Union's WFD, adopted in 2000, is a comprehensive legislative framework aimed at achieving good water quality across Europe's rivers, lakes, coastal waters, and groundwater. The directive establishes standards and objectives for water quality, requires member states to develop river basin management plans, and promotes integrated water management approaches. The WFD has led to improvements in water quality and ecosystem health in many European water bodies.

6.8.2 River Restoration Projects

Numerous river restoration projects have been undertaken across Europe to improve river ecosystems, enhance biodiversity, and restore natural hydrological processes. For example, the restoration of the Danube River's floodplains in Austria has helped mitigate flood risks, create habitat for wildlife, and improve water quality.

6.8.3 Water Reuse and Recycling

Several European countries have implemented water reuse and recycling initiatives to alleviate water scarcity and reduce dependence on freshwater sources. For instance, the Spanish city of Barcelona has implemented a reclaimed water system that treats wastewater to high standards and uses it for non-potable purposes such as irrigation, industrial processes, and street cleaning.

6.8.4 Integrated Coastal Zone Management (ICZM)

Coastal areas face unique water management challenges due to factors such as erosion, pollution, and climate change. The European Union promotes ICZM approaches to balance economic development with environmental protection in coastal regions. Projects like the Wadden Sea Conservation Area in Germany, Denmark, and the Netherlands demonstrate successful collaboration and management of coastal ecosystems.

6.8.5 Flood Risk Management

Europe has experienced significant flooding events in recent years, prompting investments in flood risk management infrastructure and strategies. The Dutch Delta Works, a system of dams, dikes, and storm surge barriers, is a renowned example of successful flood protection measures. In addition, initiatives like the Room for the River program in the Netherlands prioritize floodplain restoration and natural water retention measures.

6.8.6 Water Efficiency and Conservation Programs

Many European countries have implemented water efficiency and conservation programs to reduce water consumption and minimize waste. These initiatives include public awareness campaigns, incentives for water-saving technologies, and regulations promoting water-efficient practices in agriculture, industry, and households.

6.8.7 Transboundary Water Cooperation

European countries collaborate on transboundary river basins to manage shared water resources effectively. For example, the International Commission for the Protection of the

Danube River (ICPDR) facilitates cooperation among Danube Basin countries to address water quality, navigation, and ecological concerns.

6.9 Future Directions and Recommendations

Ensuring water security and availability in North America and Europe requires proactive strategies that address current challenges and anticipate future needs. Here are some future directions and recommendations for water management in these regions:

6.9.1 Climate Resilience Planning

Both North America and Europe should prioritize climate resilience planning to mitigate the impacts of climate change on water resources. This includes investing in infrastructure that can withstand extreme weather events, implementing adaptive water management strategies, and integrating climate projections into long-term water planning efforts.

6.9.2 Promotion of Water Conservation and Efficiency

Continued promotion of water conservation and efficiency measures is essential to reduce water demand and minimize waste. Governments, utilities, and businesses should incentivize water-saving technologies and practices in agriculture, industry, and urban areas. Public education campaigns can also raise awareness about the importance of water conservation.

6.9.3 Investment in Infrastructure Upgrades

Aging water infrastructure in both regions requires significant investment to ensure reliability and resilience. Governments should prioritize funding for infrastructure upgrades, including repairs to leaking pipes, modernization of treatment plants, and expansion of water storage and distribution systems.

6.9.4 Integrated Water Management Approaches

Embracing integrated water management approaches that consider the entire water cycle, from source to tap to treatment and back to the environment, can optimize water use and minimize conflicts between competing water users. Collaboration among stakeholders, including governments, communities, industries, and environmental organizations, is essential for implementing integrated water management plans effectively.

6.9.5 Enhanced Water Quality Protection

Measures to protect and improve water quality should be strengthened to safeguard public health and ecosystem integrity. This includes reducing pollution from agricultural runoff, industrial discharge, and urban stormwater, as well as controlling emerging contaminants like pharmaceuticals and microplastics. Monitoring programs should be expanded to track water quality trends and identify emerging threats.

6.9.6 Investment in Research and Innovation

Continued investment in research and innovation is critical for developing new technologies and approaches to address water challenges. Research areas of focus may include water reuse and recycling, advanced water treatment methods, sustainable agriculture practices, and climate-resilient infrastructure design. Governments, academia, and the private sector should collaborate to advance scientific knowledge and translate research findings into practical solutions.

6.9.7 Transboundary Cooperation and Governance

Given the interconnected nature of water resources, transboundary cooperation, and governance mechanisms are essential for managing shared water bodies effectively. North American and European countries should strengthen existing cooperation frameworks and develop new agreements as needed to collaboratively address transboundary water issues.

6.9.8 Community Engagement and Equity

Meaningful engagement with local communities, including Indigenous peoples and marginalized groups, is essential for developing water management policies and projects that are equitable and socially responsible. Governments and water utilities should prioritize community involvement in decision-making processes and ensure access to clean and affordable water for all residents.

6.9.9 Preparedness for Water Scarcity and Drought

With increasing water scarcity and drought risks, proactive planning and preparedness are essential to minimize impacts on water users and ecosystems. This includes developing drought contingency plans, implementing water rationing measures, promoting drought-resistant crop varieties, and diversifying water supplies through measures like rainwater harvesting and desalination.

6.9.10 Policy Alignment and Harmonization

Harmonizing water management policies and regulations across jurisdictions can facilitate more effective and coordinated responses to water challenges. Governments at all levels should work together to align policies related to water quality standards, water rights, land use planning, and environmental protection to ensure consistency and coherence in water management efforts.

By adopting these future directions and recommendations, North America and Europe can enhance water security, promote sustainable water management practices, and ensure the availability of clean and reliable water for current and future generations.

References

Asok Raja, S.K., Panday, D.P., and Kumar, M. (2023). Decoding the enigma of 100-year record-breaking rainfall over Tamil Nadu using wavelet analysis. *Groundwater for Sustainable Development* 101007. https://doi.org/10.1016/j.gsd.2023.101007.

CDP *CDP Global Water Report 2020. A Wave of Change: The Role of Companies in Building a Water-Secure World*, vol. 2020, 39. London: CDP Worldwide.

CRED – Centre for Research on the Epidemiology of Disasters; UNISDR – UN Office for Disaster Risk Reduction (2015). *The Human Cost of Weather-Related Disasters 1995–2015*. Geneva: UNISDR.

Das, A., Kumar, M., Jha, P.K. et al. (2022). Isotopic and hydrogeochemical tracking of dissolved nutrient dynamics in the Brahmaputra River System: a source delineation perspective. *Chemosphere* 307: 135757. https://doi.org/10.1016/j.chemosphere.2022.135757.

European Commission (2019). *Blueprint to Safeguard Europe's Water Resources*. European Commission.

European Commission (2020). *European Green Deal: Communication from the Commission to the European Parliament*. European Commission.

European Environment Agency (2018). European waters – Assessment of status and pressures.

FAO *The State of Food and Agriculture 2020. Overcoming Water Challenges in Agriculture*, 2020. Rome: FAO.

Gleick, P.H. (2018). Water security in a changing world: assessment, indicators, and implications for policy. *Water Security* 1: 1–8.

Global Water Partnership (2018). *Integrated Urban Water Management: Humid Tropics*. Global Water Partnership.

Kumar, M., Panday, D.P., Bhagat, C. et al. (2023a). Demystifying the decadal shift in the extent of groundwater in the coastal aquifers of Gujarat, India: a case of reduced extent but increased magnitude of seawater intrusion. *Science of the Total Environment* 165451. https://doi.org/10.1016/j.scitotenv.2023.165451.

Kumar, M., Mazumder, P., Silori, R. et al. (2023b). Prevalence of PPCPs, microplastics and co-infecting microbes in the post-COVID-19 era and its implications on antimicrobial resistance and potential endocrine disruptive effects. *Science of the Total Environment* 166419. https://doi.org/10.1016/j.scitotenv.2023.166419.

Makarigakis, A.K. and Jimenez-Cisneros, B.E. (2019). UNESCO's contribution to face global water challenges. *Water* 11: 388.

Neog, N., Goswami, R., Panday, D.P. et al. (2024). Arsenic contamination in the groundwater of Northeastern India: critical understandings on geotectonic controls and the need for intervention. *Current Opinion in Environmental Science & Health* 100539. https://doi.org/10.1016/j.coesh.2024.100539.

OECD (2008). *Costs of Inaction on Key Environmental Challenges*. OECD: Paris.

OECD (2013). *Water Security for Better Lives*, vol. 12. OECD: Paris.

Organisation for Economic Co-operation and Development (OECD) (2019). *OECD Environmental Outlook to 2050: The Consequences of Inaction*. Organisation for Economic Co-operation and Development (OECD).

Panday, D.P. and Kumar, M. (2022). Climate indices and hydrological extremes: deciphering the best-fit model. *Environmental Research* 215: 114301. https://doi.org/10.1016/j.envres.2022.114301.

Panday, D.P. and Kumar, M. (2023). Application of remote sensing techniques to deal with scale aspects of GRACE data to quantify groundwater levels. *MethodsX* 102108. https://doi.org/10.1016/j.mex.2023.102108.

Panday, D.P., Khosa, R., Maheswaran, R. et al. (2021a). Game-theoretic-based modelling of Krishna waters dispute: equilibrium solutions by Metagame Analysis. *The European Physical Journal B* 94: 101. https://doi.org/10.1140/epjb/s10051-021-00107-w.

Panday, D.P., Khosa, R., Maheswaran, R. et al. (2021b). Game theoretic-based modelling of Krishna waters dispute: equilibrium solutions by hypergame analysis. *The European Physical Journal B* 94: 131. https://doi.org/10.1140/epjb/s10051-021-00135-6.

Panday, D.P., Singh, A., Gaur, S. et al. (2023). Probable maximum precipitation analysis of high rainfall regions in India. *Groundwater for Sustainable Development* 100893. https://doi.org/10.1016/j.gsd.2022.100893.

Romano, O. and Akhmouch, A. (2019). Water governance in cities: current trends and future challenges. *Water* 11: 500.

U.S. Environmental Protection Agency (2020). *Clean Watersheds Needs Survey 2012: Report to Congress*. U.S. Environmental Protection Agency.

UN General Assembly (2015). *Transforming Our World: The 2030 Agenda for Sustainable Development*. New York: UN.

UN-DESA (2019). *World Population Prospects 2019: Highlights*. New York: United Nations Department of Economic and Social Affairs—Population Division.

UNESCO World Water Assessment Programme (2019). *The United Nations World Water Development Report 2019: Leaving No One Behind*, 186. UNESCO: Paris ISBN 9789231003097.

United Nations (2019). *World Population Prospects 2019: Highlights*. United Nations.

United Nations Water (2019). *World Water Development Report 2019: Leaving No One Behind*. United Nations Water.

United Nations World Water Assessment Programme (2016). *The United Nations World Water Development Report 2016: Water and Jobs*. UN-WWAP: Paris.

UN-Water *SDG 6 Synthesis Report 2018 on Water and Sanitation*, vol. 2018, 195. New York: United Nations.

Water Europe. WssTP (2017). Water Vision 2030. The Value of Water. Multiple Waters, for Multiple Purposes and Users-Towards a Future Proof Model for a European Water-Smart Society; WssTP-The European Technology Platform for Water, Brussels, Belgium, p. 32.

World Bank (2017). *High and Dry: Climate Change, Water, and the Economy*. World Bank.

WRI – World Resources Institute (2021). 17 countries, home to one-quarter of the world's population, face extremely high water stress. https://www.wri.org/insights/17-countries-home-one-quarter-worlds-population-face-extremely-high-water-stress (accessed 2 November 2021).

7

Indigenous Technology and Modern Practices in Asia

Aanchal Kumari[1], *Kanchan D. Bahukhandi*[1], *and Manish Kumar*[1,2,3]

[1] *Sustainability Cluster, School of Advanced Engineering, UPES, Dehradun, Uttarakhand, India*
[2] *Escuela de Ingeniería y Ciencias, Tecnológico de Monterrey, Monterrey, Nuevo León, Mexico*
[3] *Research and Development Initiative, Chuo University, Korakuen Campus 1-13-27, Kasuga, Bunkyo-ku, Tokyo, Japan*

7.1 Introduction

The specific conditions of women and men indigenous to a particular geographic area are deeply intertwined with unique, traditional, local knowledge (Grenier 1998). According to Gorjestani (2004), indigenous knowledge serves as "the foundation for community-level decision-making in areas pertaining to food security, natural resource management, health, education, and other crucial socioeconomic activities." It is a crucial component of the poor's social capital and serves as their principal resource in their fight for personal autonomy (Gorjestani 2004). Therefore, indigenous knowledge is an accumulated collection of practices, expressions, and "know how" that has been upheld and developed by population who have had extensive connection with their natural surroundings. For many people, especially the impoverished, it serves as the foundation for local decision-making and offers community-based approaches to solving problems. Indigenous knowledge contributes to the sustainability of development projects because its inclusion fosters mutual learning and adaptation, which ultimately helps local communities gain more power (Hiwasaki et al. 2014). Utilizing indigenous knowledge makes sense economically since efficiency, efficacy, and sustainability are important factors in determining the quality of development efforts. Early signs suggest that if indigenous knowledge is given leverage in the form of modem technology, the quality of development initiatives will significantly improve. There shouldn't be any misunderstanding between the phrase "knowledge" and "information" or "technology" and "practice." Information can just be a collection of data about a single scenario or experiment, but knowledge is linked to cumulative collective experience (Baul and McDonald 2014). To create beneficial technology, knowledge and information are required. Over time, people have created novel technologies and methods that are based on native knowledge. Even now, this still takes place (Baul and McDonald 2015).

Water Scarcity Management: Enabling Technologies, First Edition. Edited by Kanchan D. Bahukhandi, Manish Kumar, Durga P. Panday, and Tushara G.G. Chaminda.

Indigenous knowledge is a never-ending source of creative concepts that are transformed into technologies and practices (Toussaint et al. 2005). When employed in other situations or at different eras, the technologies and practices established might not be as useful. Even if indigenous knowledge has stood the test of time, there are some situations in which scientific verification may be needed before a practice based on indigenous knowledge is spread outside of its original context and place. Thoughtful solutions to a society's challenges should be developed with the aid of indigenous knowledge (Ayre and Mackenzie 2013). Existing contextual arrangements might not be suitable for the particulars of indigenous knowledge that have been adapted, necessitating such creative solutions. Indigenous knowledge typically develops gradually rather than in sharp steps and is transmitted by oral tradition and demonstration (Gautam 2014). Human settlement sustainability is unquestionably largely dependent on water. Thousands of people worldwide migrate every year as a result of the lack of water. The prolonged water crisis has put considerable strain on both ecological and human rights concerns. By 2025, according to Arnell's (2004) prediction, two-thirds of the global population will experience problems with the availability of water. In addition, water scarcity is increasingly posing a threat to culture in many regions of the world. FAO (2012) reports that a lack of agricultural water supplies will put a strain on agricultural production. Both mainstream and indigenous people have been harmed by the lack of water and continuous depletion, but the consequences on indigenous agricultural villages are more severe than on other communities. The United Nations Declaration on the Rights of Indigenous Peoples emphasized the importance of giving these populations' water rights special consideration; nevertheless, there aren't any sufficient interventions or policies in place anywhere in the globe (Warren 1991). Indigenous knowledge has many benefits for managing water resources. Indigenous practices are beneficial, but there hasn't been much research on the subject up until now. According to Oweis (2003), current development practices have increased the pressure on the traditional usage of water for subsistence (Saxena et al. 2021). According to numerous studies, indigenous communities' water management practices are generally regarded as sustainable.

In addition, indigenous practices are regarded as cost- and energy-effective strategies. Due to the large number of indigenous communities who call Nepal home, the country is home to many different and distinctive cultural styles (Gautam 2014). Among these distinctive cultural elements, Nepal's medieval civilization gave rise to the water management system of the Kathmandu Valley (Becker-Ritterspach 1990). Even though the first stone spouts appeared in Nepal in the sixth century as "Pranali," the majority of the current stone spouts were built during the Middle Ages. Later, these stone spouts were known as "Kirti," and in the Middle Ages, the term "Hiti" was created to refer to the same stone spouts (Sahoo 2020). came to the same conclusion regarding the royal canals, also known as "Rajkulo." The Rio+20 Conference's conclusions in 2012 are when the Sustainable Development Goals (SDGs) concept first appeared. The SDGs, which consist of 17 global objectives and 169 targets that must be met by 2030, were agreed by the 193 world leaders gathered at the UN General Assembly in September 2015 following three years of talks. The UN said that the 2030 Agenda was the most inclusive document it had ever created (Pun 2001). Indigenous population have been incorporated into the political declaration of the SDGs and the follow-up and review section that calls for their participation as a result of their active participation in the 2030 Agenda process. Two SDGs specifically mention indigenous population as part of their aim by 2030. The first is Goal 2 section 2.3, which focuses on increasing the agricultural productivity and income of small-scale producers, particularly indigenous population (Nair 2016) and other marginalized groups, including through secure and equal access to land, other productive resources and inputs, knowledge, financial services, markets, and opportunities for value addition and nonfarm employment (Shiva 2016).

In general, the second goal seeks to eradicate hunger, ensure food security, enhance nutrition, and advance sustainable agriculture. The second is Goal 4 section 4.5, which calls for eradicating gender gaps in education and ensuring that those who are most vulnerable, such as individuals with disabilities, indigenous population, and children in vulnerable situations, have equal access to all levels of education and vocational training (Das et al. 2022; Kumar et al. 2023a, b). The fourth SDG aspires to guarantee inclusive and equitable quality education and advance opportunities for lifelong learning for everyone. Through baseline and benchmarking studies, the governments of Bangladesh, India, Indonesia, Japan, Malaysia, Nepal, Philippines, and Thailand adopted the SDG framework and started its implementation at the national level (Rose 2004). The UN SDG goal-setting process is an innovative approach because it gives member states a great deal of implementation latitude. In order to prioritize national goals, targets, and strategies within the framework of global goals, the government's role is crucial (Murthy et al. 2022). Indigenous communities protect Asia's largest surviving natural resource (Rerkasem et al. 2009; Neog et al. 2024). Hence, the idea of incorporating indigenous knowledge systems into local and regional policies needs to be rethought. The experiences of various countries, like Australia, Canada, Mexico, and many others, can be used to investigate some forms of indigenous knowledge integration in environmental governance (Levidow et al. 2014). In addition, the International Panel on Climate Change (IPCC) even requests the inclusion of indigenous knowledge systems in international reports, underlining their significance for research, public policy, and international politics (Codjoe et al. 2014). As national governments and the global community require knowledge production, policy analysis, and expert assessments, academic researchers play a crucial role in defining the research agenda (Asok Raja et al. 2023). Due to these factors, collaborations between indigenous populations' organizations, university researchers, and national governments are necessary in order to execute and achieve the goals of the 2030 Agenda (Fadhili et al. 2002).

Table 7.1 highlights native techniques for water conservation and gives an overview of traditional water management systems found in different parts of Asia. In India, methods like step wells (Jhalaras and Talabs in Rajasthan), rainwater storage systems (Johads and Taankas) and Virda wells in Gujarat are widely used. The Khadin embankments and Nadis in Rajasthan are particularly notable for their ability to collect and retain rainfall. Ahar Pynes pipes in Bihar facilitate irrigation, and Panam Keni wells in Kerala store water by means of decomposing palm stems. Maharashtra manages surface and groundwater using Bhandara Phad and the Ramtek Model, among other approaches. Ladakh and Himachal Pradesh rely on glacier-fed systems such as Zing and Kuhls; in the northeast, the bamboo drip irrigation of Meghalaya and the Zabo of Nagaland are noteworthy. Other instances are the Eris tanks in Tamil Nadu, the Jackwells in the Great Nicobar Islands, and the Katta in Karnataka. The Qanat method in West Asia and North Africa sustainably draws water from

Table 7.1 Some of the traditional methods of water conservation in different regions of Asia.

State	Traditional water management system	Traditional technology for water conservation	References
India			
Gujarat	Virda	Virdas: nomadic Maldharis shallow wells in Banni grasslands for water supply.	Gautam et al. (2018), Bishnoi et al. (2017)
Rajasthan	Jhalara	Rectangular step wells with tiered steps, collecting water for religious, royal, and community use.	Khan and Narain (2005), Saxena (2017a, b), Yan et al. (2020), Sharma et al. (2023), Hussain et al. (2014)
	Talab/Bandhi	Talabs store water for household use, while a medium-sized lake is called a bandhi.	
	Bawari	Reservoirs with layered steps conserve water and recharge aquifers through percolation.	
	Taanka	Taankas: underground pits that store rainwater, ensuring water security in arid regions and reducing daily water-fetching burdens.	
	Johads	Johads, ancient earthen dams capturing rainwater, conserving groundwater, and connecting to rivers or streams.	
	Khadin	Khadin/dhora: hillside embankment with sluices to drain excess water, enabling crop production on water-saturated land.	
	Nadi	Nadis: village ponds storing rainwater from catchment areas, prone to siltation due to sandy sediments.	
Bihar	Ahar Pynes	Ahars: reservoirs with three-sided embankments, fed by pynes (artificial rivulets) for dry season irrigation and paddy cultivation.	Koul et al. (2012)
Kerela	Panam keni	Kuruma tribe's panam keni: wells made from rotting toddy palm stems, storing water in groundwater springs.	Sojka et al. (2016), Yan et al. (2020)
Maharashtra	Bhandara Phad	Bhandhara, kalvas, Sandams, and charis form a system of check dams, canals, and outlets for water distribution in fields.	Yan et al. (2020), Belsare (2017)
	Ramtek Model	Malguzar-led system interconnected tanks and canals for groundwater and surface water management.	
Ladakh	Zing	Can bring the melting glacier water into the tanks.	Nüsser et al. (2019), Pani et al. (2021)

Table 7.1 (Continued)

State	Traditional water management system	Traditional technology for water conservation	References
Himachal Pradesh	Kuhls	The channels carry glacial waters from rivers and streams into the fields.	Kumar and Katara (2020), Sharma et al. (2015)
Nagaland	Zabo	Forest-runoff channels collect rainwater and cattle waste, and direct it to terraced ponds and paddy fields.	(2016), Kithan (2014)
Meghalaya	Bamboo Drip irrigation	Tribal farmers divert spring water to terrace fields using bamboo pipes, delivering precise irrigation to crops.	Cajee (2018)
Great Nicobar Islands	Jackwells	Split bamboo channels collect runoff water, directing it to interconnecting jackwells for efficient water collection.	Yan et al. (2020)
Madhya Pradesh	Pat System	Diversion bunds across streams lined with teak leaves and mud create leak-proof channels for irrigation water distribution.	Xu et al. (2020)
Tamil Nadu	Eri	Eris: interconnected tanks fed by river channels or rainwater, ensuring water access and balance between villages.	Xu et al. (2020)
Karnataka	Katta	Katta: temporary tone bund across streams, storing water, and replenishing nearby wells through gradual seepage.	Xu et al. (2020)
Uttarakhand	Naula and dhara Gadhera, khal, chal, gul	Naulas are crucial water sources in Uttarakhand's hill regions, appearing as stone structures with stairs, supplying drinking water to villages. Gadhera, gul, chal, and khal also serve as water collection and supply systems.	
Asia			
West Asia and North Africa (WANA)	Qanat	Qanats tap groundwater using gravity, without pumping to transport water to the surface sustainably for domestic and agricultural use.	Oweis et al. (2004), Guttmann-Bond (2010)
Bangladesh	Jhurjhuri	Jhurjhuri is built at the base of hills, providing clear and cool water through natural filtration and underground flow.	Kabir and Faisal (1999), Pidou et al. (2007)
	Phour	Godha stores water for household and irrigation, utilizing bamboo pipes to transport water to lower-elevation fields and aiding in bamboo transportation.	
	Thagalok-Kum	To save travel, water is collected using split bamboo (Thagalok), directing it to an earthen pitcher (kum) from a nearby stream fed by seepage from surrounding hills.	

(Continued)

Table 7.1 (Continued)

State	Traditional water management system	Traditional technology for water conservation	References
	Thelya-Thok	Similar to a Godha, this structure serves the same purposes, utilizing bamboo and wooden pegs to support the earthen dams.	
	Godha	This is a kind of cross-dam that is constructed across a small hill creek to collect water for the dry season.	
Nepal	Pukhu	Pukhus has a surface drainage system for storing surplus water used for cattle, irrigation, and pottery industries.	Gautam et al. (2014), Gautam et al. (2018)
	Hiti and Tunche	Hiti and Tunche are water discharging through stone spouts, primarily used for drinking, cooking, washing, cleaning, animal feeding, and small-scale farming.	
Sri Lanka	Ralapanawa	Ralapanawa is a structure made to prevent bunt dam from erosion. It is constructed by using granite plates and shaped like a staircase	Das et al. (2022), Jayatilleke et al. (2020)
	Thawulla	The area which marked the maximum water level of a wewa	
	Pita wana	Pitawana is used to displace excessive water from wewa.	
	Athulu wana	Water inlet of a wewa is athul wana.	
	Kulu wewa	Kulu wewa is a small wewa that settles mud particles before water flows into the main reservoir and it also replenishes water resources for paddy fields.	
	Karahana	This is a device used to distribute water to paddy fields.	
	Yoda Ela	It supplies water to multiple tank cascade systems while receiving excess drainage water from upstream tank cascades.	

groundwater, whereas Bangladesh and Nepal employ buildings such as Jhurjhuri and Pukhu to collect water. In Sri Lanka, buildings like Yoda Ela and Ralapanawa help distribute water and avoid erosion in paddy fields.

Figure 7.1 illustrates the many traditional water-saving techniques used in Asia according to the table. Important methods like qanats, step wells, drip irrigation using bamboo, and linked tank systems are demonstrated.

Kuhls (Himachal): The channels carry glacial waters from rivers and streams.

Taanka (Rajasthan): Underground pits that store rainwater, ensuring water security in arid regions and reducing daily water-fetching burdens.

Bawari (Rajasthan): Reservoirs with layered steps conserve water and recharge aquifers through percolation.

Johads (Rajasthan): Ancient earthen dams capturing rainwater, conserving groundwater, and connecting to rivers or streams.

Bamboo Drip irrigation (Meghalaya): Tribal farmers divert spring water to terrace fields using bamboo pipes, delivering precise irrigation to crops.

Katta (Karnataka): Temporary tone bund across streams, storing water and replenishing nearby wells through gradual seepage.

Godha (Bangladesh): This is a kind of cross-dam that is constructed across a small hill creek to collect water for the dry season.

Qanat (West Asia North Africa): Tap groundwater using gravity, without pumping to transport water to the surface sustainably for domestic and agricultural use.

Figure 7.1 Shows the different Indigenous technology in Asia. *Source:* Adapted from Modern India Can Learn a Lot from These 20 Traditional Water Conservation Systems, 2015/https://www.thebetterindia.com/61757/traditional-water-conservation-systems-india/, last accessed 4 June 2024.

7.1.1 Positive Aspects of Indigenous Water Conservation Technology on Environment Sustainability

Indigenous water conservation technology offers several positive aspects that contribute to environmental sustainability. Indigenous water conservation techniques are frequently created depending on the unique requirements and circumstances of a certain location. They are made with the local environment in mind, considering things like soil types, rainfall patterns, and available natural resources (David and Ploeger 2014; Panday et al. 2021a, b; Panday and Kumar 2023). The sustainability of water supplies over the long term is enhanced by this local adaptation. Keeping the ecological balance of the surrounding ecosystems is a top priority for indigenous water conservation techniques. They frequently stress the need to protect natural water sources like rivers, streams, and springs since they are essential for maintaining biodiversity and serving as habitats for different species (Jackson 2005). Indigenous practices contribute to the sustainability and health of the environment overall by protecting these ecosystems (Senanayake 2006). Indigenous groups have a wealth of information about the local ecosystems in their area, particularly the water sources and how they interact with the environment (Sharma et al. 2016). This expertise, which includes efficient water management methods like rainwater gathering, refilling aquifers, and irrigation systems that reduce water waste, is frequently passed down through generations. Indigenous water conservation systems provide creative and long-lasting solutions by integrating this ancient knowledge into contemporary practices (Shrestha and Shrestha 2008; Panday and Kumar 2022).

Native American water conservation techniques typically have little to no environmental impact. They don't require complex infrastructure made of energy-intensive materials or extensive water extraction because they use straightforward, locally accessible materials. These innovations encourage the wise use of water supplies while reducing pollution and ecological impact (Gadgil et al. 1993). Indigenous water conservation systems frequently entail significant community ownership and participation. Local communities actively participate in the design, implementation, and administration of these technologies, encouraging a feeling of group accountability and stewardship. This community-centered strategy guarantees the long-term upkeep and durability of water-saving measures. Over centuries of adaptation to shifting climatic conditions, such as alterations in rainfall patterns and water availability, indigenous water conservation systems have developed. These technologies are frequently made to withstand and lessen climate change's effects (Chibememe et al. 2014). Communities can increase their resistance to water scarcity, droughts, and other climate-related problems by depending on indigenous water conservation techniques (Gupta and Agrawal 2015). In general, indigenous water conservation technology highlights the importance of local customs and traditional knowledge in fostering environmental sustainability (Van Dijk 1997). We can take advantage of these technologies' advantages to build a more sustainable future for water resources and ecosystems by integrating them into larger water management strategies (Pradhan 1990).

7.2 Modern Technology

In both rich and developing countries, effective water management is essential for achieving sustainable crop production, which in turn boosts social wealth and economic growth. It has been suggested that many different regions around the world may experience water-related conflicts in the coming ensuing decades. According to Malekian Jabali et al. (2017), the Middle East is the main location that could experience such a conflict for political, demographic, and economic reasons (Anup Das et al. 2012). Saudi Arabia and other countries in the area cannot afford to ignore the severe water deficit and the careless usage of subterranean water reserves since the region is plagued by significant water scarcity (Jackson 2005). A strain is put on the Middle East's resources, particularly its water supply, as a result of the region's limited water supplies and ongoing population expansion (Ahmed and Farooq 2022). A national policy for utilizing brackish water and recycling wastewater (such as agricultural drainage, industrial wastewater, and municipal wastewater) must be adopted in order to deal with limited water resources. Water use efficiency must also be maximized. In irrigation, agriculture has traditionally focused on water management from the demand side, and while in drainage, the focus has been on managing excess water. The use of brackish and recycled water in agriculture is one example of the kind of tactical management that will be seen more frequently (Barlow and Covenant 2007). Aquifer storage and groundwater "banking" will also be utilized more to their full potential. The primary obstacles to the development of agriculture in Saudi Arabia are now water resource scarcity and degradation (Critchely et al. 1994; Panday et al. 2023).

One of the key elements that has a significant impact on the agricultural industry is irrigation water, which can come from a variety of sources, including rainwater, treated wastewater, and groundwater. The Ministry of Agriculture is aware of how important it is to conserve water, and since the amount of water used for irrigation is continuing to rise, the government has encouraged agricultural businesses and various types of farmers to implement integrated and sustainable irrigation water management in an effort to achieve the best possible irrigation water use (Bhattacharya et al. 2011). More specifically, the ministry urged them to implement the concept of "More Crop per Less Drop," which would increase crop production by using less water, rather than the concept of "More Crop per Drop," which would increase crop production by using more water (Barlow 2014). The ministry is making a lot of effort in this situation to use contemporary irrigation techniques, such as drip irrigation or pivot sprinklers, rather than more antiquated ones, like flood irrigation, and to link agricultural licenses with the requirement of contemporary irrigation techniques.

A few contemporary techniques for conserving water that are employed in different Indian states are shown in Table 7.2. Using low-pressure systems with point or line sources, drip irrigation has been used in Gujarat to improve crop quality and productivity while conserving water (Alcon et al. 2011; Kaarthikeyan and Suresh 2019). In a similar vein, areas like Madhya Pradesh, Southern Haryana, and portions of Punjab and Rajasthan are heavily dependent on spray irrigation. With success rates ranging from 70% to 80%, this technique improves water efficiency (Debnath et al. 2020; Carrión et al. 2001). Water collecting techniques are not new to Tamil Nadu; in locations like Delhi and Burhanpur, sustainable water supplies are supported by subterranean tunnels and rainwater harvesting (Bhattacharya 2015; Chowdhury and Biswas 2016). Inspired by past practices, the Jal Jeevan Mission (JJM) is a large-scale water conservation initiative that is being implemented in states such as Goa, Telangana, Gujarat, and Haryana (Lucas-Borja et al. 2021). Greywater recycling, which is frequently utilized in urban and agricultural contexts for commercial, recreational, and environmental purposes, has been embraced by Madhya Pradesh (Islam et al. 2007; Li et al. 2009). The Eco-Water project in Noida and Delhi assesses eco-efficiency metrics in an effort to enhance knowledge of the socio-technical dynamics surrounding water consumption (Gautam 2014). Also, Gujarat has created check dams, which are little buildings that control erosion, store water, and regulate the flow of sediment (Lucas-Borja et al. 2021; Abbasi et al. 2019). Reservoirs are utilized to store water, produce hydroelectric power, and offer recreational opportunities in states such as Tamil Nadu, Karnataka, Andhra Pradesh, Kerala, Orissa, and Maharashtra (HB More 2020). These contemporary approaches are a part of a larger, dynamic approach to managing water resources, as seen in the Figure 7.2 that goes with it. This diagram most likely shows the various types of water conservation strategies and their geographic distribution across the nation. Every system reflects the fusion of cutting-edge technologies with customary methods as a regional response to particular water-related issues.

Table 7.2 Some of the modern technique methods of water conservation.

State	Modern water management system	Modern technology for water conservation	References
Gujarat	Drip irrigation	Drip irrigation conserves water, improves crop quality, and increases production with point/line sources and low pressure.	Alcon et al. (2011), Kaarthikeyan and Suresh (2019)
Madhya Pradesh, Southern Region of Haryana, and North Eastern Part of Rajasthan and Parts of Punjab	Sprinkler irrigation	Sprinkler irrigation system releases pressurized water through nozzles, improving efficiency with 70–80% effectiveness.	Debnath et al. (2020), Carrión et al. (2001)
Tamil Nadu	Water harvesting for domestic use	Delhi and Burhanpur historically used rainwater harvesting and underground tunnels for sustainable water supply.	Bhattacharya (2015), Chowdhury and Biswas (2016)
Goa, Telangana, Gujarat, and Haryana	Jal Jeevan Mission (JJM)	India's Jal Jeevan Mission (JJM) Ministry of Water Resources focus on nationwide water conservation, drawing from historical practices.	Lucas-Borja et al. (2021)
Madhya Pradesh	Greywater recycling	Wastewater recycling finds common use in agriculture, but urban reuse options include industrial, recreational, and environmental applications.	Islam et al. (2007), Li et al. (2009)
Noida Sector 2, Noida, Delhi	Eco-Water project	The project assesses ecoefficiency indicators and analyzes factors influencing adoption of innovating practices for improved understanding of sociotechnical dynamics.	Gautam (2014)
Gujarat	Check dams	Small structures across rivers controlling erosion, storing water, and managing sediment flow.	Lucas-Borja et al. (2021), Abbasi et al. (2019)
Tamil Nadu, Karnataka, Andhra Pradesh, Kerala, Orissa, and Maharashtra	Reservoirs	Reservoirs are large storage areas formed by dams, providing water supply, hydroelectric power, and recreational opportunities.	More (2020)

7.2.1 Positive Aspect of Modern Water Conservation Technology on Environment Sustainability

There are many advantages to modern water conservation technology that support environmental sustainability. Modern water-saving technology includes sensor-based faucets, smart irrigation systems, and low-flow fixtures that encourage effective water use. These technologies help save water resources by minimizing water waste, especially in places that

Sprinkler Irrigation: Sprinkler irrigation system releases pressurized water through nozzles, improving efficiency with 70-80% effectiveness. https://www.google. com/search

Drip Irrigation: Drip irrigation conserves water, improves crop quality and increases production with point/line sources and low pressure. https://www.google.com/search

Check dam: Small structures across rivers controlling erosion, storing water and managing sediment flow. https://aamjanata.com/check-dams

Reservoir: Reservoirs are large storage areas formed by dams, providing water supply, hydroelectric power and recreational opportunities. https://www.google.com/search

Figure 7.2 Shows the different modern practices images.

are experiencing water scarcity or drought. By ensuring that water resources are used effectively, this efficient use eases the burden on natural water sources (Nikouei et al. 2012). Technologies that conserve water are essential for maintaining natural ecosystems. Treated wastewater can be reused for a variety of uses, such as irrigation or industrial activities, by putting in place measures like water recycling and reclamation systems. Lowering the demand for freshwater from rivers, lakes, and aquifers preserves the natural balance and safeguards aquatic habitats (Ballogou et al. 2014).

Energy savings and water conservation methods frequently coexist. For instance, water-efficient fixtures and appliances use less energy to heat and distribute the water they use. These solutions indirectly cut the energy needed for water treatment and pumping by reducing water demand, which lowers carbon emissions and has a positive influence on the environment. Technologies that conserve water can help to lessen pollution. In order to reduce the demand for chemically treated water, rainwater harvesting systems, for instance, gather and store rainwater for non-potable uses like gardening. This results in enhanced water quality and less pollution in rivers and oceans by reducing the flow of contaminants from conventional water sources and easing the burden on wastewater treatment facilities.

7.2.2 Drawback of Modern Water Conservation Technology on Environment Sustainability

While trying to alleviate water scarcity and promote sustainability, modern water conservation technologies can have certain negative effects on environmental sustainability. Modern water conservation methods like desalination plants, water treatment facilities, and pumping systems all use a lot of energy to keep things running. The use of fossil fuels frequently can increase greenhouse gas emissions and the effects of climate change, negating any potential environmental advantages of water conservation measures (Huang et al. 2017). Significant habitat alteration and ecosystem disruption are possible as a result of several contemporary water conservation practices, such as large-scale dam construction and water diversion projects. These initiatives may damage wetlands, disrupt fish migration patterns, split river systems, and reduce biodiversity across the region (Yan et al. 2020). Some water conservation techniques, such as wastewater recycling and artificial groundwater recharge, can increase the number of pollutants or toxins in water sources. If these practices are not effectively controlled, dangerous chemicals, diseases, or an overabundance of nutrients may be released into the environment, harming aquatic ecosystems and endangering public health (Strang 2001).

Constructing infrastructure like reservoirs, pipelines, and treatment facilities may be necessary as part of the application of contemporary water-saving methods, which could result in significant land use. Deforestation, habitat damage, and the eviction of local groups or indigenous inhabitants can all come from this. In addition, if not carried out properly, the production and upkeep of water conservation devices could result in a greater ecological footprint. Natural flow patterns and hydrological cycles of rivers, lakes, and groundwater systems are altered by several contemporary water conservation practices. For instance, excessive groundwater extraction can result in land subsidence and the depletion of aquifers, which will affect the water resources available to ecosystems and future generations. It is crucial to consider the environmental effects of contemporary water conservation technologies during their design, implementation, and operation phases in order to reduce these downsides. This can be done by incorporating sustainable practices, like the use of renewable energy sources for water treatment and pumping, thorough environmental impact assessments, the use of ecofriendly materials and construction techniques, and putting policies in place to reduce the release of pollutants into water sources. In addition, a holistic strategy that incorporates indigenous water management techniques and traditional knowledge can aid in finding a balance between environmental sustainability and water conservation (Warren 1991).

7.3 Conclusion

A detailed view of the sustainable management of this priceless resource is provided by the comparison of Asian indigenous water conservation technologies and contemporary practices. Indigenous people have created inventive water management techniques over the course of the area's long history, frequently blending in with their natural environment. These age-old customs are a tribute to the knowledge of earlier generations and give us important information about how to respond to the modern problems of water scarcity and climate change. On the other hand, contemporary water conservation techniques put cutting-edge science and technology at the forefront. They provide effective responses to the escalating needs of societies that are urbanizing and industrializing quickly. Integrated water management systems, sophisticated irrigation methods, and wastewater treatment facilities are just a few of the numerous developments that have increased the efficiency of water resource use. It is vital to understand the value of combining these two strategies as we advance, though. The importance of holistic and sustainable water management is one way that indigenous knowledge can influence current practices. This includes considering how our activities may affect the environment, safeguarding traditional knowledge, and encouraging community involvement in water conservation initiatives. In addition, working together with indigenous communities, decision-makers, scientists, and engineers can result in the creation of novel solutions that incorporate the best of both cultures. We can work toward a more resilient, equitable, and sustainable water future for everyone in Asia and beyond by honoring and combining ancient practices with modern technology. By doing this, we respect the history of indigenous population and guarantee that the priceless teachings of the past continue to direct our path toward good water stewardship.

References

Abbasi, N.A., Xu, X., Lucas-Borja, M.E. et al. (2019). The use of check dams in watershed management projects: Examples from around the world. *Science of the Total Environment* 676: 683–691.

Ahmed, A., and Farooq, H. (2022). Indigenous Knowledge and Sustainable Agricultural Development: A Case History of the Indus Valley, Pakistan.

Alcon, F., de Miguel, M.D., and Burton, M. (2011). Duration analysis of adoption of drip irrigation technology in southeastern Spain. *Technological Forecasting and Social Change* 78 (6): 991–1001.

Anup Das, A.D., Ramkrushna, G.I., Choudhury, B.U. et al. (2012). Natural resource conservation through indigenous farming systems: wisdom alive in North East India. *Indian Journal of Traditional Knowledge* 11: 505–513.

Arnell, N.W. (2004). Climate change and global water resources: SRES emissions and socio-economic scenarios. *Global Environmental Change* 14 (1): 31–52.

Asok Raja, S.K., Panday, D.P., and Kumar, M. (2023). Decoding the enigma of 100-year record-breaking rainfall over Tamil Nadu using wavelet analysis. *Groundwater for Sustainable Development* 101007. https://doi.org/10.1016/j.gsd.2023.101007.

Ayre, M. and Mackenzie, J. (2013). "Unwritten, unsaid, just known": the role of Indigenous knowledge (s) in water planning in Australia. *Local Environment* 18 (7): 753–768.

Ballogou, V.Y., Soumanou, M.M., Toukourou, F., and Hounhouigan, J.D. (2014). Indigenous knowledge on landraces and fonio-based food in Benin. *Ecology of Food and Nutrition* 53 (4): 390–409.

Barlow, M. (2014). *Blue Future: Protecting Water for People and the Planet Forever*. The New Press.

Barlow, M. and Covenant, B. (2007). *The Global Water Crisis and the Coming Battle for the Right to Water*. Toronto: McClelland & Stewart.

Baul, T.K. and McDonald, M.A. (2014). Agro-biodiversity management: using indigenous knowledge to cope with climate change in the Middle-Hills of Nepal. *Agricultural Research* 3: 41–52.

Baul, T.K. and McDonald, M. (2015). Integration of indigenous knowledge in addressing climate change.

Becker-Ritterspach, R. (1990). Dhunge-dharas in the Kathmandu Valley: an outline of their architectural development. *Ancient Nepal* 116 (118): 1–9.

Belsare, S. (2017). Participatory irrigation management in Maharashtra – achievements, challenges and way ahead. *IWRA (India) Journal (Half Yearly Technical Journal of Indian Geographical Committee of IWRA)* 6 (1): 31–34.

Bhattacharya, S. (2015). Traditional water harvesting structures and sustainable water management in India: a socio-hydrological review. *International Letters of Natural Sciences* 37.

Bhattacharya, S., Dasgupta, A., Mahansaria, R.et al. (2011). Traditional rainwater harvesting in India: historical perspectives, present scenario and future prospects.

Bishnoi, S., Prateek, G., Sahay, N., and Gupta, A. (2017). Traditions of water conservation, sanitation and augmentation: insights from Gujarat. *Plurimondi* 8.

Cajee, L. (2018). Water management during the dry season in the betel leaf plantations of the southern slopes of The Meghalaya Plateau. *International Journal of Research in Management & Social Science* 6 (2): 44.

Carrión, P., Tarjuelo, J., and Montero, J. (2001). SIRIAS: a simulation model for sprinkler irrigation: I. Description of model. *Irrigation Science* 20: 73–84.

Chibememe, G., Muboko, N., Gandiwa, E. et al. (2014). Embracing indigenous knowledge systems in the management of dryland ecosystems in the Great Limpopo Transfrontier Conservation Area: the case of Chibememe and Tshovani communities, Chiredzi, Zimbabwe. *Biodiversity* 15 (2–3): 192–202.

Chowdhury, P. and Biswas, S. (2016). Biodiversity conservation envisioned to livelihood development in Nagaland. *International Journal in Management & Social Science* 4 (2): 294–314.

Codjoe, S.N.A., Owusu, G., and Burkett, V. (2014). Perception, experience, and indigenous knowledge of climate change and variability: the case of Accra, a sub-Saharan African city. *Regional Environmental Change* 14: 369–383.

Critchley, W.R.S., Reij, C., and Willcocks, T.J. (1994). Indigenous soil and water conservation: a review of the state of knowledge and prospects for building on traditions. *Land Degradation & Development* 5 (4): 293–314.

Das, A., Kumar, M., Jha, P.K. et al. (2022). Isotopic and hydrogeochemical tracking of dissolved nutrient dynamics in the Brahmaputra River System: a source delineation perspective. *Chemosphere* 307: 135757. https://doi.org/10.1016/j.chemosphere.2022.135757.

David, W. and Ploeger, A. (2014). Indigenous Knowledge (IK) of water resources management in West Sumatra, Indonesia. *Future of Food: Journal on Food, Agriculture and Society* 2: 52–60.

Debnath, S., Adamala, S., and Palakuru, M. (2020). An overview of Indian traditional irrigation systems for sustainable agricultural practices. *International Journal of Modern Agriculture* 9: 12–22.

Fadhili, B., Liwa, E., and Shemdoe, R. (2002). Integrating geospatial decision support systems with indigenous knowledge in natural resource management. *Forest* 1998 (2).

Gadgil, M., Berkes, F., and Folke, C. (1993). Indigenous knowledge for biodiversity conservation. *Ambio* 22: 151–156.

Gautam, D. (2014). *Water Management Through Indigenous Knowledge: A Case of Historic Settlement of Bhaktapur City*. Nepal: Anchor Academic Publishing (aapverlag).

Gautam, D., Thapa, B.R., and Prajapati, R.N. (2018). Indigenous water management system in Nepal: cultural dimensions of water distribution, cascaded reuse and harvesting in Bhaktapur City. *Environment, Development and Sustainability* 20: 1889–1900.

Gorjestani, N. (2004). Indigenous knowledge for development. *Protecting and Promoting Traditional Knowledge: Systems, National Experiences And International Dimensions* 265.

Grenier, L. (1998). *Working with Indigenous Knowledge: A Guide for Researchers*. idrc.

Gupta, R.K. and Agrawal, R.K. (2015). Rainwater harvesting in ancient times and its sustainable modern techniques. *ICID* 2015: 1–5.

Guttmann-Bond, E. (2010). Sustainability out of the past: how archaeology can save the planet. *World Archaeology* 42 (3): 355–366.

Hiwasaki, L., Luna, E., and Shaw, R. (2014). Process for integrating local and indigenous knowledge with science for hydro-meteorological disaster risk reduction and climate change adaptation in coastal and small island communities. *International Journal of Disaster Risk Reduction* 10: 15–27.

https://aamjanata.com/check-dams-a-practical-alternative-to-our-large-dam-fetish/.

https://www.google.com/search?q=reservior&tbm=isch&ved=2ahUKEwigqq6qptGAAxUVoekKHR4ICLoQ2.

https://www.google.com/search?sca_esv=555356848&rlz=1C1VDKB_enIN1047IN1047&sxsrf=AB5stBhhwEhjdpXPcDl5w3JJdZBMiAgCHA:1691641460658&q=sprinkler+irrigation&.

https://www.google.com/search?sca_esv=555356848&rlz=1C1VDKB_enIN1047IN1047&sxsrf=AB5stBiYdEuEqh5t4OJFZfbTtLVltWniYg:1691641739669&q=drip++irrigation&tbm=isch&source=lnms&sa=X&ved=2ahUKEwjYmOvHoNGAAxUAe2wGHQ9pAukQ0pQJegQIDRAB&biw=1366&bih=619&dpr=1.

https://www.thebetterindia.com/61757/traditional-water-conservation-systems-india/.

Huang, Q., Wang, J., and Li, Y. (2017). Do water saving technologies save water? Empirical evidence from North China. *Journal of Environmental Economics and Management* 82: 1–16.

Hussain, J., Husain, I., and Arif, M. (2014). Water resources management: traditional technology and communities as part of the solution. *Proceedings of the International Association of Hydrological Sciences* 364: 236–242.

Islam, M.M., Kabir, M.R., and Chou, F.N. (2007). Feasibility study of rainwater harvesting techniques in Bangladesh. *Rainwater and Urban Design* 2007: 726–733.

Jabali, M.M., Okhravi, S., Eslamian, S., and Gohari, S. (2017). Water conservation techniques. In: *Handbook of Drought and Water Scarcity*, 501–520. CRC Press.

Jackson, S. (2005). Indigenous values and water resource management: a case study from the Northern Territory. *Australasian Journal of Environmental Management* 12 (3): 136–146.

Jayatilleke, Y. D., Mudalige, C., and Dewapura, D. (2020). An Anthropological Study Of Traditional Irrigation, Technological Wisdom and Ecocivilization in Anuradhapura, Sri Lanka.

Kaarthikeyan, G.M. and Suresh, A. (2019). A study on understanding the adoption of water saving technology: a case study of drip irrigation. *International Journal of Recent Technology and Engineering* 7 (6): 1123–1130.

Kabir, M.R. and Faisal, I.M. (1999). Indigenous practice for water harvesting in Bangladesh. *Proceedings of the Regional Workshop on Traditional Water Harvesting System.*

Khan, M.A. and Narain, P. (2005). Mainstreaming rainwater harvesting as resource for conjunctive use. *12th International Conference on Rainwater Catchment Systems, Mainsteaming Rainwater Harvesting,* New Delhi, India, pp. 1–9.

Kithan, L.N. (2014). Indigenous system of Paddy cultivation in Terrace and Jhum fields among the Nagas of Nagaland. *International Journal of Scientific and Research Publications* 4 (3): 1–4.

Koul, D.N., Singh, S., Neelam, G., and Shukla, G. (2012). Traditional water management systems – an overview of Ahar-pyne system in South Bihar plains of India and need for its revival.

Kumar, S. and Katara, S. (2020). Traditional methods of water management in India. *Himachal Pradesh University Journal* 8 (1): 53–70.

Kumar, M., Panday, D.P., Bhagat, C. et al. (2023a). Demystifying the decadal shift in the extent of groundwater in the coastal aquifers of Gujarat, India: a case of reduced extent but increased magnitude of seawater intrusion. *Science of the Total Environment* 165451. https://doi.org/10.1016/j.scitotenv.2023.165451.

Kumar, M., Mazumder, P., Silori, R. et al. (2023b). Prevalence of PPCPs, microplastics and co-infecting microbes in the post-COVID-19 era and its implications on antimicrobial resistance and potential endocrine disruptive effects. *Science of the Total Environment* 166419. https://doi.org/10.1016/j.scitotenv.2023.166419.

Levidow, L., Zaccaria, D., Maia, R. et al. (2014). Improving water-efficient irrigation: prospects and difficulties of innovative practices. *Agricultural Water Management* 146: 84–94.

Li, F., Wichmann, K., and Otterpohl, R. (2009). Review of the technological approaches for grey water treatment and reuses. *Science of the Total Environment* 407 (11): 3439–3449.

Lucas-Borja, M.E., Piton, G., Yu, Y. et al. (2021). Check dams worldwide: objectives, functions, effectiveness and undesired effects. *Catena* 204: 105390.

More, H.B. (2020). A review of Indian traditional method of rain water harvesting. *International Journal of Innovative Research in Science, Engineering and Technology* 9 (4): 2080–2086.

Murthy, R.N.S., Srikonda, R., and Kasinath, I.V. (2022). Traditional Water Management Systems of India. *ISVS e-Journal* 9 (2): 61–77.

Nair, G.U. (2016). Traditional wisdom in harvesting water. *Journal of Traditional and Folk Practices* 4 (1): 50–53.

Neog, N., Goswami, R., Panday, D.P. et al. (2024). Arsenic contamination in the groundwater of Northeastern India: Critical understandings on geotectonic controls and the need for intervention. *Current Opinion in Environmental Science & Health* 100539. https://doi.org/10.1016/j.coesh.2024.100539.

Nikouei, A., Zibaei, M., and Ward, F.A. (2012). Incentives to adopt irrigation water saving measures for wetlands preservation: an integrated basin scale analysis. *Journal of Hydrology* 464: 216–232.

Nüsser, M., Dame, J., Kraus, B. et al. (2019). Socio-hydrology of "artificial glaciers" in Ladakh, India: assessing adaptive strategies in a changing cryosphere. *Regional Environmental Change* 19: 1327–1337.

Oweis, T. (2003). Indigenous water management knowledge: From traditional to modern technology.

Oweis, T., Hachum, A., and Bruggeman, A. (2004). The role of indigenous knowledge in improving present water-harvesting practices. *Indigenous Water-Harvesting Systems in West Asia and North Africa* 1–20.

Panday, D.P. and Kumar, M. (2022). Climate indices and hydrological extremes: deciphering the best-fit model. *Environmental Research* 215: 114301. https://doi.org/10.1016/j.envres.2022.114301.

Panday, D.P. and Kumar, M. (2023). Application of remote sensing techniques to deal with scale aspects of GRACE data to quantify groundwater levels. *MethodsX* 102108. https://doi.org/10.1016/j.mex.2023.102108.

Panday, D.P., Khosa, R., Maheswaran, R. et al. (2021a). Game-theoretic-based modelling of Krishna waters dispute: equilibrium solutions by Metagame Analysis. *The European Physical Journal B* 94: 101. https://doi.org/10.1140/epjb/s10051-021-00107-w.

Panday, D.P., Khosa, R., Maheswaran, R. et al. (2021b). Game theoretic-based modelling of Krishna waters dispute: equilibrium solutions by hypergame analysis. *The European Physical Journal B* 94: 131. https://doi.org/10.1140/epjb/s10051-021-00135-6.

Panday, D.P., Singh, A., Gaur, S. et al. (2023). Probable maximum precipitation analysis of high rainfall regions in India. *Groundwater for Sustainable Development* 100893. https://doi.org/10.1016/j.gsd.2022.100893.

Pani, A., Ghatak, I., and Mishra, P. (2021). Understanding the water conservation and management in India: an integrated study. *Sustainable Water Resources Management* 7: 1–16.

Pidou, M., Memon, F.A., Stephenson, T. et al. (2007, September). Greywater recycling: treatment options and applications. *Proceedings of the Institution of Civil Engineers-Engineering Sustainability*, vol. 160, no. 3, 119–131. Thomas Telford Ltd.

Pradhan, R. (1990). Dhunge Dhara: A case study of the three cities of Kathmandu Valley. *Ancient Nepal: Journal of the Department of Archaeology* 116: 10–14.

Pun, S. (2001). Role of gender in Sali-Nadi (Shanku Raj Kulo) irrigation management: a case study.

Rerkasem, K., Yimyam, N., and Rerkasem, B. (2009). Land use transformation in the mountainous mainland Southeast Asia region and the role of indigenous knowledge and skills in forest management. *Forest Ecology and Management* 257 (10): 2035–2043.

Rose, D.B. (2004). Fresh water rights and biophilia: Indigenous Australian perspectives. *Academy of the Social Sciences in Australia: Dialogue* 22 (3): 35–44.

Sahoo, P. (2020). Study of traditional ahar-pyne system and community governance methods in Asarganj Block, Munger District, Bihar, India. *Bonfring International Journal of Industrial Engineering and Management Science* 3: 192–195.

Saxena, D. (2017a). Water conservation: traditional rain water harvesting systems in Rajasthan. *International Journal of Engineering Trends and Technology* 52 (2): 91–98.

Saxena, D. (2017b). Water conservation: traditional rain water harvesting systems in Rajasthan. *International Journal of Engineering Trends and Technology* 52 (2): 91–98.

Saxena, A., Raghuvanshi, M.S., and Suna, T. (2021). Traditional water management of subsistence agriculture system in cold arid Ladakh: a review.

Senanayake, S.G.J.N. (2006). Indigenous knowledge as a key to sustainable development. *Journal of Agricultural Sciences–Sri Lanka* 2 (1).

Sharma, A.K., Sharma, K.D., and Prakash, B. (2015). Death of Kuhl irrigation system of Kangra Valley of Himachal Pradesh: institutional arrangements and technological options for revival. *Indian Journal Of Agricultural Economics* 70 (902-2016-68394): 350–364.

Sharma, A., Sharma, K.D., and Prakash, B. (2016). Himachal Pradesh: institutional arrangements and technological options for revival. *Indian Journal Of Agricultural Economics* 70: 350–364.

Sharma, P.K., Srivastava, S., and Chandauriya, M. (2023). Indigenous knowledge and traditional practices for water resource management in Rajasthan, India. In: *Traditional Ecological Knowledge of Resource Management in Asia*, 137–157. Cham: Springer International Publishing.

Shiva, V. (2016). *Water Wars: Privatization, Pollution, and Profit.* North Atlantic Books.

Shrestha, S. and Shrestha, B.K. (2008). The influence of water in shaping culture and modernisation of the Kathmandu Valley. In: *Water and Urban Development Paradigms*, 201–206. CRC Press.

Sojka, S., Younos, T., and Crawford, D. (2016). Modern urban rainwater harvesting systems: design, case studies, and impacts. *Sustainable Water Management in Urban Environments* 209–234.

Strang, V. (2001). Poisoning the rainbow: mining, pollution and Indigenous cosmology in far north Queensland. In: *Mining and Indigenous Lifeworlds in Australia* (ed. A.J. Weiner), 208–225.

Toussaint, S., Sullivan, P., and Yu, S.(2005). Water ways in Aboriginal Australia: an interconnected analysis. *Anthropological Forum*, vol. 15, no. 1, 61–74. Taylor & Francis Group.

UNEP, U. D. FAO, 2012. SIDS-FOCUSED *Green* Economy: An Analysis of Challenges and Opportunities www.unep.org/greeneconomy and www.unep.org/regionalseas.

Van Dijk, J.A. (1997). Indigenous soil and water conservation by teras in eastern Sudan. *Land Degradation & Development* 8 (1): 17–26.

Warren, D. M. (1991). Using indigenous knowledge. *Agricultural Development World Bank Discussion Papers,* no. 127, p. 46.

Xu, Z.X., Li, S.Y., Li, B.F. et al. (2020). A review of development methods and EOR technologies for carbonate reservoirs. *Petroleum Science* 17: 990–1013.

Yan, H., Hui, X., Li, M., and Xu, Y. (2020). Development in sprinkler irrigation technology in China. *Irrigation and Drainage* 69: 75–87.

8

Indigenous and Modern Practices for Water Conservation and Management in Africa

Pushpa Sharma, Uttamasha B. Borah, and Saumya Patel

Department of Petroleum Engineering and Earth Sciences (Energy Cluster), School of Advanced Engineering, UPES, Dehradun, Uttarakhand, India

8.1 Introduction

After population expansion, climate change, and rising energy and agriculture demands have devalued water's crucial role, a worldwide water crisis is imminent. Sustainable water conservation is needed (Kowenje et al. 2022). With more than 20% of renewable water resources currently in use, Sub-Saharan Africa (SSA) is where the effects of climate change are most noticeable. Millions of people are at risk due to the extreme water stress this region is under, with withdrawals above 40%. Unsafe water sources are another factor in deaths, and vulnerability is increased by deforestation, agriculture, and greenhouse gas emissions that disturb the water cycle. Water scarcity is made worse by the difficulties faced by climate-dependent industries, most notably agriculture (Ngoran et al. 2015).

Indigenous knowledge, which is marked by accumulated wisdom handed down through generations, comes forth as a significant strength in handling these issues. This information system is flexible, adapting to changes coming from within as well as outside sources. Rainwater harvesting (RWH) techniques, which are based on Indigenous wisdom and involve the use of Zai pits, are examples of sustainable practices because of their adaptability, continual existence, and ongoing evolution (Flavier et al. 1995; Oweis and Hachum 2006). Riparian buffer zones, or the vegetated areas beside rivers and streams, are also becoming crucial components of modern water resource conservation. These zones function as organic filters, filtering out pollutants, nutrients, and sediments before they reach water bodies. Riparian buffer zones need to be maintained in order to enhance water quality and ecological balance while lessening the damaging effects of urbanization, industrialization, and agriculture on aquatic ecosystems (Olokeogun and Kumar 2022). Solar photovoltaic (PV) technologies have high-potential approaches to rural electrification and water extraction in the particular setting of SSA. Solar PV technologies, such as the creation of solar PV pumps, have gained popularity, especially as the cost of solar panels continues to decrease. By offering a sustainable and effective method for both electrification and water extraction for irrigation

Water Scarcity Management: Enabling Technologies, First Edition. Edited by Kanchan D. Bahukhandi, Manish Kumar, Durga P. Panday, and Tushara G.G. Chaminda.

and residential use, these inventions meet important demands for smallholder farmers (Singh et al. 2021).

Water resource conservation takes a complete and robust approach when these contemporary technologies are combined with traditional methods like riparian buffer zones and Zai pits. The effectiveness of PV pumps combined with the low cost of solar panels has the potential to revolutionize African rural communities by improving access to clean water for home and agricultural usage in addition to providing sustainable energy solutions. The combination of traditional knowledge, cutting-edge technology, and ecological preservation produces a harmonious balance that effectively tackles the complex issues brought about by shifting environmental conditions and social expectations.

In addition to preserving the delicate environmental balance, this comprehensive strategy improves community livelihoods and general well-being. Through the preservation of natural buffer zones, the advancement of contemporary technologies, and the appreciation of Indigenous wisdom, society may create comprehensive solutions that are flexible enough to adjust to the changing trends in water resource conservation. Moreover, this kind of approach can be used as a template for international cooperation, recognizing the interdependence of problems related to water and the necessity of group efforts to guarantee a fair and sustainable future. It is anticipated that water-related issues will grow increasingly complicated, posing challenges to communications, energy, agriculture, transportation, environment, health, and rural and regional development, among other areas (Molobela and Sinha 2011).

The ultimate aim is to conserve water, either in terms of quantity or quality, through various methods such as collecting and storing rainwater, capturing and holding onto rainwater through planting holes, utilizing alternative energy sources like solar power for irrigation, or utilizing protective vegetated areas near water bodies to reduce pollution, improve water quality, etc. by functioning as organic filters as well buffers against runoff via nearby lands. This can be accomplished by learning about the issues pertaining to water that a particular community in a particular region – in this study, SSA – faces, conducting surveys, and implementing water conservation techniques into action while evaluating their efficacy, adaptability, and socioeconomic ramifications.

8.2 Materials and Methods

Mekelle City in northern Ethiopia is the subject of the first case study for RWH. The city is critically short of water, which forces residents to look for other sources of water. Due to poor design and underutilization, only 32 of the 67 RWH systems currently in use are dependable. A number of factors, including inadequate surface area and exposed roof parts, add to the area's RWH systems' unreliability. Taffere et al. (2016) did this investigation. Zai farming successfully stops land degradation in the second case study by Schuler et al. (2016), enhancing livelihoods and ensuring food security. When compared to traditional methods, it produces noticeably larger economic gains per hectare in cereal production – Zai millet surpassing the gross margin by three times. The labour-intensive process of preparing the soil and implementing planting pits, however, prevents the Zai method from being widely adopted.

In the third scenario, limited information about better farming methods restricts possibilities for rural smallholder households, which is made worse by erratic rainfall and depleted soil nutrients. Food shortages are exacerbated by deteriorated agricultural land caused by intensive agriculture and erosion. A lack of consistent access to subterranean water is detrimental to food security and agriculture. Urban growth and development are impacted by the difficulty of installing Solar-Based Irrigation Systems (SBIS) nearer to populated regions; vast distances make investment locations impossible from an economic standpoint. Birhanu et al. conducted this investigation in 2023 (Birhanu et al. 2023). In the most recent case study by Petersen et al. (2020), agricultural practices and the invasion of alien plant species are causing land-derived consequences in the study area, including as nutrient and sediment contamination.

8.2.1 Study Area

SSA, or Sub-Sahara or Non-Mediterranean Africa, is the area and regions of the continent of Africa that lie south of the Sahara. These include Central Africa, East Africa, Southern Africa, and West Africa. The study areas for the different case studies practising various water conservation practices are spread across SSA. Figure 8.1 is a map pinpointing the locations.

Figure 8.1 shows that Indigenous practices like RWH are carried out in Mekelle city, Tigray regional state in Ethiopia, and Zai pits for water conservation are practised in five Northern Burkina Faso states, respectively.

And, modern practices like SBIS are carried out in Bougouni and Koutiala districts, the Sikkaso region in Mali, as well as in the riparian buffer zones of the Breede-Gouritz Water Management Area (WMA), on the southern coast of South Africa.

Water may be everywhere, but its use has always been constrained in terms of availability, quantity, and quality (Biswas 2004). Despite water's role in human development, most individuals and places have a water crisis and water demand exceeds resources in some SSA countries (Mahlangu and Garutsa 2014).

8.3 Water Conservation Practices

8.3.1 Rainwater Harvesting (RWH): An Indigenous Practice

RWH collects, stores, and uses rainwater from rooftops, pavements, and landscapes. Rainwater is collected in tanks, cisterns, or reservoirs and used for irrigation, domestic use, washing, and drinking after purification. This practice conserves water, reduces usage of traditional water sources, and reduces stormwater runoff, making water conservation more sustainable. This method collects rainwater and stores it to minimize water waste and dependence on traditional water sources. Rainwater collection systems can be simple, elaborate, or highly advanced. RWH captures and maximizes rainwater near its source. Examples include improving local food security, passively cooling cities in summer, lowering living costs and energy use, managing erosion, preventing flooding, reviving dead streams, decreasing water pollution, establishing communities, celebrating, and more (Denison and Wotshela 2009). This practice increases climate resilience and local water security as climate patterns change. RWH combines ecological responsibility with practical water needs, making it essential for sustainable living (Raimondi et al. 2023).

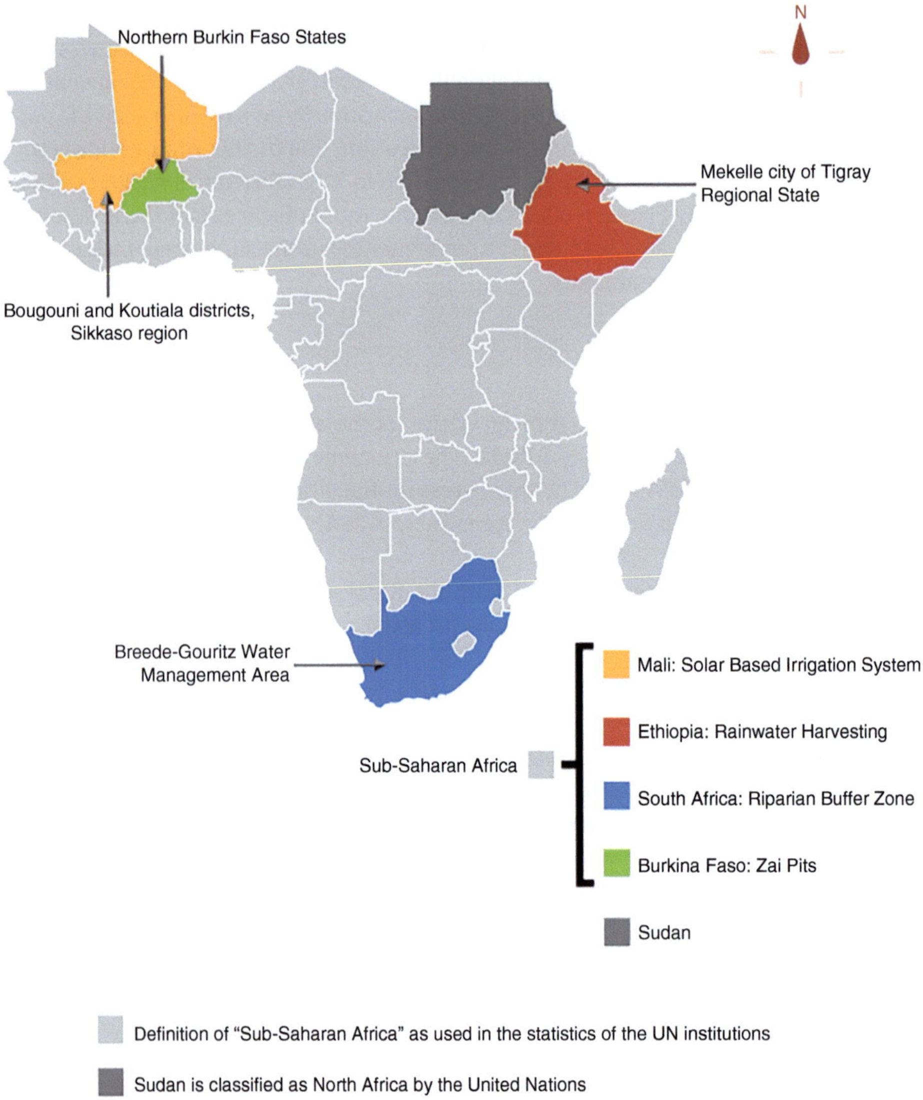

Figure 8.1 A map of Sub-Saharan Africa showcasing the different study areas practising various water conservation practices. *Source:* Adapted from Wikipedia Contributors (2024, January 20). Sub-Saharan Africa.

8.3.2 Types of RWH Systems

RWH encompasses various techniques tailored to diverse environments and needs (Kahinda and Taigbenu 2011; Jakhar et al. 2018).

1) **Rooftop catchment systems**: This is a frequently used technique for gathering rainwater from building rooftops. Rainwater collects on the roof's surface and is guided into the storage tanks or cisterns by gutters and downspouts. This method offers a simple and economical way to collect rainwater for home consumption, irrigation, and a variety of non-potable uses. It works especially well for small-scale and household applications.

2) **Micro catchment systems**: An increasingly site-specific method of RWH is represented by this kind. This technique involves the strategic design of small-scale structures, such as trenches, bunds, or berms, to collect rainwater in a limited space. In arid and semiarid areas, these micro catchments increase soil moisture, encouraging the growth of flora and making agriculture easier. This method works well for maintaining localized ecosystems and assisting agricultural activities in water-scarce regions by optimizing rainwater consumption at the micro level.
3) **Macro catchment systems**: These, on the other hand, work on a bigger scale and frequently entail building contour trenches, check dams, or embankments to collect and channel rainfall across wide areas. In hilly or sloping terrain, where rainwater runoff may be effectively directed and retained for agricultural uses, these systems are very useful. Macro catchment systems enhance rainwater usage for large-scale irrigation projects, minimize soil erosion, and conserve soil by utilizing the natural environment.
4) **Floodwater systems**: This kind of RWH offers a distinctive method of RWH in areas vulnerable to seasonal flooding. These devices use extra rainfall collected during floods to power them instead of traditional catchment techniques. Man-made constructions, like basins or storage reservoirs, are positioned in specific ways to collect and hold floodwater for future use. Floodwater systems can be reliable sources of water throughout dry seasons and help lessen the disastrous effects of flooding by diverting excess water for useful purposes, even if they need careful planning to prevent negative environmental effects.

8.3.3 Zai Pits: An Indigenous Practice

Zai, an ancient agricultural technique from Burkina Faso, involves excavating prior soil to create small planting trenches known as Zais. These trenches, typically 20–30 cm wide and 10–20 cm deep, spaced 60–80 cm apart, play a crucial role in breaking surface crusts and enhancing water infiltration (Nuraddeen et al. 2015). Farmers in the northern region of Burkina Faso use Zais to rehabilitate degraded drylands and restore soil fertility, responding to recurrent droughts and harvest failures. This traditional land rehabilitation technology aims to reclaim severely degraded farmland with hardened surfaces (Ehiakpor et al. 2019). The Zai technique involves constructing planting pits filled with organic matter during the dry season to foster soil moisture retention and nutrient availability. Executed properly, it serves as a conventional soil rehabilitation conservation practice (Nyamekye et al. 2018).

8.3.4 Types of Zai Pits

1) **Classical Zai pits**: Traditional Zai pits are tiny basins that are used by farmers for sowing seeds of seasonal or perennial crops. It is common practice to fill the pits with a combination of mulch, fertilizer, or enriched soil in order to concentrate the nutrients and moisture near the roots of the crop. Because of this method, water is retained more effectively, and the conditions that are ideal for plant growth are created. The ancient method of exploiting these trenches demonstrates a deep awareness of sustainable

agriculture, particularly in areas where there is a lack of water or where the soil is degraded (Nuraddeen Danjuma and Mohammed 2015).

2) **Half-moon Zai pits**: These pits, which are used to address concerns related to land degradation, are shaped like a crescent moon and are used to implement solutions. These half-moon depressions are used by farmers to cultivate consecutive rows of sorghum and legume plants. Through the utilization of this design, water is more effectively captured and directed, resulting in water and soil conservation. This technique not only helps to prevent soil erosion, but also encourages a mutually beneficial interaction between the various types of crops (Nyamekye et al. 2018).
3) **Micro catchment Zai pits**: The micro catchment area is known as Zai pits. Zai pits are bigger pits that are ringed by basins made specifically for the purpose of collecting water from a wider region. This method is especially useful in arid locations, where effective water conservation is essential for the continuation of agricultural production. The design of the micro catchment system makes it possible to collect water over a wider surface area, which is beneficial to the growth of plants and contributes to the overall efforts designed to conserve water (Kimaru-Muchai et al. 2021).
4) **Double Zai pits**: It is a type of pit that consists of two pits that are next to one another and are designed to maximize the amount of water that can be held. In dry locations, where conservation of water is essential for effective crop production, this design is especially advantageous since it allows for more efficient use of water. A reliable water supply for plants during dry seasons is provided by the dual-pit system, which also contributes to greater agricultural resilience. The technology helps maximize water retention (Mati 2006).
5) **Tassa Zai pits**: These are a modified variant of Zai pits that feature stone bunds for the purpose of preventing erosion. Tassa Zai pits are the seventh type of pit. The ability of Zai pits to conserve soil is further improved by this adaptation, particularly in regions that are vulnerable to soil erosion. The incorporation of stone bunds into the frame of the pits helps to strengthen them, reducing the negative effects of water flow and preventing soil erosion (Nuraddeen Danjuma and Mohammed 2015).

8.4 Solar-Based Irrigation Systems (SBIS): A Modern Practice

Within the field of resource-efficient innovations and agricultural sustainability, SBIS have surfaced as a game-changing way to tackle the problems associated with traditional irrigation techniques. It is extremely challenging to extend the electric grid to every site where it is required and obtain dues from every farmer in developing countries, which are often made up of multiple communities and farmers. In addition, there are still problems with water scarcity in many emerging nations, particularly in rural regions (Ramos and Ramos 2009). A more effective, economical, and ecologically friendly method of watering crops and landscapes is made possible by SBIS, which use the power of the sun to revolutionize irrigation practices in response to the world's growing need for agricultural output and water conservation. SBIS use solar energy, which is a plentiful and clean resource, to power water pumping mechanisms, improving water availability and efficiency for horticultural and agricultural uses. This is an environmentally friendly method of producing energy that, after an initial expenditure, yields free energy. Solar-powered solutions are necessary for sustainable development (Suresh et al. 2014). Solar energy is essential for achieving SDG 7, which is to guarantee that everyone has access to modern, cheap, sustainable, and dependable energy sources. The implementation of SBIS holds great promise as resource utilization needs to be optimized and global climate patterns grow more unpredictable. In addition to addressing the energy-water nexus, this technology helps achieve food security in an environmentally responsible way, lessen greenhouse gas emissions, and mitigate climate change (Wikipedia Contributors 2023).

8.4.1 Types of SBIS

1) **Solar photovoltaic water pumping systems (SPWPS)**: Use PV panels to capture solar energy and turn it into electricity, which powers water extraction pumps. Solar PV panels, a motor, and a pump are the standard parts of the system; charge regulators and storage batteries are optional extras. In areas with plenty of sunlight, SPWPS provide an effective and low-maintenance solution. However, obstacles to their wider adoption include their high initial cost and worries about the adverse environmental effects of PV panel production (Gopal et al. 2013;Sontake and Kalamkar 2016; Muhsen et al. 2017).
2) **Solar thermal water pumping systems (STWPS)**: Use concentrated surface collectors or flat-plate solar collectors. The heat is subsequently converted into mechanical work that powers pumps or generates electricity in an engine by means of the Stirling, Brayton, or Rankine cycles. While poor thermal efficiency and greater initial costs are disadvantages of STWPS, new developments, such as medium-to-low temperature Stirling engines, offer promise in mitigating these problems. In some situations, the lack of mechanical moving parts, less maintenance needs, and reduced expenses render STWPS a desirable substitute (Pytilinski 1978; Gopal et al. 2013).

8.4.2 Riparian Buffer Zones: A Modern Practice

Riparian zones are the plains and surroundings of a river, or a different body of water. Riparian buffers that border the floodplains are also considered to be part of the riparian zone, in addition to floodplains. Riparian zones provide recreational benefits to groundwater, downstream land regions, streams, and offer several environmental advantages (Parkyn 2004). The term "riparian" originates from the Latin word "riparius," which means "of or associated with the shore of a river," is where (Naiman and Décamps 1997). As the interfaces between aquatic and terrestrial systems, riparian zones are dynamic, complex habitats with a multitude of uses and advantages. An ecological complex known as riparian zones is created when wetland areas and floodplains are next to rivers or other bodies of water (Parson 1991). These areas are essential for protecting forest ecosystems and enhancing biodiversity (Akay et al. 2012). Riparian zones provide several hydro-ecological benefits, such as soil stabilization, water quality protection, erosion control, blocking the flow of sediment and pollutants into streams, and lowering groundwater flow rates (Dindaroğlu et al. 2015).

8.4.3 Types of Riparian Buffer Zones

There are many different forms of riparian buffers, each meeting a distinct environmental need. Riparian buffers are crucial for maintaining biological balance and water quality (Cunningham et al. 1914).

1) **Grass buffer**: This kind, which is made up entirely of grasslands and forbs, is placed strategically alongside tiny streams and uses native grasses to effectively filter silt and offer important habitat for wildlife. These buffers, which are typically 20–30 ft wide, need to be periodically maintained in order to prevent invasive species and restore the grass.
2) **Three-zone forest buffer system**: This kind provides flexibility to meet landowner aims and water quality standards. This buffer balances timber extraction and aesthetic preferences with Zone 1 having stream-edge forests for durability and habitat, Zone 2 screening sediment and soaking nutrients, and Zone 3 being a grass strip to reduce run-off. A revised version places a small tree and shrub zone between Zones 2 and 3, improving the area's attraction to animals. Depending on the slope, a minimum net buffer width of 50–100 ft is advised.
3) **Two-zone forest buffer**: This kind creates controlled and unmanaged forest zones by eliminating the grass zone from the three-zone buffer. Although it may be desirable to certain landowners, the grass buffer's value and environmental role could be lost. The trade-offs between ecological concerns and land conservation preferences are highlighted by this sort of barrier.
4) **Wildlife buffers**: These are similar to three-zone buffers but place more of an emphasis on wildlife-beneficial vegetation to increase biodiversity near streams. Widening to up to 300 ft in width, wildlife buffers serve as both travel routes and connecting points among larger forest tracts due to their unique ecological features and multiple layers of vegetation.
5) **Urban buffers**: These are essential for capturing pollutants and runoff from populated regions. These buffers use larger planting material, which provides an instant appearance and resistance to human use. They have been created to withstand human impacts. Urban buffers have two functions: they can be recreational paths with greenways along waterways, and they can help with environmental education.
6) **Naturalized buffer**: The kind, which was created from naturally scattered grass, tree, and shrub seeds, is an affordable method. This buffer, whose recommended width is between 50 and 100 ft, efficiently captures runoff while promoting the growth of native plants. This buffer type is a good choice for environmentally responsible land conservation because of its affordability.

8.5 Discussion on Case Studies

8.5.1 Rainwater Harvesting (RWH): An Indigenous Practice

Water scarcity in SSA requires alternative solutions for sustainable home water delivery, and RWH has emerged as a feasible solution, according to this case study by Taffere et al. (2016). In particular, the case study focuses on the dependability of RWH systems in

Mekelle, the capital of the Tigray Regional State in northern Ethiopia, where RWH is made possible by the annual rainfall (Abera et al. 2022). On the other hand, doubts about the efficacy of RWH devices in the area prompted a thorough reliability analysis.

Evaluating the dependability of Mekelle City's current RWH systems is the study's main goal (Imteaz et al. 2013). In order to complete this assessment, which takes into account variables like roof area, monthly water usage, and storage capacity, a mix of field surveys, data gathering, and statistical analysis are used. Considering 90.6% of the total RWH systems within the research area being deemed unreliable, the results raised serious concerns. The roof area is one of the main factors affecting the dependability of RWH systems in Mekelle. It was discovered that a significant portion of homes had additional roof space that was left exposed below the gutter, resulting in ineffective RWH. This emphasizes how crucial it is to maximize roof area use when designing and putting into place RWH systems. The study also found a significant disparity between the project implementers' design criteria and the mean daily water use in the studied area. This discrepancy suggests that the design criteria should be reevaluated in order to better reflect the region's real water demand and usage patterns. The study's ramifications are extensive, going beyond Mekelle to highlight more general difficulties in areas with limited water resources. The study area's underutilization of RWH schemes highlights a systemic problem and the necessity of specialized water policies and strategies for RWH optimization.

Several suggestions are put forward according to the study's findings in order to raise the region's RWH systems' level of dependability. These include maximizing roof area usage, adjusting design requirements to account for changes in household features and locations, and raising the storage capacity for RWH systems to correspond with mean daily water consumption. In conclusion, the dependability analysis of the RWH systems in Mekelle City offers insightful information on the prospects and difficulties related to RWH in SSA's semiarid regions. The study emphasizes the importance of considering local parameters, such as household size, water use, and environmental conditions, when designing and implementing RWH systems, highlighting the need for customized approaches. These variables can be taken into consideration when optimizing RWH systems to efficiently augment the supply of potable water in water-scarce areas. This will support resilient water conservation practices and sustainable water conservation (Taffere et al. 2016).

8.5.2 Zai Pits: An Indigenous Practice

The Zai farming practice, an old conservation technique in Burkina Faso, has acquired popularity for its role in minimizing land degradation and enhancing agricultural output, as demonstrated in this case study by Schuler et al. (2016). In order to better understand the socioeconomic effects of Zai farming in the Ouahigouya area, this case study looks at adoption rates, household features, and potential economic effects. The Zai technique, which was developed in reaction to the destruction of land brought about by the drought of the 1970s, entails digging planting trenches and adding organic matter to improve the fertility of the soil and retention of water (Roose et al. 1999). This approach is commonly used in villages like Tangaye, Salla, Somyaga, and Ziga.

Surveys and interviews were used to gather data for the study from 101 families as well as 16 farms that were revisited. Then, in order to examine the labour and material inputs, yields, and financial gains related to the production of Zai grain, the OLYMPE economical farm system was utilized (Le Grusse et al. 2006). The results showed profitable results for farmers and fields alike, with Zai cultivators showing larger revenue margins per hectare than those using other techniques. With the Zai method, cereal crops – especially sorghum and millet – showed increased yields and productivity. On the other hand, issues including expensive labour and scarce supply were noted as possible roadblocks to Zai farming's further growth. When examining the characteristics of the home, an average farm community in the focus area consists of 13 people, of whom 5 are actively involved in farming. Farm managers are mostly men, and 99% of them are wedded. Based on size, home farms were divided into three categories: small, medium-sized, and big. Particularly, 65% of the owners of households are members of farmers' groups, suggesting that farmers take a considerable interest in the community.

Utilizing the OLYMPE model for economic analysis, information was provided regarding labour expenses, input costs for crops and livestock, and prospective revenue generation. The research showed that Zai farming has improved farm-level lives and food security, highlighting its socioeconomic importance in the area. In summary, Zai farming in Burkina Faso has a significant socioeconomic impact, resulting in favourable economic consequences and increased agricultural output. In addition to reducing soil degradation, the Zai method has improved farming households' standard of living. Potential obstacles to the broad adoption of Zai farming include labour prices and availability, underscoring the need for ongoing study and assistance. This case study emphasizes how important it is for traditional agricultural methods to be used in Burkina Faso to meet current environmental and social issues (Schuler et al. 2016).

8.5.3 Solar-Based Irrigation Systems (SBIS): A Modern Practice

The agricultural landscape of SSA is mostly dependent on rainfed systems, as noted in this case study by Birhanu et al. (2023). However, inconsistent precipitation patterns and poor soils have resulted in low agricultural production and land degradation (Birhanu et al. 2019). In order to overcome these obstacles, a case study concentrated on applying SBIS in the southern Mali region of Sikkaso, specifically in the districts of Bougouni and Koutiala. Due to irregular rainfall patterns and deteriorated soils, SSA has difficulties that lead to low agricultural production and worsening food scarcity. The subterranean hydrological system has limited access to dependable water supplies, which exacerbates the problems with agricultural and nutritional security. The study used Boolean (Shahabi et al. 2014) and fuzzy approaches to select appropriate regions for SBIS deployment and to understand farmers' perspectives based on economic and social information gathered from 112 farm households. The results showed that 20% of the two districts' total land area is viable for investments in solar-powered irrigation. The respondents stressed the necessity of village council-level education initiatives on SBIS functioning and suggested policies for long-term consumption of water and system maintenance. For sustainable agricultural development, the study emphasizes the significance of combining these findings with soil and water conservation strategies. Encouraging gender and youth involvement in SBIS use requires fostering an environment that supports sustainable finance mechanisms.

The effective application of SBIS has great potential to enhance agricultural practices in SSA. Through the provision of a dependable and sustainable irrigation water source, SBIS can improve agricultural output and aid in the conservation of natural resources. Among the difficulties noted are the requirements for policies and training courses to guarantee long-term use and expand the technology into more appropriate environments. The report suggests including women and youth in the implementation of SBIS and recommending sustainable funding methods. The case study concludes by highlighting the promise of SBIS to revolutionize agricultural methods in SSA, especially in southern Mali. The multi-criteria decision-making technique, which identifies appropriate locations for SBIS deployment based on farmers' opinions, offers insightful information about irrigation practices. Future studies should examine the economic and social effects of SBIS on agricultural yields and the utilization of natural resources in SSA, with a particular emphasis on a cost-benefit analysis aimed at encouraging rural farmers to adopt new technologies (Uyan 2013). The effective application of SBIS has the ability to completely transform agricultural practices, resulting in increased agricultural productivity, efficient use of water, and improved food security in the area. Policymakers, scholars, and stakeholders interested in SSA's sustainable agricultural growth and water resource conservation will find this case study to be a useful resource (Birhanu et al. 2023).

8.5.4 Riparian Buffer Zones: A Modern Practice

The assessment of riparian buffer zones to minimize land-derived impacts on water quality in the Southern Cape region of South Africa's Duiwe River catchment is presented in this case study by Petersen et al. (2020). This study provides significant insights into the dynamics of nutrient fluxes and river water quality (Blanchè 2002; Macfarlane et al. 2014). Located in the Breede-Gouritz water conservation area, the study compares the riparian vegetation and land cover/use in the Klein Keurbooms River, a tributary of the Duiwe River, along a downstream gradient. Researchers monitored rainfall, gathered runoff water samples, and investigated water quality variables across native forests, semi-Indigenous riparian areas, and deteriorated zones characterized by cultivation and foreign plant species. The catchment area of the study was 33.82 km^2, with minimal seasonal variation in rainfall. The findings show a distinct gradient of anthropogenic effect that is determined by land use and cover. Because of their closed canopy, greater plant water absorption, and improved transpiration, Indigenous forests showed reduced runoff volumes and higher total nitrogen (TN) concentrations (APHA (American Public Health Association) 2012). On the other hand, nutrient loads were more significantly impacted by cultivation and foreign riparian vegetation, with heavier rainfall events leading to higher concentrations of nutrients. Water quality of discharge and sediment was greatly impacted by species composition variations in riparian buffer zones.

The importance of riparian buffer zones in reducing the effects of land-derived pollutants on water quality is discussed, with particular attention paid to the function that riparian thickness and vegetation type play throughout the downstream river slope. The study provides knowledge about riparian zone conservation and restoration initiatives by highlighting the necessity of particular sites and buffer lengths to maximize nutrient removal rates. In summary, this case study adds insightful information about how well riparian buffer zones mediate nitrogen fluxes and improve the quality of river water. It highlights the value of native forests and clarifies how cultivation and foreign plant invasions affect the quality of the water in South Africa's Southern Cape. The results have consequences for the sustainable conservation of water resources, offering a basis for well-informed choices and focused actions in the restoration and conservation of riparian zones (Petersen et al. 2020).

8.6 Impact

In Ethiopia's semiarid areas, RWH is becoming more and more well-known as a workable way to supplement the scarce surface water and groundwater sources. Taffere et al.'s 2016 study (Taffere et al. 2016) in Mekelle City evaluates the reliability of current RWH systems and the contribution of rainfall to the water supply, offering important information for future adaptation plans in semiarid areas. According to Schuler et al. (2016), Zai farming significantly improves lifestyle and food security by reversing land degradation. When compared to traditional methods, it produces noticeably larger financial advantages per hectare in cereal production – Zai millet surpassing the gross margin by three times. The labour-intensive process of preparing the soil and implementing planting pits, however, prevents the Zai method from being widely adopted.

The financial well-being of small rural farm households is directly impacted by low livestock and crop productivity, which exacerbates the increased food shortages brought on by degraded and less productive agricultural land. Options for agricultural and nutritional security are further limited by unreliable availability of water resources, as noted in the case study by Birhanu et al. (2023). Furthermore, SBIS's close proximity to populated regions could be a barrier to further development and the growth of metropolitan areas. According to a study by Petersen et al. (2020), riparian buffer zones are essential for reducing problems with water quality in South Africa's farming rivers. They offer essential functions such as nutrient absorption, aquifer replenishment, and flood control. When it comes to lowering nutrient and sediment loads, riparian buffer zones perform better than larger regions overrun by alien plant species. Superior ecological benefits are provided by small buffers using native or hybrid plants. Riparian zones differ in their physiographic contexts, which influences their effectiveness and width. Plant invasions and other changes may make it more difficult for them to carry out vital ecological tasks.

Based on the case studies, here is a table comparing key aspects of Indigenous and modern water conservation practices:

To summarize Table 8.1, traditional methods provide decentralized solutions that are ideal for smallholder farms in rural areas. However, modern systems necessitate the integration of technology, markets, financing, and capacity building to achieve widespread acceptance and optimize potential benefits. All things considered, an integrated approach incorporating both traditional and modern techniques enables communities to approach water conservation issues holistically.

Table 8.1 Comparison of key aspects of Indigenous and modern water conservation practices.

Parameter	Rainwater harvesting systems	Zai pits	Solar-based irrigation systems	Riparian buffer zones
Description	Capture and store rainwater from surfaces	Small planting pits to trap rainfall	Use solar energy to power water pumps	Vegetated zones along waterways
Traditional or modern	Traditional	Traditional	Modern	Modern
Setup costs	Low	Low	High equipment costs	Low if allowing natural regrowth
Maintenance needs	Low	Labor intensive	Medium, some technical skills required	Low if protected from impacts
Crop productivity impact	Supplements rain-fed farming	Enhances soil fertility	Enables year-round irrigation	Stabilizes soil, water supplies
Environmental impact	Minimal if well-managed	Risk of soil erosion	PV panel production concerns	Protects water quality, wildlife
Socioeconomic impact	Reduces water expenses	Increases yields and incomes	Requires financing and markets	Enhances ecosystem services
Climate resilience	Rainfall dependent	Soil moisture retention reduces risk	Solar energy offsets rainfall variability	Buffers extreme weather impacts

Source: Schuler et al. (2016), Taffere et al. (2016), Petersen et al. (2020) and Birhanu et al. (2023).

8.7 Mitigative Measures

1) Promoting sustainable irrigation systems that maximize food production without increasing water use and with environmental benefits is crucial. Sustainable irrigation methods could boost agricultural output and reduce environmental impacts like energy use (Nhamo et al. 2023).
2) Investments in water infrastructure are vital for socioeconomic growth, particularly in developing nations with variable water resources. These services include access, storage, regulation, and conservation of water resources (Grey and Sadoff 2007). In SSA, climate-related problems comprise droughts, floods, and erratic rainfall. The region's food, water, and economic security are all at risk due to the economic and human capacity constraints in water resource conservation, which restrict access to clean water (Conway et al. 2015; Affoh et al. 2022). The provision of water infrastructure and investment has been given top priority by several SSA governments in response to the UN's Millennium Development Goals (MDG) and Sustainable Development Goals (SDGs) (Dangui and Jia 2022).

3) Developing biologically and chemically safe wastewater treatment systems. Effective wastewater collection and treatment system should have low implementation costs, established technology, ease of operation, and flexibility for future upgrades (Bahri et al. 2008).

For cooling and boiler feed, treated wastewater is a sustainable choice; however, food businesses need potable water for cleaning (i). Impurities, scaling, corrosion, biological growth, fouling, and foaming can be removed by advanced wastewater treatment (ii). Low hardness, salt, sediments, organic matter, nutrients, and pathogenic microbe elimination are goals for industrial water treatment systems (Peletz et al. 2018; Gulamussen et al. 2019; Fils 2021).

1) Education is essential for water conservation awareness and home water conservation. Clean water boosts health and education. Waterborne infections and water collection time might impact school performance. Due to reduced water collection burdens, better access to water increases engagement and academic performance, according to studies. Promoting water and sanitation services reduces healthcare expenses, empowers communities, especially women, and creates a more sustainable future (Panday et al. 2021; Das et al. 2022; Kumar et al. 2023a, b; Neog et al. 2024).
2) Water quality monitoring is vital for public health, particularly in low-income nations where unsafe drinking water causes diseases and fatalities, especially among children. Over 500,000 people die from waterborne illnesses annually. The right water quality data assist in recognizing dangerous water sources, guide treatment, and evaluate sanitation initiatives. Operations and surveillance water quality monitoring are usually required by regulations, but limited finances and enforcement often prevent compliance. Improving facilities, providing equipment, deploying mobile data conservation apps, and educating staff are suggestions. Qualitative Comparative Analysis (QCA) helps identify institutional factors that affect water quality monitoring in order to establish effective improvement measures (Panday et al. 2021, 2023; Panday and Kumar 2022, 2023; Asok Raja et al. 2023).

8.8 Conclusion

African water challenges are addressed in a variety of ways, as evidenced by the comparison of traditional and modern approaches to water conservation on the continent. Indigenous techniques such as RWH and Zai pits show how adaptive and resilient societies can be when it comes to using their natural resources. These methods have their roots in customs and conventional wisdom. These techniques significantly increase agricultural productivity and promote sustainable utilization of water.

Conversely, newer techniques like SBIS and riparian buffer zones emphasize the value of cutting-edge science and technology when it comes to solving modern water issues. The integration of modern advances and Indigenous wisdom presents a holistic and dynamic approach to water conservation. This strategy recognizes the value of customs while also emphasizing the need for technological advancements to guarantee Africa's access to water in the future. Collaboration between antiquated customs and contemporary technology is crucial for paving a resilient and sustainable path forward for water conservation on the continent; this is wherein Integrated Water Resource Management (IWRM) enters into play.

References

Abera, K.A., Gebreyohannes, T., Abrha, B. et al. (2022). Vulnerability mapping of groundwater resources of Mekelle City and surroundings, Tigray Region, Ethiopia. *Water* 14 (16): 2577.

Affoh, R., Zheng, H., Dangui, K., and Dissani, B.M. (2022). The impact of climate variability and change on food security in sub-Saharan Africa: perspective from panel data analysis. *Sustainability* 14 (2): 759.

Akay, A.E., Wing, M.G., and Sessions, J. (2012). Estimating structural properties of riparian forests with airborne lidar data. *International Journal of Remote Sensing* 33 (22): 7010–7023.

APHA (American Public Health Association) (2012). *Standard Methods for the Examination of Water and Wastewater*, 22e. American Public Health Association.

Asok Raja, S.K., Panday, D.P., and Kumar, M. (2023). Decoding the enigma of 100-year record-breaking rainfall over Tamil Nadu using wavelet analysis. *Groundwater for Sustainable Development* 101007. https://doi.org/10.1016/j.gsd.2023.101007.

Bahri, A., Drechsel, P., andBrissaud, F. (2008). Water reuse in Africa: challenges and opportunities. RePEc: research papers in economics. AgEcon search.

Birhanu, B.Z., Traoré, K., Gumma, M.K. et al. (2019). A watershed approach to managing rainfed agriculture in the semiarid region of southern Mali: integrated research on water and land use. *Environment, Development and Sustainability* 21 (5): 2459–2485.

Birhanu, B.Z., Sanogo, K., Traore, S.S. et al. (2023). Solar-based irrigation systems as a game changer to improve agricultural practices in sub-Sahara Africa: a case study from Mali. *Frontiers in Sustainable Food Systems* 7: 1–14.

Biswas, A.K. (2004). Integrated water resources management: a reassessment. *Water International* 29 (2): 248–256.

BlanchèC. (2002). The use of riparian buffer zones for the attenuation of nitrate in agricultural landscapes.MSc thesis.University of Natal, South Africa.

Conway, D., van Garderen, E.A., Deryng, D. et al. (2015). Climate and southern Africa's water–energy–food nexus. *Nature Climate Change* 5 (9): 837–846.

Cunningham, K., Stuhlinger, C., &Liechty, H. (1914). Riparian buffers: types and establishment methods [fact sheet]. Agriculture and Natural resources.

Dangui, K. and Jia, S. (2022). Water infrastructure performance in sub-Saharan Africa: an investigation of the drivers and impact on economic growth. *Water* 14 (21): 3522.

Das, A., Kumar, M., Jha, P.K. et al. (2022). Isotopic and hydrogeochemical tracking of dissolved nutrient dynamics in the Brahmaputra River System: a source delineation perspective. *Chemosphere* 307: 135757. https://doi.org/10.1016/j.chemosphere.2022.135757.

Denison, J.andWotshela, L. (2009). Indigenous Water Harvesting and Conservation Practices: Historical Context, Cases and Implications.Report No. TT 392/09.Water Research Commission.

Dindaroğlu, T., Reis, M., Akay, A.E., and Tonguç, F. (2015). Hydroecological approach for determining the width of riparian buffer zones for providing soil conservation and water quality. *International Journal of Environmental Science and Technology* 12 (1): 275–284.

Ehiakpor, D.S., Danso-Abbeam, G., Dagunga, G., and Ayambila, S.N. (2019). Impact of Zai technology on farmers' welfare: Evidence from northern Ghana. *Technology in Society* 59: 101189.

Fils, L.R.N. (2021). Linking water access and education in the sustainable development goals in sub-Saharan Africa. *Asian Journal of Empirical Research* 11 (6): 50–58.

Flavier, J.M., De Jesús, A., and Navarro, C.S. (1995). *The Regional Program for the Promotion of Indigenous Knowledge in Asia (REPPIKA).* Practical Action Publishing.

Gopal, C., Mohanraj, M., Chandramohan, P., and Chandrasekar, P. (2013). Renewable energy source water pumping systems—a literature review. *Renewable and Sustainable Energy Reviews* 25: 351–370.

Grey, D. and Sadoff, C.W. (2007). Sink or swim? Water security for growth and development. *Water Policy* 9 (6): 545–571.

Gulamussen, N.J., Arsénio, A.M., Matsinhe, N.P., and Rietveld, L.C. (2019). Water reclamation for industrial use in sub-Saharan Africa – a critical review. *Drinking Water Engineering and Science* 12 (2): 45–58.

Imteaz, M.A., Ahsan, A., and Shanableh, A. (2013). Reliability analysis of rainwater tanks using daily water balance model: variations within a large city. *Resources, Conservation and Recycling* 77: 37–43.

Jakhar, S., Chakhiya, R., and Dhankar, M. (2018). Rain water harvesting and its methods. *International Journal of Engineering Research and Technology* 3 (10): 1–4.

Kahinda, J.M. and Taigbenu, A.E. (2011). Rainwater harvesting in South Africa: challenges and opportunities. *Physics and Chemistry of the Earth, Parts a/B/C* 36 (14–15): 968–976.

Kimaru-Muchai, S.W., Ngetich, F.K., Mucheru- Muna, M.W., and Baaru, M. (2021). Zai pits for heightened sorghum production in drier parts of Upper Eastern Kenya. *Heliyon* 7 (9): e08005.

Kowenje, C., Haarstrick, A., Biswick, T. et al. (2022). *From Traditional to Modern African Water Management*, 1e. Springer International Publishing.

Kumar, M., Panday, D.P., Bhagat, C. et al. (2023a). Demystifying the decadal shift in the extent of groundwater in the coastal aquifers of Gujarat, India: a case of reduced extent but increased magnitude of seawater intrusion. *Science of the Total Environment* 165451. https://doi.org/10.1016/j.scitotenv.2023.165451.

Kumar, M., Mazumder, P., Silori, R. et al. (2023b). Prevalence of PPCPs, microplastics and co-infecting microbes in the post-COVID-19 era and its implications on antimicrobial resistance and potential endocrine disruptive effects. *Science of the Total Environment* 166419. https://doi.org/10.1016/j.scitotenv.2023.166419.

Le Grusse, P., Belhouchette, H., Le Bars, M. et al. (2006). Participative modelling to help collective decision-making in water allocation and nitrogen pollution: application to the case of the Aveyron-Lere Basin. *International Journal of Agricultural Resources, Governance and Ecology* 5 (2/3): 247.

Macfarlane, DM.,Bredin, IP., Adams, JB., Zungu, MM., Bate, GC., &Dickens, CWS. (2014). *Preliminary Guideline for the Determination of Buffer Zones for Rivers, Wetlands and Estuaries*. Report No. TT 610/14.Water Research Commission.

Mahlangu, M. and Garutsa, T.C. (2014). Application of indigenous knowledge systems in water conservation and management: the case of Khambashe, Eastern Cape South Africa. *Academic Journal of Interdisciplinary Studies* 3 (4): 151–160.

Mati, B.M. (2006). *Overview of Water and Soil Nutrient Management UnderSmallholder Rain-Fed Agriculture in East Africa*. IWMI.

Molobela, I.P. and Sinha, P.K. (2011). Management of water resources in South Africa: a review. *African Journal of Environmental Science and Technology* 5 (12): 993–1002.

Muhsen, D.H., Khatib, T., and Nagi, F. (2017). A review of photovoltaic water pumping system designing methods, control strategies and field performance. *Renewable and Sustainable Energy Reviews* 68: 70–86.

Naiman, R.J. and Décamps, H. (1997). The ecology of interfaces: Riparian Zones. *Annual Review of Ecology and Systematics* 28 (1): 621–658.

Neog, N., Goswami, R., Panday, D.P. et al. (2024). Arsenic contamination in the groundwater of Northeastern India: critical understandings on geotectonic controls and the need for intervention. *Current Opinion in Environmental Science & Health* 100539. https://doi.org/10.1016/j.coesh.2024.100539.

Ngoran, S.D., Dogah, K.E., and Xue, X. (2015). Assessing the impacts of climate change on water resources: the sub-Saharan Africa perspective. *Journal of Economics and Sustainable Development* 6 (1): 185–193.

Nhamo, L., Mpandeli, S., Liphadzi, S. et al. (2023). *Advances in Water Research: Enhancing Sustainable Water Use in Irrigated Agriculture in South Africa*. Elsevier.

Nuraddeen Danjuma, M. and Mohammed, S. (2015). Zai pits system: a catalyst for restoration in the drylands. *IOSR Journal of Agriculture and Veterinary Science (IOSR-JAVS)* 8: 2319–2372.

Nyamekye, C., Thiel, M., Schönbrodt-Stitt, S. et al. (2018). Soil and water conservation in Burkina Faso, West Africa. *Sustainability* 10 (9): 3182.

Olokeogun, O.S. and Kumar, M. (2022). Indicator-based vulnerability assessment of riparian zones in Nigeria's Ibadan region due to urban settlement pressure. *Environmental Challenges* 7: 100501.

Oweis, T. and Hachum, A. (2006). Water harvesting and supplemental irrigation for improved water productivity of dry farming systems in West Asia and North Africa. *Agricultural Water Management* 80 (1–3): 57–73.

Panday, D.P. and Kumar, M. (2022). Climate indices and hydrological extremes: deciphering the best-fit model. *Environmental Research* 215: 114301. https://doi.org/10.1016/j.envres.2022.114301.

Panday, D.P. and Kumar, M. (2023). Application of remote sensing techniques to deal with scale aspects of GRACE data to quantify groundwater levels. *MethodsX* 102108. https://doi.org/10.1016/j.mex.2023.102108.

Panday, D.P., Khosa, R., Maheswaran, R. et al. (2021). Game-theoretic-based modelling of Krishna waters dispute: equilibrium solutions by metagame analysis. *The European Physical Journal B* 94: 101. https://doi.org/10.1140/epjb/s10051-021-00107-w.

Panday, D.P., Khosa, R., Maheswaran, R. et al. (2021). Game theoretic-based modelling of Krishna waters dispute: equilibrium solutions by hypergame analysis. *The European Physical Journal B* 94: 131. https://doi.org/10.1140/epjb/s10051-021-00135-6.

Panday, D.P., Singh, A., Gaur, S. et al. (2023). Probable maximum precipitation analysis of high rainfall regions in India. *Groundwater for Sustainable Development* 100893. https://doi.org/10.1016/j.gsd.2022.100893.

Parkyn, S., (2004). Review of Riparian Buffer Zone effectiveness. Publication no. MAF technical paper no: 2004/05.

Parson, A. (1991). *Conservation and Ecology of Riparian Tree Communities in the Murray-Darling Basin, New South Wales: Literature Review*. New South Wales: NSW National Parks and Wildlife Service.

Peletz, R., Kisiangani, J., Bonham, M. et al. (2018). Why do water quality monitoring programs succeed or fail? A qualitative comparative analysis of regulated testing systems in sub-Saharan Africa. *International Journal of Hygiene and Environmental Health* 221 (6): 907–920.

Petersen, C., Jovanovic, N., and Grenfell, M. (2020). The effectiveness of riparian zones in mitigating water quality impacts in an agriculturally dominated river system in South Africa. *African Journal of Aquatic Science* 45 (3): 336–349.

Pytilinski, J.T. (1978). Solar energy installations for pumping irrigation water. *Solar Energy* 21 (4): 255–262.

Raimondi, A., Quinn, R., Abhijith, G.R. et al. (2023). Rainwater harvesting and treatment: state of the art and perspectives. *Water* 15 (8): 1518.

Ramos, J.S. and Ramos, H.M. (2009). Solar powered pumps to supply water for rural or isolated zones: a case study. *Energy for Sustainable Development* 13 (3): 151–158.

Roose, E., Kabore, V., and Guenat, C. (1999). Zai practice: a West African traditional rehabilitation system for semiarid degraded lands, a case study in Burkina Faso. *Arid Soil Research and Rehabilitation* 13 (4): 343–355.

Schuler, J., Voss, A.K., Ndah, H.T. et al. (2016). A socioeconomic analysis of the zaï farming practice in northern Burkina Faso. *Agroecology and Sustainable Food Systems* 40 (9): 988–1007.

Shahabi, H., Keihanfard, S., Ahmad, B., and Amiri, M.J.T. (2014). Evaluating Boolean, AHP and WLC methods for the selection of waste landfill sites using GIS and satellite images. *Environmental Earth Sciences* 71 (9): 4221–4233.

Singh, D.B., Mahajan, A., Devli, D. et al. (2021). A mini review on solar energy-based pumping system for irrigation. *Materials Today Proceedings* 43: 417–425.

Sontake, V.C. and Kalamkar, V.R. (2016). Solar photovoltaic water pumping system - A comprehensive review. *Renewable and Sustainable Energy Reviews* 59: 1038–1067.

Suresh, H., Kumar, R., Umasankar, V., and Thulasi, V. (2014). Solar powered smart irrigation system. *Research India Publications* 4 (4): 341–346.

Taffere, G.R., Beyene, A., Vuai, S.A.H. et al. (2016). Reliability analysis of roof rainwater harvesting systems in a semi-arid region of sub-Saharan Africa: case study of Mekelle, Ethiopia. *Hydrological Sciences Journal* 61 (6): 1135–1140.

Uyan, M. (2013). GIS-based solar farms site selection using analytic hierarchy process (AHP) in Karapinar region, Konya/Turkey. *Renewable and Sustainable Energy Reviews* 28: 11–17.

Wikipedia Contributors (2023, November 6). Sustainable development goal 7. Wikipedia.

Wikipedia Contributors (2024, January 20). Sub-Saharan Africa. Wikipedia.

9

Water Security in South America

Sushila Arya[1], *Ganesh Datt Bhatt*[2], *Minakshi Pandey*[3], *and Divya Chaudhary*[4]

[1] *Department of Agriculture, Dev Bhoomi Uttarakhand University, Dehradun, Uttarakhand, India*
[2] *Department of Agro-Forestry, College of Agriculture Sciences, An ICAR Accredited College, Teerthanker Mahaveer University, Moradabad, Uttar Pradesh, India*
[3] *Department of Chemistry, School of Science, IFTM University, Moradabad, Uttar Pradesh, India*
[4] *Department of Genetics and Plant Breeding, G.B. Pant University of Agriculture and Technology, Udham Singh Nagar, Uttarakhand, India*

9.1 Introduction

Water is the essence of life and an invaluable natural resource that sustains all living beings and ecosystems on Earth. The distribution of water resources is not uniform across the globe, owing to the planet's geographical diversity (Perry and Vanderklein 2009). From arid deserts to lush rainforests, and from towering mountains to expansive plains, the Earth's varied landscapes directly influence the availability and accessibility of water.

Water is a fundamental resource that sustains life, ecosystems, and economies. Water is a lifeline for agriculture, powering hydroelectricity, supporting industries, and maintaining vital ecosystems. Water stress situation can have serious consequences on societal well-being, economic development, and ecosystem health. It often results in competition among sectors for limited water resources, affecting agriculture, industries, and communities and can even lead to conflicts over water allocation (Cosgrove and Loucks 2015). Several regions across the world are currently facing varying degrees of water stress. This underscores the urgency to adopt strategies that enhance water security and alleviate water stress. Water security is a critical global concern that revolves around ensuring reliable access to clean and sufficient water resources for both human populations and the environment. Water is essential for various human activities, from drinking and sanitation to agriculture, industry, and energy production (Cook and Bakker 2012). However, the increasing demands for water, coupled with factors like population growth, urbanization, climate change, and pollution, have led to water stress becoming a significant issue. WHO views water security as the assurance of safe and sufficient water for personal and domestic uses. This includes access to safe drinking water, proper sanitation facilities, and hygiene practices to prevent waterborne diseases and improve public health (Howard et al. 2003).

Water Scarcity Management: Enabling Technologies, First Edition. Edited by Kanchan D. Bahukhandi, Manish Kumar, Durga P. Panday, and Tushara G.G. Chaminda.

Whereas FAO focuses on water security's connection to agriculture and food production. It highlights the need to manage water resources for agricultural purposes to ensure food security while also protecting ecosystems (Mabrey and Vittorio 2018). In the context of South America, water plays a pivotal role in shaping the region's environment, society, and development. This chapter explores the intricate web of water security, availability, and the challenges faced by countries across South America. It explores the diverse landscapes, the significance of water resources, the pressures threatening water security, and potential strategies for a sustainable water future. South America's vast expanse encompasses a remarkable variety of ecosystems, from the Amazon rainforest to the Andes mountains and the arid Atacama Desert. This diversity leads to distinct water availability patterns, with countries like Brazil and Colombia possessing abundant water resources, while nations like Chile and Peru experience chronic water scarcity. The Andes mountain range serves as a significant water source, hosting glaciers and rivers that provide water for agriculture, energy production, and domestic use across the continent. The water scarcity level and mean water crowding index 10-year return period (m^3 per capita/year) in South America are represented in Figure 9.1.

9.2 Water Security in South America

9.2.1 South America

South America is the fourth-largest continent, a land of breath-taking landscapes, cultural diversity, and historical significance. Bordered by the Atlantic Ocean to the east, the Pacific Ocean to the west, and the Caribbean Sea to the north, South America is a continent of contrasts, featuring everything from dense Amazon rainforests to towering Andes mountains. The continent boasts a wide range of geographical features, the Amazon Rainforest, located in South America, is home to the Amazon River, which is the world's largest river by discharge volume (Crane and Fletcher 2017). The Amazon Basin is rich in water resources and plays a crucial role in regulating the global climate and biodiversity. Despite having vast water resources in certain areas, South America also experiences water stress in various regions. Water stress occurs when demand for water exceeds the available supply (Hoekstra et al. 2012). Factors contributing to water stress in the region include population growth, urbanization, agricultural expansion, and changing climate patterns.

9.2.2 Factor Affecting Water Security in South America

9.2.2.1 Geographical Diversity

South America exhibits significant geographical diversity in terms of water security due to variations in climate, geography, hydrology, and human factors. Water security refers to the availability, access, and sustainable management of water resources to meet societal needs. Here is an overview of the geographical diversity of water security in South America:

Amazon basin: The Amazon River and its vast basin cover a substantial portion of South America, providing abundant freshwater resources. However, despite the apparent abundance, water security can be a concern due to uneven distribution and potential pollution from human activities, including deforestation and mining (de Alkmim 2015).

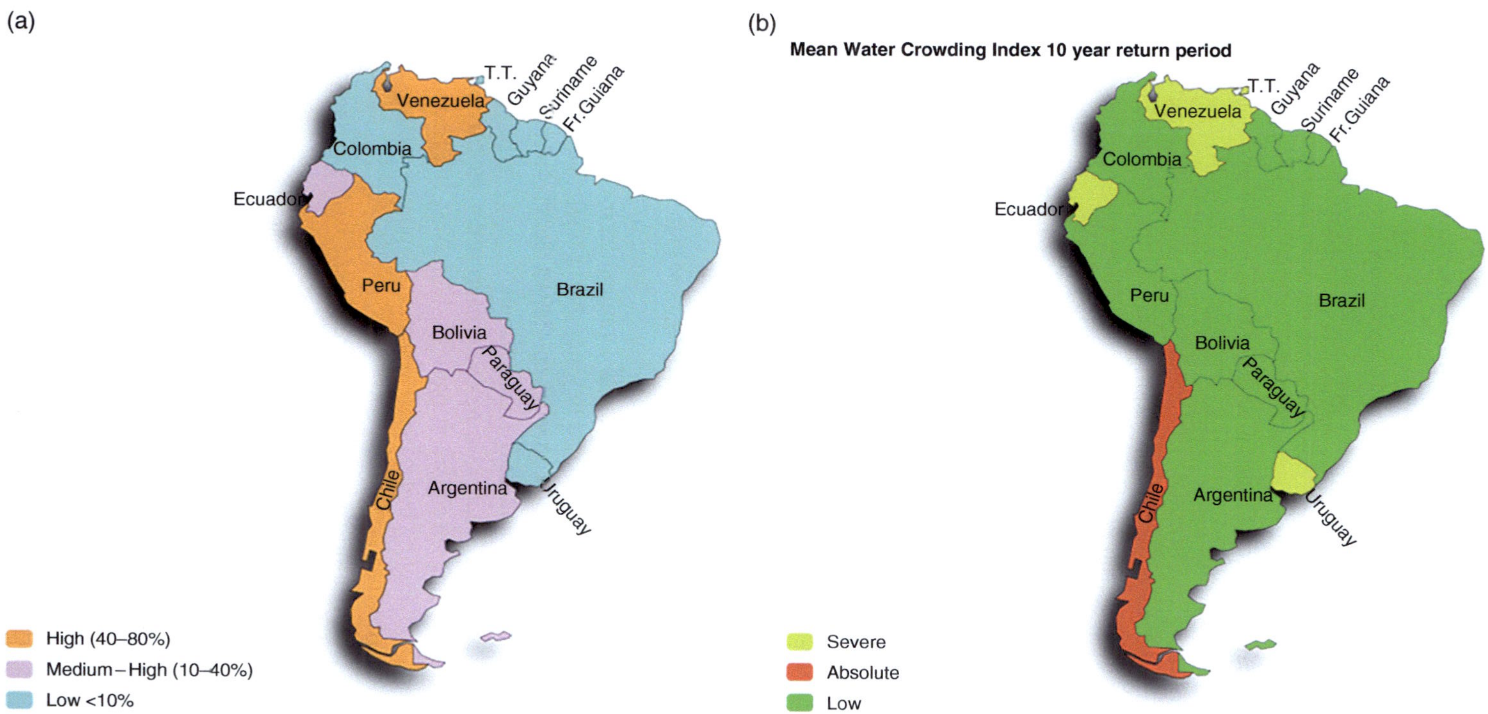

Figure 9.1 (a) Water scarcity level and (b) mean water crowding index 10-year return period (m^3 per capita/year) in South America.

Andean region: The Andes Mountains stretch across several countries in South America. Glaciers and high-altitude ecosystems in this region play a critical role in providing water resources to downstream communities (Johansen et al. 2018). However, the accelerated melting of glaciers due to climate change threatens long-term water availability, leading to water security challenges for communities that depend on glacial meltwater.

Atacama desert: Located along the western coast, the Atacama Desert is one of the driest places on Earth. Water scarcity is a constant challenge for the populations in this region, leading to the development of innovative water management strategies, such as desalination and water recycling.

Brazilian highlands: The Brazilian Highlands include the vast Cerrado biome, which serves as a crucial water source for major rivers like the Paraná and São Francisco. Agriculture and urbanization in this region can impact water quality and availability, affecting both rural and urban communities (Maliva and Missimer 2012).

Pampas and Chaco: These grassland ecosystems cover parts of Argentina, Uruguay, Paraguay, and southern Brazil. Water security in these areas is closely linked to sustainable land-use practices to prevent soil degradation and maintain water quality (Liu et al. 2012).

Coastal cities: Coastal cities like Rio de Janeiro, São Paulo, and Lima face water security challenges due to urbanization, pollution, and increasing demand. These cities often need to manage their water resources carefully to ensure a consistent and safe water supply for their populations.

Hydropower and water governance: South America heavily relies on hydropower for electricity generation, making water availability crucial for energy security. Efficient water governance, transboundary agreements, and sustainable development are essential for managing shared water resources, such as those in the Paraná and La Plata basins (Mugagga and Nabaasa 2016).

Indigenous communities: Many Indigenous communities in South America have traditional knowledge and practices related to water management. However, their access to clean and reliable water sources is often compromised due to external factors, including infrastructure development and land encroachment.

Climate change impact: Climate change exacerbates water security challenges across South America, affecting precipitation patterns, increasing the frequency of extreme weather events, and altering hydrological systems. Adaptation strategies are crucial to mitigate these impacts.

Transboundary water management: Several rivers and basins in South America cross international borders, necessitating cooperation and agreements among neighboring countries to ensure equitable water sharing and sustainable management (Giordano and Wolf 2003).

9.2.3 Role of Topography

The Earth topography, characterized by its mountains, valleys, plateaus, and plains, plays a pivotal role in shaping water availability and distribution. Mountains, for instance, act as natural barriers to the movement of water, giving rise to phenomena like the rain shadow effect. As moist air rises against the windward side of a mountain range, it cools and condenses, resulting in rainfall. Conversely, the leeward side, or rain shadow, experiences drier conditions due to descending air masses. The Himalayas and the Andes are classic examples of such topographically induced water disparities. The topography of South America plays a significant role in shaping water security across the continent. Here is how topography influences water security in South America:

Mountainous regions and watershed management: The Andes Mountains run along the western edge of South America, creating a complex network of watersheds and river systems. These mountains are the source of many major rivers, such as the Amazon, Orinoco, and Paraná. The topography influences the distribution of water resources, as precipitation in the highlands feeds into these river systems (Winemiller et al. 2008). Effective watershed management is crucial to ensure equitable access to water resources downstream and to prevent water scarcity in regions that depend on these rivers for their water supply.

Glacial meltwater: Many Andean glaciers provide a significant portion of the water supply for downstream communities. However, due to climate change, these glaciers are melting at an accelerated rate, affecting the timing and availability of water resources (Mark et al. 2010). This impacts the reliability of water supply for both human consumption and agriculture. Changes in glacier melt patterns require adaptive water management strategies to ensure continued water security in the region.

Rainfall patterns and droughts: The topography of South America also affects rainfall patterns. The east side of the Andes, for instance, receives heavy rainfall due to orographic effects. In contrast, the western slopes are much drier due to the rain shadow effect. Variability in rainfall can lead to periods of drought in certain regions, affecting water availability for agriculture and other purposes. Water storage infrastructure, such as reservoirs, can help mitigate the impacts of drought by storing excess water during wet periods.

Flooding and erosion: The steep slopes of the Andes can lead to rapid runoff during heavy rainfall, causing flash floods and soil erosion (Houston 2002). Effective land-use planning, soil conservation measures, and flood management strategies are essential to minimize the negative impacts of such events on water security and local communities.

Transboundary water management: The Andean region spans multiple countries and many of its major rivers cross international borders. This creates challenges for equitable water allocation and management among neighboring countries (Albrecht et al. 2021). The topography influences the movement of water across borders, and collaborative efforts are required to prevent conflicts over shared water resources.

Urbanization and water quality: Urban areas in South America often develop in valleys and plains, where topography can influence drainage patterns and water quality (Lowrance et al. 1997). Poor urban planning and inadequate wastewater treatment can lead to pollution of water sources, impacting both human health and ecosystem integrity. Proper wastewater management and the protection of water sources are essential for maintaining water security.

Hydropower generation: The steep terrain of the Andes offers excellent opportunities for hydropower generation. While hydropower can contribute to energy security, the construction of dams can also impact local ecosystems and alter water flow patterns, affecting downstream water availability.

9.2.4 Rivers and Watersheds

Rivers serve as lifelines for civilizations, but their presence and vitality are heavily influenced by geographical diversity. The intricate networks of rivers and their associated watersheds are directly shaped by factors such as geological formations, soil types, and vegetation cover. Rivers and watersheds play a crucial role in determining water security in South America. Water security refers to the availability of sufficient, safe, and reliable water resources for various purposes, including drinking, agriculture, industry, and ecosystem health. In South America, several factors related to rivers and watersheds contribute to the region's water security challenges:

Geographic diversity and climate variability: South America encompasses diverse ecosystems, from tropical rainforests to arid deserts and high-altitude Andean regions. The distribution of water resources varies significantly across these regions due to varying precipitation patterns, temperature, and altitude. Some areas experience heavy rainfall and flooding, while others face prolonged droughts.

Andean glaciers and snowmelt: The Andes mountain range stretches along the western edge of South America and is a significant source of water for downstream regions. Glaciers and snowpack store water that is released slowly over time, contributing to the flow of major rivers like the Amazon and the Orinoco (Urrutia and Vuille 2009). However, the rapid melting of glaciers due to climate change threatens the reliability of this water source.

Transboundary rivers: Many major rivers in South America are transboundary, flowing through multiple countries. This can lead to complex challenges related to water allocation, sharing, and management. Disputes over water usage and dam construction can strain diplomatic relations and impact water security for downstream nations.

Deforestation and land use: The Amazon Rainforest, often referred to as the "lungs of the Earth," plays a critical role in regulating the water cycle in South America. Deforestation and land-use changes can disrupt this cycle, leading to reduced rainfall, altered precipitation patterns, and increased soil erosion (Oyague and Cooper 2020). These changes affect both local water availability and regional climate patterns.

Urbanization and pollution: Rapid urbanization in South American cities has led to increased demand for water resources. Unplanned growth can strain existing water supply systems, leading to water scarcity in urban areas. In addition, inadequate wastewater treatment and pollution from industries and agriculture can contaminate water sources, impacting both water quality and availability (Edokpayi et al. 2017).

Hydropower and infrastructure: South America relies heavily on hydropower for electricity generation. While hydropower contributes to energy security, large dams and reservoirs can alter river ecosystems, affect fish migration, and displace communities (Kuriqi et al. 2021). The management of these structures is essential to balance energy needs with environmental and social considerations.

Climate change: The changing climate is expected to have significant impacts on water resources in South America. Changes in precipitation patterns, increased temperatures, and more frequent extreme weather events can exacerbate existing water security challenges (Rangecroft et al. 2013). Melting glaciers, altered river flows, and shifting ecosystems all contribute to uncertainties in future water availability.

Inequitable access: Water security challenges in South America often highlight inequities in access to water resources. Indigenous communities, rural populations, and marginalized groups may have limited access to clean and reliable water sources, leading to issues of social justice and human rights.

9.2.5 Arid and Semiarid Regions

Arid and semiarid regions can have a significant impact on water security in South America due to their inherent low levels of precipitation and high evaporation rates. These regions often experience water scarcity and limited access to freshwater resources, which can lead to various challenges for both human populations and ecosystems. Here is how these regions affect water security in South America:

Limited water availability: Arid and semiarid regions receive limited rainfall, resulting in reduced surface water availability. This scarcity of water resources can make it difficult for local communities to meet their domestic, agricultural, and industrial water needs (Falkenmark et al. 1989).

Agricultural challenges: Agriculture is a major water-consuming sector, and in arid and semiarid regions, water scarcity can hamper agricultural productivity (Alam 2015). Limited water availability may lead to decreased crop yields, reduced agricultural output, and even desertification in some cases.

Economic impact: Water scarcity can have economic implications, particularly in regions where agriculture plays a vital role. Decreased agricultural output can lead to income loss for farmers and affect the overall economic stability of the region.

Ecosystem stress: Arid and semiarid ecosystems are adapted to low water availability, but excessive water extraction can disrupt the delicate balance and harm local flora and fauna (Bortolini et al. 2018). Wetlands, rivers, and other water-dependent ecosystems may suffer from reduced water flows, impacting biodiversity and ecosystem health.

Urban water supply: Cities located in arid and semiarid regions face challenges in providing sufficient water to their growing populations. Urban water supply systems may struggle to meet the demands of residents, leading to water rationing and potential conflicts over water resources.

Water quality issues: Limited water availability can lead to increased concentrations of pollutants in water bodies, as there is less water available to dilute and cleanse contaminants. This can result in degraded water quality and pose health risks to communities that rely on these water sources (Stewart et al. 2006).

Interregional water conflicts: In some cases, water scarcity in arid and semiarid regions can lead to conflicts between neighboring regions or countries that share water resources. Competing demands for limited water supplies can strain diplomatic relations and hinder cooperation for sustainable water management.

Climate change impact: Climate change can exacerbate the challenges faced by arid and semiarid regions. Altered precipitation patterns, increased temperatures, and more frequent droughts can intensify water scarcity and further stress water resources (Misra 2014).

To address these challenges and enhance water security in arid and semiarid regions of South America, a comprehensive approach is needed. This may involve implementing efficient water management practices, promoting water conservation, investing in water storage and distribution infrastructure, adopting drought-resistant crops, and fostering international cooperation for transboundary water management. In addition, sustainable development and environmental conservation efforts can help maintain the health of ecosystems and ensure the availability of water resources for future generations.

9.2.6 Coastal Zones and Estuaries

Coastal zones, characterized by their proximity to oceans and seas, possess a dynamic interplay of saltwater and freshwater resources. Estuaries, where rivers meet the sea, serve as vital ecosystems, providing habitat for diverse species and serving as nurseries for marine life (Cicin-Sain et al. 1998). Managing water resources in these regions requires addressing the challenges of saltwater intrusion, coastal erosion, and the impacts of rising sea levels.

Coastal zones and estuaries play a crucial role in water security in South America, as they are often densely populated and ecologically diverse areas that provide a range of ecosystem services. However, they also face various challenges that can impact water security in the region. Here are some ways in which coastal zones and estuaries affect water security in South America:

Freshwater availability and salinity intrusion: Estuaries are transitional zones where freshwater from rivers meets saltwater from the ocean. Changes in sea level rise, river flow patterns, and human activities can lead to saltwater intrusion into estuaries, contaminating freshwater resources and affecting water quality (Wang et al. 2021). This intrusion can reduce the availability of freshwater for drinking, agriculture, and industrial use.

Erosion and infrastructure impact: Coastal erosion and rising sea levels can threaten infrastructure, including water supply and sanitation systems. As erosion advances, it can compromise water distribution networks, sewage treatment facilities, and other critical water-related infrastructure, leading to disruptions in water supply and sanitation services.

Biodiversity and ecosystem services: Coastal zones and estuaries support diverse ecosystems that provide essential services like water filtration, nutrient cycling, and habitat for fisheries (Zhao et al. 2004). Human activities such as urban development, industrialization, and pollution can degrade these ecosystems, compromising their ability to regulate water quality and maintain ecological balance, which can, in turn, impact water security for both human and aquatic life.

Pollution and contamination: Coastal zones often serve as receptacles for pollutants from inland areas, which can include agricultural runoff, industrial discharges, and untreated sewage. These pollutants can contaminate both surface and groundwater, affecting the safety of drinking water sources and aquatic ecosystems.

Climate change resilience: Coastal areas are highly vulnerable to the impacts of climate change, including increased storm intensity, sea level rise, and extreme weather events. These factors can lead to flooding, coastal inundation, and the destruction of water infrastructure, disrupting water supply systems and compromising water security (Adeyemo 2003).

Population pressure: Many of South America's major cities and urban centers are located along the coast, placing significant demands on local water resources. The concentration of population and economic activity can strain water supply systems, especially during periods of drought or water scarcity.

Conflict over water resources: Competing demands for water resources in coastal zones, such as agriculture, industry, and domestic use, can lead to conflicts over water allocation. These conflicts can exacerbate water scarcity and threaten the overall water security of the region.

Inadequate governance and management: In some cases, inadequate governance, lack of effective regulations, and poor management of water resources can contribute to overextraction of freshwater, habitat destruction, and pollution in coastal zones and estuaries (Newton et al. 2012).

To address these challenges and enhance water security in South America's coastal zones and estuaries, integrated coastal zone management, sustainable development practices, pollution control measures, climate resilience strategies, and effective water governance are essential. Collaborative efforts between governments, local communities, industries, and environmental organizations are crucial to ensure the sustainable use and conservation of these vital resources.

9.2.7 Urbanization and Water Demand

Geographical diversity also encompasses the urban–rural divide, where the concentration of population and industrialization places stress on water resources. Urban areas require vast amounts of water for domestic, industrial, and commercial purposes. As cities expand, they often rely on distant sources, leading to concerns about water transfer, ecological disruption, and competition with rural communities.

Urbanization and water demand have significant impacts on water security in South America, as in many other regions around the world. Here is how these factors interact and contribute to water security challenges in this context:

Urbanization: Urbanization refers to the process of population growth and migration from rural to urban areas, resulting in the expansion of cities and towns. South America has experienced rapid urbanization in recent decades, with a significant proportion of the population now living in urban centers (Novotný 2016). This trend has several implications for water security:

1) **Increased water demand**: Urban areas tend to have higher population densities and greater water consumption per capita compared to rural areas. As more people move to cities, the demand for water for domestic, industrial, and commercial purposes increases, putting stress on local water resources.
2) **Infrastructure challenges**: Rapid urbanization often outpaces the development of necessary infrastructure, including water supply, sanitation, and wastewater treatment systems (Ichimura 2003). This can lead to inadequate access to clean and safe water, increasing the risk of waterborne diseases and contamination.

3) **Land-use changes**: Urbanization can lead to changes in land-use patterns, such as deforestation and increased impervious surfaces (concrete, asphalt), which can alter natural water cycles, reduce groundwater recharge, and increase the risk of flooding.

Water demand: The growing water demand driven by urbanization and industrialization has several implications for water security:

1) **Competing demands**: Urbanization often leads to competition for water resources between various sectors, including agriculture, industry, and domestic use (Bao and Fang 2012). This can result in conflicts over water allocation, especially during periods of water scarcity.
2) **Groundwater depletion**: Increased water demand can lead to overexploitation of groundwater resources, causing aquifer depletion. This can lead to subsidence of land, reduced water availability, and deteriorated water quality.
3) **Ecological impacts**: Diverting water for urban use can reduce the flow of water to natural ecosystems such as rivers, wetlands, and lakes, affecting aquatic biodiversity and ecosystem health.

Water security: Water security refers to the availability of adequate and safe water resources to meet various human and environmental needs. Urbanization and increased water demand can challenge water security in several ways:

1) **Scarcity and stress**: Rapid urbanization and rising water demand can lead to water scarcity, where demand exceeds available supply. This can result in water stress, impacting the well-being of both urban and rural populations (Du Plessis and du Plessis 2019).
2) **Water quality**: As urban areas expand, improper wastewater disposal and industrial pollution can contaminate water sources, reducing water quality and posing health risks.
3) **Vulnerability to climate change**: Urbanization, combined with changing climate patterns, can exacerbate water security challenges. More frequent and intense rainfall events can lead to flooding, while droughts can further stress water supplies.

Addressing challenges: To address these challenges and enhance water security in the face of urbanization and increased water demand, it is important to implement comprehensive strategies:

1) **Integrated water management**: Implementing integrated water management approaches that consider the needs of various sectors and ecosystems can help optimize water use and allocation (Jønch-Clausen and Fugl 2001).
2) **Infrastructure development**: Investment in water supply, sanitation, and wastewater treatment infrastructure is crucial to ensure access to clean and safe water for growing urban populations.
3) **Conservation and efficiency**: Promoting water conservation practices and improving water use efficiency in urban settings can help reduce demand and mitigate scarcity (Hamdy et al. 2003).
4) **Ecosystem protection**: Protecting and restoring natural ecosystems like watersheds, wetlands, and rivers can help maintain water quality and regulate water flows.
5) **Policy and governance**: Effective water governance, regulations, and policies that consider both urbanization and water demand are essential for sustainable water management (Choi et al. 2017).

9.2.8 Climate Change and Adaptation

Climate change is a cross-cutting concern that intersects with geographical diversity and water resources. Altered precipitation patterns, melting glaciers, and changing sea levels are already affecting water availability and distribution. Geographical diversity necessitates localized adaptation strategies, from building resilient infrastructure to implementing efficient water management practices that consider the unique challenges posed by each landscape. Climate change is having a significant impact on water security in South America, with various regions facing challenges related to changes in precipitation patterns, rising temperatures, melting glaciers, and shifts in hydrological cycles. These changes are affecting both the availability and quality of water resources, which in turn has implications for ecosystems, agriculture, economies, and human well-being. Here are some key ways in which climate change is impacting water security in South America:

Changing precipitation patterns: Climate change has led to altered precipitation patterns across the continent. Some areas experience more intense rainfall events, leading to an increased risk of flooding, erosion, and landslides (Mirza 2003). Others may face extended periods of drought, reducing the availability of freshwater resources.

Glacier melting: The Andes mountain range in South America is home to many glaciers, which are an important source of freshwater for downstream communities. However, these glaciers are rapidly melting due to rising temperatures, impacting water supply during the dry season. This can lead to water scarcity in regions that heavily rely on glacial meltwater.

Shifts in hydrological cycles: Climate change is causing shifts in the timing and intensity of hydrological cycles. This can affect the timing of river flows, making it difficult for communities and ecosystems to adapt to changing water availability.

Sea level rise: Coastal areas in South America are vulnerable to sea level rise, which can lead to saltwater intrusion into freshwater sources, making them unsuitable for consumption and agriculture (Ashrafuzzaman et al. 2022).

Ecosystem impact: Changes in water availability can disrupt ecosystems and biodiversity. Wetlands, for instance, might dry up due to decreased water flow, affecting the habitats of various species.

Agriculture and food security: Changes in water availability can impact agricultural productivity. Droughts can lead to crop failures, while floods can damage crops and infrastructure. These changes have implications for food security and livelihoods.

Human health: Changes in water availability can also affect sanitation and water quality. Increased temperatures can promote the growth of waterborne diseases, impacting human health (Ashrafuzzaman et al. 2022).

Economic implications: Sectors like energy production, industry, and tourism that rely on water resources can also be negatively affected by changing water availability.

To address these challenges and ensure water security in the face of climate change, adaptation strategies are crucial. Some strategies include:

Water management and infrastructure: Improved water storage and distribution infrastructure can help manage the impacts of changing water availability.

Rainwater harvesting and conservation: Promoting rainwater harvesting and efficient water use practices can help conserve water during times of scarcity.

Ecosystem-based adaptation: Protecting and restoring ecosystems, such as wetlands and forests, can help regulate water flows and enhance resilience.

Diversification of water sources: Exploring alternative water sources, such as treated wastewater and desalination, can reduce dependence on traditional sources.

Climate-resilient agriculture: Implementing agricultural practices that are more resilient to water stress can help ensure food security.

Early warning systems: Developing effective early warning systems for floods and droughts can help communities prepare and respond.

Cross-border cooperation: Many rivers in South America are transboundary, so cooperation among countries is essential for effective water resource management.

Overall, the complex interactions between climate change, water security, and adaptation strategies highlight the need for interdisciplinary efforts involving governments, communities, scientists, and policymakers to build resilience and ensure a sustainable water future for South America. The various factors affecting the water security in South America are shown in Figure 9.2.

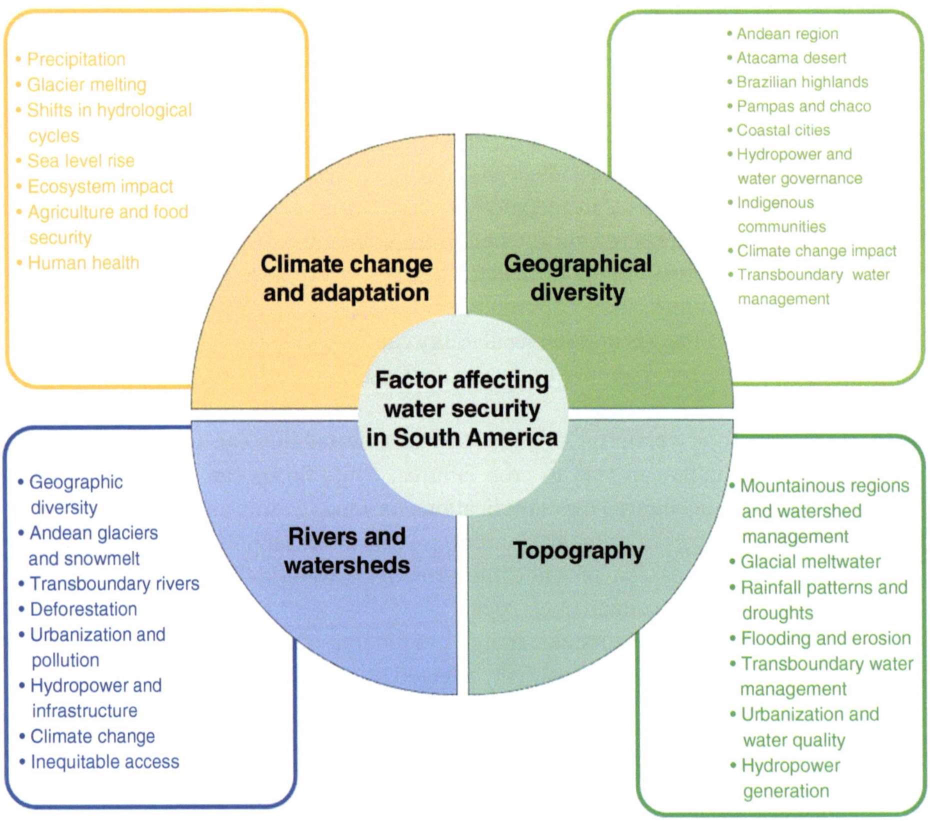

Figure 9.2 Various factors affecting the water security in South America.

9.3 Opportunities and Challenges in Achieving Water Security in South America

Water security in South America, as in many other parts of the world, is a critical issue that involves ensuring reliable access to clean water for all aspects of human development while maintaining the ecological health of aquatic ecosystems. South America faces a unique set of opportunities and challenges in achieving water security:

9.3.1 Opportunities

Abundant water resources: South America is endowed with abundant freshwater resources, including major rivers like the Amazon, Paraná, and Orinoco, as well as numerous lakes and aquifers. Proper management of these resources can provide a strong foundation for water security (Roggeri 2013).

Hydropower potential: The continent's extensive river systems provide significant potential for hydropower generation, which can contribute to clean energy production and economic development.

Biodiversity and ecosystem services: South America's rich biodiversity and intricate ecosystems provide valuable ecosystem services, such as water purification, flood control, and habitat for aquatic species. Protecting these ecosystems supports water security (Baron et al. 2002).

International cooperation: Transboundary river basins in South America, like the Amazon basin, offer opportunities for collaboration among countries to jointly manage water resources, mitigating conflicts and enhancing security.

Community involvement: Many local communities in South America have a deep connection to water resources. Involving these communities in decision-making processes can lead to more sustainable and inclusive water management practices.

9.3.2 Challenges

Uneven distribution: Despite the abundance of water resources, South America experiences significant variability in water distribution. Some regions face water scarcity, while others are prone to floods. Uneven distribution can lead to conflicts and disparities in access (Pereira et al. 2009).

Climate change: Changing precipitation patterns, melting glaciers, and more frequent extreme weather events due to climate change can disrupt water availability and exacerbate water-related challenges.

Urbanization and pollution: Rapid urbanization often outpaces infrastructure development, leading to pollution of water sources from industrial discharge, sewage, and improper waste disposal.

Agricultural demands: Agriculture is a major consumer of water in South America. Ensuring food security while managing water resources sustainably requires careful planning and efficient irrigation practices.

Deforestation: Large-scale deforestation in regions like the Amazon basin can disrupt local and regional hydrological cycles, impacting water availability and quality.

Transboundary conflicts: Several major river systems span multiple countries, leading to potential conflicts over water usage, dam construction, and resource allocation.

Limited access to clean water: Many rural and marginalized communities lack access to clean and safe drinking water and sanitation facilities, leading to health issues and social inequalities.

Inadequate governance: Inadequate regulatory frameworks, lack of enforcement mechanisms, and insufficient data collection can hinder effective water management.

9.3.3 Some Specific Regions in South America Face Notable Water Stress Challenges

Chile: Northern Chile is one of the driest places on Earth, facing severe water scarcity issues. The country's capital, Santiago, and other urban centers experience significant water stress (Herrera-León et al. 2022).

Peru: Parts of Peru, including its capital city Lima, face water stress due to a combination of factors, including limited water infrastructure, glacial retreat in the Andes, and population growth (Motschmann et al. 2020).

Brazil: While the Amazon Basin has abundant water resources, some parts of Brazil, especially in the semiarid Northeast region, experience water scarcity, which can lead to conflicts over water resources (De Melo Branco et al. 2005).

Bolivia: The country has faced water-related conflicts, notably the "Water War" in Cochabamba in the early 2000s, highlighting issues of water access and privatization (Bustamante 2004).

Paraguay and Uruguay: These countries also face water stress, particularly during dry periods when water levels in shared rivers decrease (Hoekstra et al. 2012).

To achieve water security in South America, countries and stakeholders need to prioritize integrated water resource management, invest in sustainable infrastructure, promote cross-sectoral cooperation, engage local communities, and incorporate climate resilience strategies (Panday et al. 2021; Das et al. 2022; Kumar et al. 2023a, b; Neog et al. 2024). International collaboration and sharing best practices can also play a crucial role in addressing the challenges and capitalizing on the opportunities presented by the region's water resources (Panday et al. 2021; Panday and Kumar 2022; Asok Raja et al. 2023; Panday and Kumar 2023; Panday et al. 2023).

9.4 Conclusion

The intricate tapestry of water security and availability in South America presents a multifaceted challenge that requires holistic and collaborative approaches for sustainable management. The diverse landscapes, ecosystems, and socioeconomic contexts across the continent underscore the need for region-specific strategies that balance ecological preservation with human development. As we have explored in this chapter, the interplay of factors such as climate change, population growth, industrialization, and agricultural expansion exerts considerable pressure on South America water resources. While the challenges are formidable, there are promising avenues for addressing these issues. Integrated water resource management, policy reforms, and investment in infrastructure hold the key to ensuring a resilient water future for the region. The cases of effective transboundary cooperation and community-driven initiatives underscore the potential for local and international collaboration in achieving sustainable water security. However, addressing water security in South America requires not only technical solutions but also awareness and education. Raising public consciousness about water conservation, pollution prevention, and the importance of equitable access to clean water can drive behavioral change and policy advocacy. Moreover, embracing traditional knowledge and Indigenous perspectives can enrich our understanding of local hydrology and contribute to culturally sensitive solutions.

References

Adeyemo, O.K. (2003). Consequences of pollution and degradation of Nigerian aquatic environment on fisheries resources. *Environmentalist* 23: 297–306.

Alam, K. (2015). Farmers' adaptation to water scarcity in drought-prone environments: a case study of Rajshahi District, Bangladesh. *Agricultural Water Management* 148: 196–206.

Albrecht, T.R., Varady, R.G., Zuniga-Teran, A.A. et al. (2021). Unraveling transboundary water security in the arid Americas. In: *Putting Water Security to Work*, 117–155. Routledge.

de Alkmim, F.F. (2015). Geological background: a tectonic panorama of Brazil. *Landscapes and Landforms of Brazil* 9–17.

Ashrafuzzaman, M., Artemi, C., Santos, F.D., and Schmidt, L. (2022). Current and future salinity intrusion in the south-western coastal region of Bangladesh. *Spanish Journal of Soil Science* 12: 10017.

Asok Raja, S.K., Panday, D.P., and Kumar, M. (2023). Decoding the enigma of 100-year record-breaking rainfall over Tamil Nadu using wavelet analysis. *Groundwater for Sustainable Development* 101007. https://doi.org/10.1016/j.gsd.2023.101007.

Bao, C. and Fang, C.L. (2012). Water resources flows related to urbanization in China: challenges and perspectives for water management and urban development. *Water Resources Management* 26: 531–552.

Baron, J.S., Poff, N.L., Angermeier, P.L. et al. (2002). Meeting ecological and societal needs for freshwater. *Ecological Applications* 12 (5): 1247–1260.

Bortolini, L., Maucieri, C., and Borin, M. (2018). A tool for the evaluation of irrigation water quality in the arid and semi-arid regions. *Agronomy* 8 (2): 23.

Bustamante, R. (2004). The water war: resistance against privatisation of water in Cochabamba, Bolivia. *Revista de Gestión Del Agua en América Latina* 1: 37–46.

Choi, I.C., Shin, H.J., Nguyen, T.T., and Tenhunen, J. (2017). Water policy reforms in South Korea: a historical review and ongoing challenges for sustainable water governance and management. *Water* 9 (9): 717.

Cicin-Sain, B., Knecht, R.W., Jang, D., and Fisk, G.W. (1998). *Integrated Coastal and Ocean Management: Concepts and Practices*. Island press.

Cook, C. and Bakker, K. (2012). Water security: debating an emerging paradigm. *Global Environmental Change* 22 (1): 94–102.

Cosgrove, W.J. and Loucks, D.P. (2015). Water management: current and future challenges and research directions. *Water Resources Research* 51 (6): 4823–4839.

Crane, R. and Fletcher, L. (2017). *Island Genres, Genre Islands: Conceptualisation and Representation in Popular Fiction*. Rowman & Littlefield.

Das, A., Kumar, M., Jha, P.K. et al. (2022). Isotopic and hydrogeochemical tracking of dissolved nutrient dynamics in the Brahmaputra River System: a source delineation perspective. *Chemosphere* 307: 135757. https://doi.org/10.1016/j.chemosphere.2022.135757.

De Melo Branco, A., Suassuna, J., and Vainsencher, S.A. (2005). Improving access to water resources through rainwater harvesting as a mitigation measure: The case of the Brazilian semi-arid region. *Mitigation and Adaptation Strategies for Global Change* 10: 393–409.

Du Plessis, A. and du Plessis, A. (2019). Current and future water scarcity and stress. *Water as an Inescapable Risk: Current Global Water Availability, Quality and Risks with a Specific Focus on South Africa* 13–25.

Edokpayi, J.N., Odiyo, J.O., and Durowoju, O.S. (2017). Impact of wastewater on surface water quality in developing countries: a case study of South Africa. *Water Quality* 10 (66561): 10–5772.

Falkenmark, M., Lundqvist, J., & Widstrand, C. (1989) Macro-scale water scarcity requires micro-scale approaches: Aspects of vulnerability in semi-arid development. In *Natural Resources Forum* (Vol. 13, No. 4, pp. 258–267). Oxford, UK: Blackwell Publishing Ltd.

Giordano, M. A., & Wolf, A. T. 2003 Sharing waters: post-Rio international water management. In *Natural Resources Forum* (Vol. 27, No. 2, pp. 163–171). Oxford: Blackwell Publishing Ltd.

Hamdy, A., Ragab, R., and Scarascia-Mugnozza, E. (2003). Coping with water scarcity: water saving and increasing water productivity. *Irrigation and Drainage: The Journal of the International Commission on Irrigation and Drainage* 52 (1): 3–20.

Herrera-León, S., Cruz, C., Negrete, M. et al. (2022). Impact of seawater desalination and wastewater treatment on water stress levels and greenhouse gas emissions: the case of Chile. *Science of the Total Environment* 818: 151853.

Hoekstra, A.Y., Mekonnen, M.M., Chapagain, A.K. et al. (2012). Global monthly water scarcity: blue water footprints versus blue water availability. *PLoS ONE* 7 (2): e32688.

Hoekstra, A.Y., Mekonnen, M.M., Chapagain, A.K. et al. (2012). Global monthly water scarcity: blue water footprints versus blue water availability. *PLoS ONE* 7 (2): e32688.

Houston, J. (2002). Groundwater recharge through an alluvial fan in the Atacama Desert, northern Chile: mechanisms, magnitudes and causes. *Hydrological Processes* 16 (15): 3019–3035.

Howard, G., Bartram, J., Water, S., and World Health Organization (2003). Domestic water quantity, service level and health.

Ichimura, M. (2003). Urbanization, urban environment and land use: challenges and opportunities. In: *Asia-Pacific Forum for Environment and Development, Expert Meeting*, vol. 23, 1–14. Global Environmental Strategies (IGES).

Johansen, K.S., Alfthan, B., Baker, E. et al. (2018). *The Andean Glacier and Water Atlas: The Impact of Glacier Retreat on Water Resources*. UNESCO Publishing.

Jønch-Clausen, T. and Fugl, J. (2001). Firming up the conceptual basis of integrated water resources management. *International Journal of Water Resources Development* 17 (4): 501–510.

Kumar, M., Panday, D.P., Bhagat, C. et al. (2023a). Demystifying the decadal shift in the extent of groundwater in the coastal aquifers of Gujarat, India: a case of reduced extent but increased magnitude of seawater intrusion. *Science of the Total Environment* 165451. https://doi.org/10.1016/j.scitotenv.2023.165451.

Kumar, M., Mazumder, P., Silori, R. et al. (2023b). Prevalence of PPCPs, microplastics and co-infecting microbes in the post-COVID-19 era and its implications on antimicrobial resistance and potential endocrine disruptive effects. *Science of the Total Environment* 166419. https://doi.org/10.1016/j.scitotenv.2023.166419.

Kuriqi, A., Pinheiro, A.N., Sordo-Ward, A. et al. (2021). Ecological impacts of run-of-river hydropower plants – current status and future prospects on the brink of energy transition. *Renewable and Sustainable Energy Reviews* 142: 110833.

Liu, X., Lee Burras, C., Kravchenko, Y.S. et al. (2012). Overview of Mollisols in the world: distribution, land use and management. *Canadian Journal of Soil Science* 92 (3): 383–402.

Lowrance, R., Altier, L.S., Newbold, J.D. et al. (1997). Water quality functions of riparian forest buffers in Chesapeake Bay watersheds. *Environmental Management* 5: 687–712.

Mabrey, D. and Vittorio, M. (2018). Moving from theory to practice in the water–energy–food nexus: an evaluation of existing models and frameworks. *Water-Energy Nexus* 1 (1): 17–25.

Maliva, R. and Missimer, T. (2012). *Arid Lands Water Evaluation and Management*. Springer Science & Business Media.

Mark, B.G., Bury, J., McKenzie, J.M. et al. (2010). Climate change and tropical Andean glacier recession: evaluating hydrologic changes and livelihood vulnerability in the Cordillera Blanca, Peru. *Annals of the Association of American Geographers* 100 (4): 794–805.

Mirza, M.M.Q. (2003). Climate change and extreme weather events: can developing countries adapt? *Climate Policy* 3 (3): 233–248.

Misra, A.K. (2014). Climate change and challenges of water and food security. *International Journal of Sustainable Built Environment* 3 (1): 153–165.

Motschmann, A., Huggel, C., Muñoz, R., and Thür, A. (2020). Towards integrated assessments of water risks in deglaciating mountain areas: water scarcity and GLOF risk in the Peruvian Andes. *Geoenvironmental Disasters* 7: 1–20.

Mugagga, F. and Nabaasa, B.B. (2016). The centrality of water resources to the realization of Sustainable Development Goals (SDG). A review of potentials and constraints on the African continent. *International Soil and Water Conservation Research* 4 (3): 215–223.

Neog, N., Goswami, R., Panday, D.P. et al. (2024). Arsenic contamination in the groundwater of Northeastern India: critical understandings on geotectonic controls and the need for intervention. *Current Opinion in Environmental Science & Health* 100539. https://doi.org/10.1016/j.coesh.2024.100539.

Newton, A., Carruthers, T.J., and Icely, J. (2012). The coastal syndromes and hotspots on the coast. *Estuarine, Coastal and Shelf Science* 96: 39–47.

Novotný, L. (2016). Urban development and migration processes in the urban region of Bratislava from the post-socialist transformation until the global economic crisis. *Urban Geography* 37 (7): 1009–1029.

Oyague, E. and Cooper, D.J. (2020). Peatlands of the central Andes puna, South America. *Wetland Science & Practice* 37 (4): 255–260.

Panday, D.P. and Kumar, M. (2022). Climate indices and hydrological extremes: deciphering the best-fit model. *Environmental Research* 215: 114301. https://doi.org/10.1016/j.envres.2022.114301.

Panday, D.P. and Kumar, M. (2023). Application of remote sensing techniques to deal with scale aspects of GRACE data to quantify groundwater levels. *MethodsX* 102108. https://doi.org/10.1016/j.mex.2023.102108.

Panday, D.P., Khosa, R., Maheswaran, R. et al. (2021). Game-theoretic-based modelling of Krishna waters dispute: equilibrium solutions by Metagame Analysis. *The European Physical Journal B* 94: 101. https://doi.org/10.1140/epjb/s10051-021-00107-w.

Panday, D.P., Khosa, R., Maheswaran, R. et al. (2021). Game theoretic-based modelling of Krishna waters dispute: equilibrium solutions by hypergame analysis. *The European Physical Journal B* 94: 131. https://doi.org/10.1140/epjb/s10051-021-00135-6.

Panday, D.P., Singh, A., Gaur, S. et al. (2023). Probable maximum precipitation analysis of high rainfall regions in India. *Groundwater for Sustainable Development* 100893. https://doi.org/10.1016/j.gsd.2022.100893.

Pereira, L.S., Cordery, I., and Iacovides, I. (2009). *Coping with Water Scarcity: Addressing the Challenges*. Springer Science & Business Media.

Perry, J. and Vanderklein, E.L. (2009). *Water Quality: Management of a Natural Resource*. Wiley.

Rangecroft, S., Harrison, S., Anderson, K. et al. (2013). Climate change and water resources in arid mountains: an example from the Bolivian Andes. *Ambio* 42: 852–863.

Roggeri, H. (2013). Tropical freshwater wetlands: a guide to current knowledge and sustainable management.

Stewart, C., Johnston, D.M., Leonard, G.S. et al. (2006). Contamination of water supplies by volcanic ashfall: a literature review and simple impact modelling. *Journal of Volcanology and Geothermal Research* 158 (3–4): 296–306.

Urrutia, R. and Vuille, M. (2009). Climate change projections for the tropical Andes using a regional climate model: temperature and precipitation simulations for the end of the 21st century. *Journal of Geophysical Research: Atmospheres* 114 (D2): 1–5.

Wang, J., Tong, Y., Feng, L. et al. (2021). Satellite-observed decreases in water turbidity in the Pearl River Estuary: potential linkage with sea-level rise. *Journal of Geophysical Research: Oceans* 126 (4): e2020JC016842.

Winemiller, K.O., López-Fernández, H., Taphorn, D.C. et al. (2008). Fish assemblages of the Casiquiare River, a corridor and zoogeographical filter for dispersal between the Orinoco and Amazon basins. *Journal of Biogeography* 35 (9): 1551–1563.

Zhao, B., Kreuter, U., Li, B. et al. (2004). An ecosystem service value assessment of land-use change on Chongming Island, China. *Land Use Policy* 21 (2): 139–148.

10

Indigenous Technology and Modern Practices in Australia

Navjot Hothi[1] *and Saumyaranjan Sahoo*[2]

[1] *Department of Physics, School of Engineering, UPES, Dehradun, Uttarakhand, India*
[2] *Department of Civil Engineering, School of Engineering, UPES, Dehradun, Uttarakhand, India*

10.1 Introduction

Australia is often referred to as a "dry land" due to its status as the driest inhabited continent globally. The country predominantly experiences a semiarid or arid climate characterized by limited rainfall and high temperatures, particularly in its interior regions (Deshpande et al. 2020). The annual rainfall map of Australia for the period of 2023–2024 in Figure 10.1 shows a relatively low annual rainfall below 600 mm in most part of the continent. Natural water distribution in Australia is predominantly scarce and variable. According to the Australian Bureau of Statistics, approximately 70% of the landmass is classified as arid or semiarid, receiving low to moderate rainfall and exhibiting high evaporation rates. These regions are typified by sparse vegetation, desert landscapes, and ecosystems susceptible to drought. Contrariwise, around 20% of the Australian landmass is categorized as having high disposal of surface and groundwater. This is majorly found in the wetter domains such as the northern tropical areas and the east coast. These regions possess extensive river basins, lush vegetation, wetlands, and rainforests. The residual 10% of the landmass is characterized as having moderate accessibility of surface and groundwater. This represents the transitional sectors between arid and wet atmospheres, such as savannas and grasslands in the northern and the southwest parts of Australia (Bark et al. 2012; Das et al. 2022, Kumar et al. 2023a, b).

Australia's climate is affected by its presence in the southern hemisphere. Most of the continent is located within the latitudes of the westerly winds accountable for rainfall in various other parts of the world. Furthermore, the Australian interior is far from the controlling effects of the ocean, and this leads to significant temperature changes. Consequently, the arid and semiarid landscapes dwarf much of Australia, surrounding vast deserts such as the Great Victoria Desert, the Great Sandy Desert, and the Simpson Desert. However, there are also regions with more moderate climates and higher rainfall,

Water Scarcity Management: Enabling Technologies, First Edition. Edited by Kanchan D. Bahukhandi, Manish Kumar, Durga P. Panday, and Tushara G.G. Chaminda.

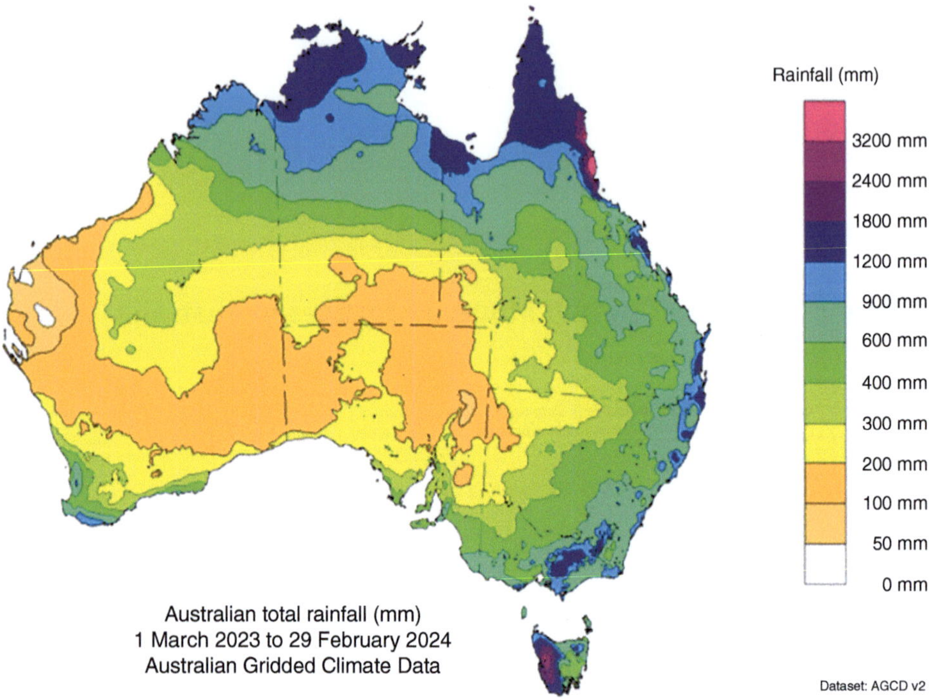

Figure 10.1 Annual rainfall map of Australia for the period 2023–2024. *Source:* Adapted from Bureau of Meteorology.

particularly in the eastern and southeastern parts of the country where major cities like Sydney and Melbourne are located (Ramírez and Briones 2017). While certain areas experience abundant rainfall, such as the tropical north and some sections of the east coast, prolonged droughts and water scarcity are prevalent across much of the country (Cosens and Chaffin 2016).

Australia has invested significantly in water infrastructure and management practices, improving access to water resources, and enhancing the resilience of water systems. Initiatives like the National Water Initiative (NWI) aim to promote sustainable water management and utilization throughout the nation. Although Australia does not possess abundant natural water resources, its investments in infrastructure and management have mitigated water scarcity and ensured access to water for various purposes. Indigenous Australians have a longstanding history of sustainable coexistence with their environment, including the management and utilization of water resources in a sustainable manner (Petzold et al. 2020). Indigenous communities in Australia have developed an array of traditional knowledge and practices for water treatment and management, passed down through generations. These traditional approaches exemplify the ingenuity and sustainability of Indigenous Australians in water resource management. The integration of this knowledge into modern water management practices holds the potential to provide innovative and sustainable solutions for water treatment and management in Australia (Pinto and Maheshwari 2010; Panday et al. 2021a, b; Panday and Kumar 2023).

10.2 Natural Water Resources in Australia

Australia is blessed with numerous rivers, among which the Murray-Darling River system stands as the longest, traversing multiple states and territories (Mooney and Tan 2012). Additional notable rivers include the Fitzroy River in Western Australia and the Mitchell River in Queensland. The country is also home to various expansive natural lakes, such as Lake Eyre, the largest lake in Australia, and Lake Torrens. Typically, these lakes exhibit low water levels or remain dry, yet they have the capacity to fill up during periods of heavy rainfall. The locations of the above-mentioned river basins and lakes are shown in a map in Figure 10.2. Crucial for agriculture, industry, and urban water supply, major rivers like the Murray-Darling Basin and the Lake Eyre Basin face considerable strain due to factors such as overuse, climate fluctuations, and environmental degradation (Panday and Kumar 2022).

Groundwater is of significant importance in Australia, particularly in its arid and semiarid regions. The groundwater resources in the continent and location of selected groundwater basins are shown in Figure 10.3. The Great Artesian Basin, Figure 10.3,

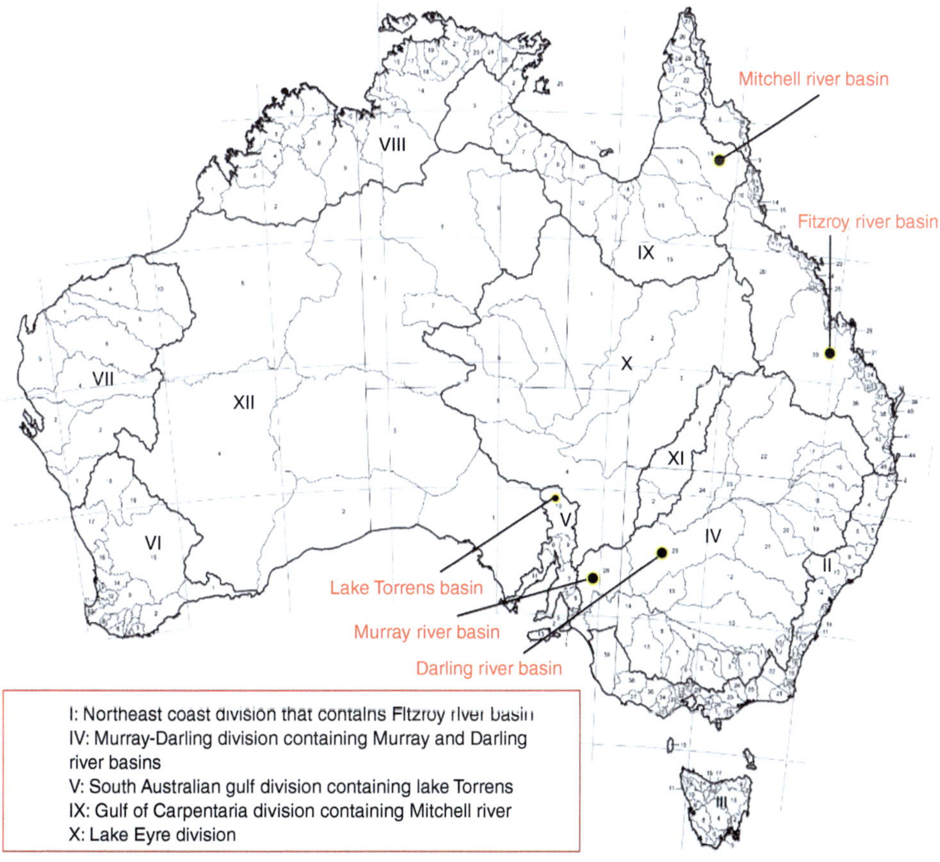

Figure 10.2 Various notable river basins and lakes in Australia. *Source:* Australian Government Bureau of Meteorology/Crown Copy Right/Public Domain.

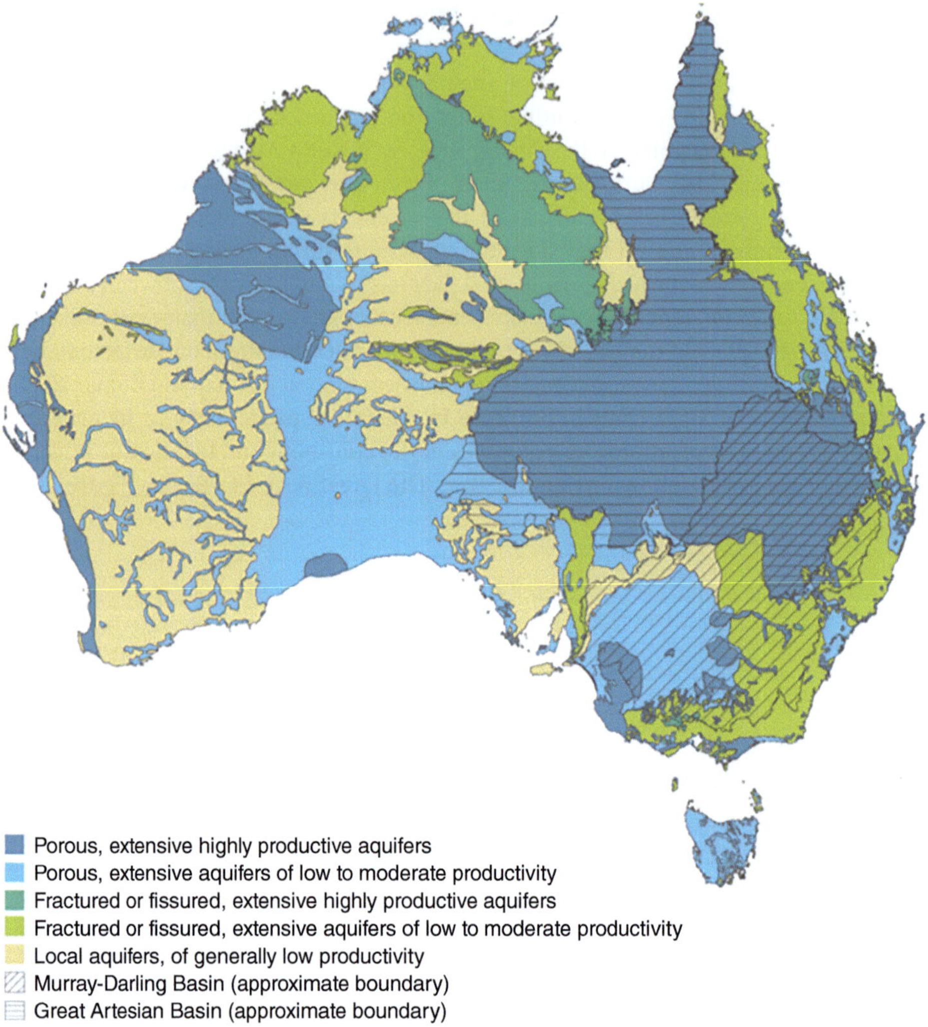

Figure 10.3 Groundwater resources of Australia. *Source:* Australia State of the Environment 2021/Commonwealth of Australia/Public Domain.

encompassing more than a quarter of the country, ranks as the largest groundwater basin globally (van Vliet et al. 2021). Groundwater resources also play a vital role in water supply across various parts of Australia, particularly in regions with arid and semiarid climates. Nevertheless, groundwater faces intensifying pressures from climate variations, pollution, and over-extraction. Australia has various significant wetlands, comprising the famous Kakadu National Park in the Northern Territory, the Macquarie Marshes in New South Wales, and the Coorong in South Australia. Furthermore, the country's widespread coastline incorporates numerous marine environments and beaches, which act as critical settings for fishing and other activities (Asok Raja et al. 2023, Panday et al. 2023).

Despite the abundance of natural water resources, water scarcity and drought periods are prevalent throughout much of Australia due to its arid and semiarid climate. Climate change and weather patterns like El Niño and La Niña further affect quality and water availability across the country. Subsequently, Australia has made considerable investments in management practices and water infrastructure to safeguard access to water for diverse purposes. The purpose may be industrial use, agriculture, and urban water supply (Ramírez and Briones 2017).

To address these water-related challenges, Australia has implemented a range of measures. These include considerable investments in water infrastructure, the employment of water recycling initiatives, and water conservation crusades. The country has also initiated thorough water management plans, such as the Murray-Darling Basin Plan, directed at guaranteeing sustainable water resource management and considering the needs of all stakeholders. However, the ongoing influence of climate change and the rising demand for water continue to present considerable tests to Australia's water resources.

10.3 Indigenous Water Treatment and Conservation Technologies in Australia

10.3.1 Indigenous Technology of Water Treatment in Australia

Over centuries, Indigenous communities in Australia have honed distinctive water treatment techniques, which remain in practice today (Ayre and Mackenzie 2013). These technologies embody the profound ecological comprehension of the Indigenous people, encompassing the land, waterways, and natural resources. Various Indigenous water treatment methods employed in Australia comprise various notable practices. They are presented in the current section (Pinto and Maheshwari 2010). Indigenous communities have long utilized sand filtration as a means of water purification. This process involves passing water through a bed of sand, effectively eliminating impurities and bacteria. Sand filtration stands as a straightforward and efficient method still widely embraced in numerous Indigenous communities today. Boiling water remains one of the most efficient methods to eliminate bacteria and pathogens present in water. Indigenous communities have traditionally relied on boiling as a water purification technique, and it continues to be extensively employed today. Charcoal possesses innate antibacterial properties that aid in purifying water by removing impurities. The Indigenous communities have harnessed charcoal filtration for centuries, and it remains a prevalent method for water purification. The Indigenous communities have ingeniously employed reeds and grasses to filter water for consumption and cooking purposes. These plants possess the capacity to absorb impurities, thereby enhancing the taste and quality of water.

Traditional water sourcing for Indigenous communities involved the utilization of natural springs, often nestled in remote and pristine locations. Spring water undergoes natural filtration as it passes through rocks and soil, rendering it one of the purest water sources available. The Indigenous Australians have long employed a diverse array of plants for their medicinal properties, including those with water purification attributes. For instance, the roots of the Mallee tree can be utilized to filter water and eliminate impurities.

Indigenous Australians have employed smoke as a means of water purification. By burning specific types of wood and plants, the resulting smoke possesses disinfectant properties capable of eliminating harmful bacteria and viruses. Indigenous Australians have also developed techniques for water storage, such as constructing underground storage tanks known as "soakage pits." These tanks effectively hold substantial amounts of water while safeguarding it against evaporation and contamination. Indigenous Australians have a wide perception of the natural environment and its complex cycles, including the water systems. They have nurtured an all-inclusive approach to water management, diagnosing the connection of the ecosystem and the cruciality of maintaining balance.

These traditional water treatment technologies not only validate usefulness but also demonstrate the reflective bond Indigenous people hold with their environment. This naturally highlights the importance of preserving natural resources. At present, many Indigenous communities in Australia continue to devote these traditional water treatment methods in concurrence with modern technologies, safeguarding the quality and safety of their water supply (Jackson et al. 2005).

10.3.2 Role of Kamilaroi Community in Indigenous Water Conservation Technology

The Kamilaroi community, also known as the Gamilaraay or Gomeroi, resides in the northwestern region of New South Wales, Australia (Moggridge et al. 2022). Their ancestral lands encompass a vast area, including towns and regions such as Moree, Narrabri, Gunnedah, Tamworth, and Walgett. For thousands of years, the Kamilaroi people have nurtured and inhabited this land, preserving their cultural heritage, knowledge systems, and spiritual practices. Today, they continue to live in and around their traditional territories while actively contributing to Australian society. The Kamilaroi people are profoundly rooted in their rich Indigenous tradition. They have established a refined and sustainable method for water conservation. Their procedure is established upon a thoughtful connection to the land and a complete understanding of the complicated relationship between water, culture, people, and country (Figure 10.4). At the core of the Kamilaroi methodology lies a profound reverence for water as a life-giving force and a recognition of its sacred nature. They view water as a living entity, deserving of the utmost respect and protection. Understanding its indispensable role in all aspects of life, from drinking to the preservation of delicate ecosystems, the Kamilaroi people transmit this wisdom through traditional knowledge and oral traditions, ensuring the continuity of their water management practices.

Essential to their method is the smart observation and understanding of natural indicators. The Kamilaroi people have complicated information of the environment, sharpened through generations, permitting them to distinguish crucial signs indicating the quality and availability of water sources. They carefully observe the behaviour of animals, such as the movements of birds or fish patterns, which provide an appreciated understanding of the presence of water. They also intensely monitor the pattern of growth of plants, as specific species flourish in distinctive moisture conditions, enlightening dynamic information about water availability in each region. In addition, the Kamilaroi people highlight the protection and preservation of water ecosystems. They reasonably identify the mutuality between water bodies and the surrounding habitats, thus safeguarding the conservation of these ecosystems. By protecting critical water catchment areas and maintaining the health of habitats, the Kamilaroi people promote the sustainability and overall well-being of water resources. They understand that upsetting the subtle balance of ecosystems can have far-flung consequences on water availability and security.

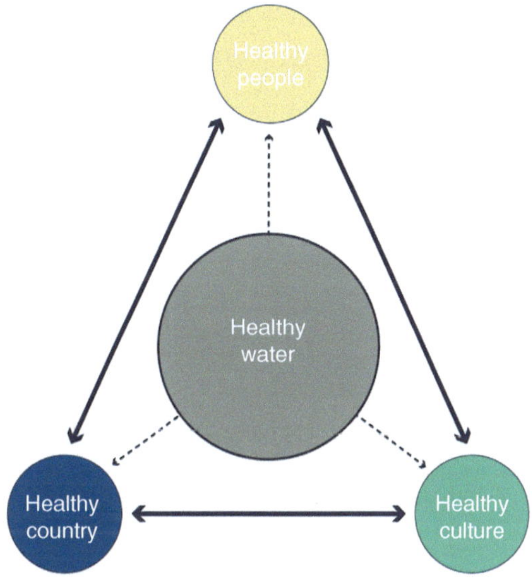

Figure 10.4 Relation between water, culture, country and people as viewed by the Indigenous communities of Australia. *Source:* The Australia state of the environment/Commonwealth of Australia Edit/CC BY 4.0.

10.4 Modern Water Treatment Practices in Australia

Australia has implemented a highly advanced water treatment system that combines natural and artificial processes to guarantee the delivery of safe and high-quality water to households and businesses (Bohnet and Kinjun 2009). The contemporary water treatment practices in Australia can be broadly classified into four key stages (Wang 2012) (Figure 10.5). The initial stage involves the collection of water from diverse sources such as rivers, lakes, and underground reservoirs. The collected water undergoes screening to eliminate large debris like leaves, twigs, and other floating materials. In this second stage, chemicals are put into the water to expedite the settling and coagulation of impurities. This procedure helps in the elimination of fine specks including sand, organic matter, and dirt. In the third stage, the passage of water through a series of filters is encompassed and is designed to remove residual impurities, such as viruses, bacteria, and other microorganisms. Various filtration techniques are utilized, including gravel, carbon, membrane filters, and sand. The last stage involves the use of chemical agents like a zone of chlorine for disinfection of water. This particular step essentially removes any remaining virus, bacteria, or other harmful microorganisms.

Apart from these four stages, modern water treatment plants in Australia often integrate advanced technologies such as ultraviolet (UV) radiation, reverse osmosis, and advanced oxidation to further remove pollutants and guarantee the water's safety for consumption. Thus, Australia's water treatment practices are categorized among the most

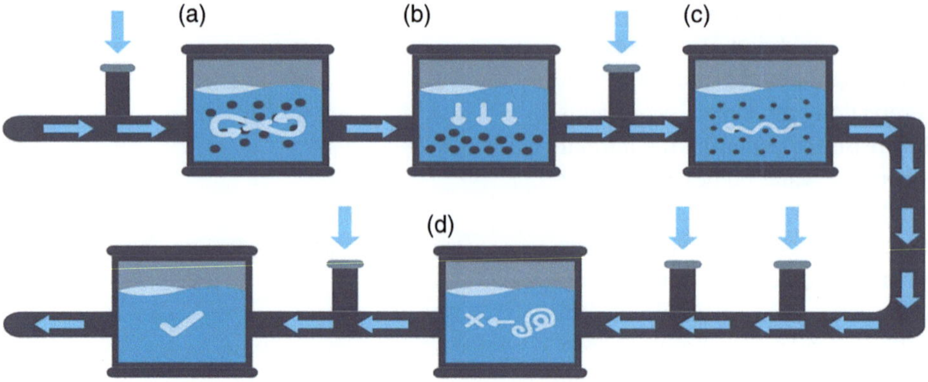

Figure 10.5 Graphical representation of various processes involved in modern water treatment practice in Australia. (a) Water collection and screening. (b) Coagulation and sedimentation. (c) Filtration. (d) Disinfection.

advanced globally, and the country has duly earned a reputation for providing its people with safe and clean drinking water (Hibbard et al. 2008).

10.5 Water Conservation and Demand Management

Figure 10.6 shows that household water consumption in Australia has increased marginally in the years 2015–2022 and the average water consumption over this period is 1.78 million megalitres. Australia being one of the driest countries in the earth requires effective water supply and management practices to reduce high-water consumption. Australia has implemented a range of strategies to promote water conservation and demand management, acknowledging the need to reduce water consumption (Scherer and Pfister 2016). These initiatives include public education campaigns, incentives such as rebates for water-efficient appliances, and widespread adoption of water-saving devices like dual-flush toilets and

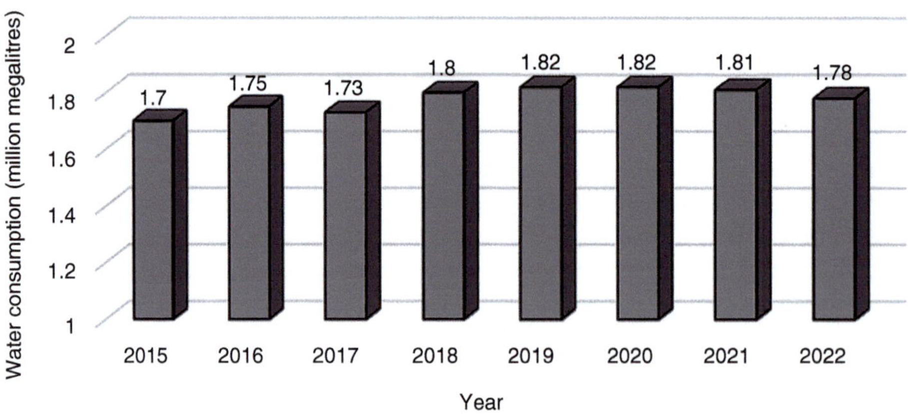

Figure 10.6 Household water consumption in Australia for the years 2015–2022. *Source:* Data taken https://www.statista.com/statistics/974952/australia-water-consumption-by-sector/.

low-flow showerheads. Furthermore, water restrictions and pricing mechanisms have incentivized individuals and businesses to decrease their water usage. These collective efforts have led to a significant decrease in per capita water consumption nationwide (Basdew et al. 2017). Australia has been at the forefront of implementing advanced systems for water recycling and reuse in urban areas (Mukherjee and Jensen 2020). Through the application of cutting-edge technologies like membrane filtration and reverse osmosis, treated wastewater is transformed into high-quality recycled water suitable for non-potable purposes. This recycled water is utilized for tasks such as irrigation of parks, sports fields, and golf courses, industrial processes, and even toilet flushing in certain buildings. By embracing water recycling, Australia has reduced its reliance on freshwater sources and minimized the discharge of treated wastewater into environmentally sensitive ecosystems (Grant et al. 2012).

In regions characterized by frequent rainfall, stormwater harvesting and management have emerged as effective measures for water conservation. Traditionally seen as waste, urban stormwater runoff is now captured, treated, and stored for various beneficial uses. This practice not only helps replenish groundwater reserves but also alleviates pressure on potable water supplies. Australia has implemented stormwater harvesting projects in cities and towns, integrating green infrastructure features like rain gardens, permeable pavements, and retention basins to enhance water quality and recharge aquifers (Domínguez et al. 2017). In response to prolonged droughts and limited freshwater resources, Australia has invested in desalination plants to augment its water supply. Desalination involves the removal of salts and impurities from seawater or brackish water to produce potable water. The establishment of large-scale desalination plants in coastal areas has provided a significant buffer against water scarcity. Although desalination is energy-rigorous, developments in knowledge, equipment, and the application of renewable energy sources have improved the sustainability and proficiency of these plants. Figure 10.7 represents the percentage of water contribution of the various desalination plants in Australia (Bark et al. 2012).

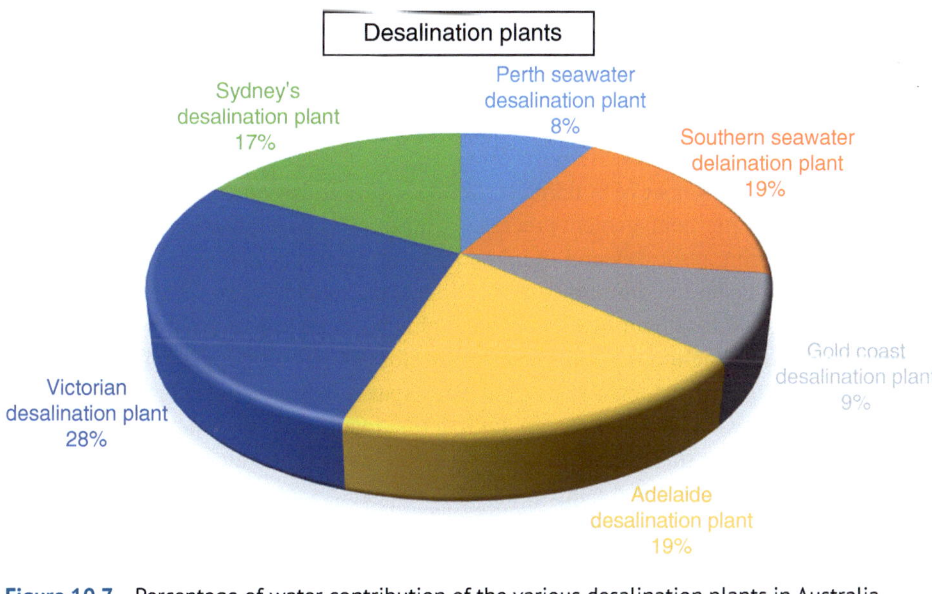

Figure 10.7 Percentage of water contribution of the various desalination plants in Australia.

Australia distinguishes the significance of a widespread approach to water management and has encompassed practices known as Integrated Water Management (IWM). The IWM includes the coordination of wastewater management, water supply, environmental considerations, and stormwater management. It aims to enhance water resources, safeguard ecosystems, and reduce demand through cooperative decision-making and planning. By amalgamating the efforts of community stakeholders, government agencies, and water utilities, Australia has attained more buoyant and sustainable water systems.

10.6 Forums and Initiatives for Implementing Water Management Technologies

10.6.1 Indigenous Water Planning Forum 2009

The Indigenous Water Planning Forum, held in Australia in 2009, was a significant event aimed at addressing the distinctive water management challenges encountered by Indigenous communities in the country. Recognizing the importance of embracing Indigenous voices in decision-making processes, the forum worked as a stage for Indigenous community members, leaders, and specialists to deliberate their detailed concerns regarding water and develop tactics that are allied with their environmental and cultural values. A major objective of the Indigenous Water Planning Forum was to honour and know the profound connection between water and Indigenous communities. Water embraces enormous spiritual, economic, and cultural importance for Indigenous people, and their customary knowledge plays a key role in sustainable water management. The forum highlighted the significance of amalgamating Indigenous perceptions and practices into conventional water planning techniques to confirm the protection of cultural environmental and heritage sustainability (Tsatsaros et al. 2018).

Through the forum, the participants engaged in deliberations regarding the challenges faced by Indigenous communities regarding water allocation, clean water access, and the impact of water-related policies on their livelihoods and lands. Indigenous leaders shared their experiences and wisdom, emphasizing the importance of all-inclusive tactics to water management that include not only economic contemplations but also ecological balance and cultural values (Pert et al. 2015). The forum underlined the requirement for augmented Indigenous participation and decision-making authority in water governance and planning. It highlighted the importance of building capability within Indigenous communities to efficiently involve in water management processes. This implicated supporting Indigenous-led creativities, providing resources and training, and nurturing partnerships among government agencies, Indigenous communities, and other stakeholders (Escott et al. 2015).

The Indigenous Water Planning Forum also highlighted the preservation and protection of Indigenous water rights. It provided a podium for discussing policy and legal frameworks that identify and endorse these rights. Participants discovered mechanisms to reinforce Indigenous water governance, including the institution of water authorities or committees that permit Indigenous communities to achieve their water resources in agreement with their traditional and cultural values. Endorsing sustainable water exercises in Indigenous communities was an additional acute aspect of the forum. Contributors shared efficacious best practices and case studies that incorporated traditional knowledge with modern water management practices. This method invigorated the adoption of water-saving technologies, the revitalization of traditional water harvesting and preservation methods, and the refurbishment of tarnished water ecosystems.

10.6.2 The National Water Initiative (NWI) in Australia: Promoting Sustainable Water Management and Engaging Indigenous Communities

Australia has long grappled with water scarcity and the need for efficient water management due to its arid climate and limited water resources. In response to these challenges, the NWI was established in 2004. This comprehensive policy framework aims to achieve sustainable water use and allocation, enhance water efficiency, and secure the long-term availability of water resources throughout Australia. The NWI is a cooperative effort involving the Australian federal government, diverse stakeholders, and territory and state governments. It endorses robust water market systems, integrated planning, efficient pricing mechanisms, and to ensure effective water management.

10.6.2.1 Key Objectives and Components of the NWI

The NWI pursues to institute well-defined and transparent systems for water allocation and entitlement that offer elasticity and certainty to water users while contemplating environmental sustainability. The proposal highlights the importance of adaptive and thorough water planning, considering issues such as population growth, competing water demands, and climate change. It inspires the expansion of water resource plans that incorporate environmental considerations. The NWI provisions the foundation of efficient water markets that expedite the trading of water privileges, permitting water to flow to areas where it holds greater values and promotes more efficient use of water. Knowing the importance of capitalizing in water infrastructure, the NWI promotes for the employment of water efficiency procedures across segments, including industry, agriculture, and urban areas, to augment water use.

10.6.2.2 The Role of Indigenous Communities in the NWI

The NWI identifies the importance of involving Indigenous communities and revering their spiritual, economic, and cultural connections to water. Indigenous practices and knowledge offer treasured perceptions of sustainable water management and the conservation of ecosystems (Dovers et al. 2005). The NWI supports the rights of Indigenous communities to cope with and manage water resources in coalition with their cultural traditions and values. It ropes the instituting of collaborative arrangements and partnerships among Indigenous communities, stakeholders, and governments, to confirm Indigenous participation in decision-making processes and water planning. The NWI encourages the establishment of Indigenous water governance bodies, such as water committees or authorities, empowering Indigenous communities to autonomously manage their water resources according to their cultural values and traditional knowledge. The initiative identifies the importance of acquiring water prerogatives for Indigenous communities, certifying their retrieve to water for cultural practices, economic activities, and environmental stewardship. The NWI endorses projects to improve Indigenous participation in water management procedures through capacity building. It aims to reinforce Indigenous involvement by providing resources, training, and confessing Indigenous knowledge and rehearses in decision-making.

10.7 Conclusions

Australia has turned out as a forerunner in innovative water management practices, driven by its parched climate and constrained water resources. Acknowledging the value of merging Indigenous practices and technology into modern water management, Australian water policymakers have accentuated the importance of leveraging Indigenous innovations and wisdom. This amalgamation is important for increasing the sustainability of water, for conserving cultural heritage, and nurturing collaborative and equitable methods to water resource management in the nation. Australia faces various challenges in safeguarding water security due to urbanization, the effects of climate change, and population growth (Klein Goldewijk et al. 2010). However, the country has incorporated technological advancements, effective policies, and community involvement to prioritize efficacy, conservation in its modern water practices and sustainability.

Australia's assurance to ecological water management is obvious in its adoption of various schemes. These comprise recycling, water conserving, stormwater harvesting, IWM, and desalination. By incorporating these practices, the country has made prominent progress in guaranteeing water security while curtailing the negative environmental influences. By decreasing water consumption, protecting development, and diversifying water sources, these preparations support both economic development and environmental preservation. As Australia renews to confront sprouting water challenges, further investment and innovation in modern water practices will be fundamental to confirm a sustainable and prosperous future for all living beings.

Statement of Declaration

The author declares that there is no potential conflict of interest to disclose.

Data Availability

There is no data to share for this manuscript.

References

Asok Raja, S.K., Panday, D.P., and Kumar, M. (2023). Decoding the enigma of 100-year record-breaking rainfall over Tamil Nadu using wavelet analysis. *Groundwater for Sustainable Development* 101007. https://doi.org/10.1016/j.gsd.2023.101007.

Ayre, M. and Mackenzie, J. (2013). "Unwritten, unsaid, just known": the role of Indigenous knowledge(s) in water planning in Australia. *Local Environment* 18 (7): 753–768. https://doi.org/10.1080/13549839.2012.665864.

Bark, R.H., Garrick, D.E., Robinson, C.J., and Jackson, S. (2012). Adaptive basin governance and the prospects for meeting Indigenous water claims. *Environmental Science & Policy* 19–20: 169–177. https://doi.org/10.1016/j.envsci.2012.03.005.

Basdew, M., Jiri, O., and Mafongoya, P.L. (2017). Integration of indigenous and scientific knowledge in climate adaptation in KwaZulu-Natal, South Africa. *Change and Adaptation in Socio-Ecological Systems* 3 (1): https://doi.org/10.1515/cass-2017-0006.

Bohnet, I.C. and Kinjun, C. (2009). Community uses and values of water informing water quality improvement planning: a study from the Great Barrier Reef region, Australia. *Marine and Freshwater Research* 60 (11): 1176. https://doi.org/10.1071/MF08329.

Cosens, B. and Chaffin, B. (2016). Adaptive governance of water resources shared with indigenous peoples: the role of law. *Water* 8 (3): 97. https://doi.org/10.3390/w8030097.

Das, A., Kumar, M., Jha, P.K. et al. (2022). Isotopic and hydrogeochemical tracking of dissolved nutrient dynamics in the Brahmaputra River System: a source delineation perspective. *Chemosphere* 307: 135757. https://doi.org/10.1016/j.chemosphere.2022.135757.

Deshpande, A., Miller-Petrie, M.K., Lindstedt, P.A. et al. (2020). Mapping geographical inequalities in access to drinking water and sanitation facilities in low-income and middle-income countries, 2000–17. *The Lancet Global Health* 8 (9): e1162–e1185. https://doi.org/10.1016/S2214-109X(20)30278-3.

Domínguez, I., Ward, S., Mendoza, J. et al. (2017). End-user cost-benefit prioritization for selecting rainwater harvesting and greywater reuse in social housing. *Water* 9 (7): 516. https://doi.org/10.3390/w9070516.

Dovers, S., Grafton, R.Q., and Connell, D. (2005). A critical analysis of the National Water Initiative. *Australasian Journal of Natural Resources Law and Policy* 10 (1): 81–107.

Escott, H., Beavis, S., and Reeves, A. (2015). Incentives and constraints to Indigenous engagement in water management. *Land Use Policy* 49: 382–393. https://doi.org/10.1016/j.landusepol.2015.08.003.

Grant, S.B., Saphores, J.-D., Feldman, D.L. et al. (2012). Taking the "waste" out of "wastewater" for human water security and ecosystem sustainability. *Science* 337 (6095): 681–686. https://doi.org/10.1126/science.1216852.

Hibbard, M., Lane, M.B., and Rasmussen, K. (2008). The split personality of planning: indigenous peoples and planning for land and resource management. *Journal of Planning Literature* 23 (2): 136–151. https://doi.org/10.1177/0885412208322922.

Jackson, S., Storrs, M., and Morrison, J. (2005). Recognition of Aboriginal rights, interests and values in river research and management: perspectives from northern Australia. *Ecological Management and Restoration* 6 (2): 105–110. https://doi.org/10.1111/j.1442-8903.2005.00226.x.

Klein Goldewijk, K., Beusen, A., and Janssen, P. (2010). Long-term dynamic modeling of global population and built-up area in a spatially explicit way: HYDE 3.1. *The Holocene* 20 (4): 565–573. https://doi.org/10.1177/0959683609356587.

Kumar, M., Panday, D.P., Bhagat, C. et al. (2023a). Demystifying the decadal shift in the extent of groundwater in the coastal aquifers of Gujarat, India: a case of reduced extent but increased magnitude of seawater intrusion. *Science of the Total Environment* 165451. https://doi.org/10.1016/j.scitotenv.2023.165451.

Kumar, M., Mazumder, P., Silori, R. et al. (2023b). Prevalence of PPCPs, microplastics and co-infecting microbes in the post-COVID-19 era and its implications on antimicrobial resistance and potential endocrine disruptive effects. *Science of the Total Environment* 166419. https://doi.org/10.1016/j.scitotenv.2023.166419.

Moggridge, B.J., Thompson, R.M., and Radoll, P. (2022). Indigenous research methodologies in water management: learning from Australia and New Zealand for application on Kamilaroi country. *Wetlands Ecology and Management* 30 (4): 853–868.

Mooney, C. and Tan, P.-L. (2012). South Australia's River Murray: social and cultural values in water planning. *Journal of Hydrology* 474: 29–37. https://doi.org/10.1016/j.jhydrol.2012.04.010.

Mukherjee, M. and Jensen, O. (2020). Making water reuse safe: a comparative analysis of the development of regulation and technology uptake in the US and Australia. *Safety Science* 121: 5–14. https://doi.org/10.1016/j.ssci.2019.08.039.

Panday, D.P. and Kumar, M. (2022). Climate indices and hydrological extremes: deciphering the best-fit model. *Environmental Research* 215: 114301. https://doi.org/10.1016/j.envres.2022.114301.

Panday, D.P. and Kumar, M. (2023). Application of remote sensing techniques to deal with scale aspects of GRACE data to quantify groundwater levels. *MethodsX* 102108. https://doi.org/10.1016/j.mex.2023.102108.

Panday, D.P., Khosa, R., Maheswaran, R. et al. (2021a). Game-theoretic-based modelling of Krishna waters dispute: equilibrium solutions by Metagame Analysis. *The European Physical Journal B* 94: 101. https://doi.org/10.1140/epjb/s10051-021-00107-w.

Panday, D.P., Khosa, R., Maheswaran, R. et al. (2021b). Game theoretic-based modelling of Krishna waters dispute: equilibrium solutions by hypergame analysis. *The European Physical Journal B* 94: 131. https://doi.org/10.1140/epjb/s10051-021-00135-6.

Panday, D.P., Singh, A., Gaur, S. et al. (2023). Probable maximum precipitation analysis of high rainfall regions in India. *Groundwater for Sustainable Development* 100893. https://doi.org/10.1016/j.gsd.2022.100893.

Pert, P.L., Hill, R., Maclean, K. et al. (2015). Mapping cultural ecosystem services with rainforest aboriginal peoples: integrating biocultural diversity, governance and social variation. *Ecosystem Services* 13: 41–56. https://doi.org/10.1016/j.ecoser.2014.10.012.

Petzold, J., Andrews, N., Ford, J.D. et al. (2020). Indigenous knowledge on climate change adaptation: a global evidence map of academic literature. *Environmental Research Letters* 15 (11): 113007. https://doi.org/10.1088/1748-9326/abb330.

Pinto, U. and Maheshwari, B.L. (2010). Reuse of greywater for irrigation around homes in Australia: understanding community views, issues and practices. *Urban Water Journal* 7 (2): 141–153. https://doi.org/10.1080/15730620903447639.

Ramírez, I.J. and Briones, F. (2017). Understanding the El Niño Costero of 2017: the definition problem and challenges of climate forecasting and disaster responses. *International Journal of Disaster Risk Science* 8 (4): 489–492. https://doi.org/10.1007/s13753-017-0151-8.

Scherer, L. and Pfister, S. (2016). Dealing with uncertainty in water scarcity footprints. *Environmental Research Letters* 11 (5): 054008. https://doi.org/10.1088/1748-9326/11/5/054008.

Tsatsaros, J., Wellman, J., Bohnet, I. et al. (2018). Indigenous water governance in Australia: comparisons with the United States and Canada. *Water* 10 (11): 1639. https://doi.org/10.3390/w10111639.

Van Vliet, M.T.H., Jones, E.R., Flörke, M. et al. (2021). Global water scarcity including surface water quality and expansions of clean water technologies. *Environmental Research Letters* 16 (2): 024020. https://doi.org/10.1088/1748-9326/abbfc3.

Wang, S.-Y.S. (ed.) (2012). *Modern Climatology*. InTech. https://doi.org/10.5772/2014.

11

Indigenous Technology and Modern Practices in North America and Europe

Ruchi Mehrotra[1], Sonali Aggarwal[1], Rajesh Tiwari[1], and Kishor Joshi[2]

[1] *Department of Management Studies, Graphic Era (Deemed to be University), Dehradun, Uttarakhand, India*
[2] *CTVS, Mira Heart Centre, Dehradun, Uttarakhand, India*

11.1 Introduction

Due to the explosive growth in the population and changes in demographic structures of rural and urban populations, there is a gap in the water demand and supply in the entire globe. Emerging market economies have seen urbanisation, population burden in urban cities, and this has led to a shortage of drinking water across the globe. Due to the presence of numerous chemical, physical, and microbiological contaminants, the natural water is unfit for human consumption. These contaminants are hazardous to health, and therefore this impure water needs purification. International organisations such as the World Health Organisation (WHO) and the Environmental Protection Agency (EPA) have set certain standards for the level of impurities in drinking water. Various water purification technologies like ultrafiltration (UF), sedimentation, reverse osmosis, etc. have been used for decades to purify the contaminated water and make it safe for human consumption (Ahmad and Azam 2019).

There has been a significant difference in the water treatment processes in the United States and Europe for nearly 50 years. In the United States, mainly conservative water treatment methods have been in practice while Europe mainly focuses on various cost-effective and efficient designs for water treatment. It is also considered that drinking water purification technologies used in the United States and Europe will be similar. The technology used in the United States has an advantage over that used in Europe as the United States considers the consultation with the water industry while developing its technology (Crowley and Packham 1993). If we study North America, especially Mexico, and consider Argentina as an emerging market economy with upward GDP growth trends. There is inequality in the access to drinking water between the high- and low-income groups in Mexico, which is a new parameter of water and income disparity to look at. One study uses economic measures of inequality, such as Lorenz Curve and Gini's Coefficient, to measure inequality. As per the results of a survey conducted in 2011, high-income households benefit more as

Water Scarcity Management: Enabling Technologies, First Edition. Edited by Kanchan D. Bahukhandi, Manish Kumar, Durga P. Panday, and Tushara G.G. Chaminda.

compared to low-income households when it comes to access to drinking water and subsidies (Morales et al. 2018).

11.2 Wastewater Treatment – Traditional Water Purification Technology

When discussing sustainability and green technology management, it is important to compare "green" and conventional methods of wastewater treatment. These methods are less non-renewable sources of energy as compared to others (Brix 1999). Even for most developing economies where clean drinking water is a limited resource and an economic calculation, it can surely limit the usage of fertilisers and the re-use of wastewater becomes an essential issue to handle. During the process of oil and gas extraction from underground reservoirs, the wastewater called produced water (PW), containing various organic, inorganic, and toxic compounds, is generated in huge quantities. Conventional processes like phase separators, coarse filters, etc. are used to treat this wastewater. Furthermore, some polishing processes are required to prevent environmental pollution. It has to be ensured that PW is treated optimally, keeping in view the environmental, social, and legal concerns. Advanced oxidation processes (AOP) and membrane techniques are being used for these processes. These technologies have to be improved and modified with time for better results (Jiménez et al. 2018).

One of the techniques that is being used for wastewater treatment is flotation. Some of the pollutants present in the wastewater are chemicals, powders, oils, residual reagents, metal ions, etc. Flotation is one of the modern techniques that treat wastewater in an efficient manner. It also generates less sludge. The technique will soon be used by mineral industries for wastewater treatment. There have also been recent advancements in wastewater treatment for heavy metals (Rubio et al. 2002). For decades now, the presence of chemical and biological contaminants in the water is deteriorating the environment and has also become a matter of concern for society and industries. Wastewater with toxic contaminants is released from most of the domestic and industrial activities, and therefore,

Table 11.1 The traditional wastewater purification technology.

S. no.	Technology name	Type/remark
1	Green filter	Chiapas, Mexico Water balance methodology Multi-stage green filter
2	Microalgae treatment process	Biological process for removal of emerging contaminants (ECs)
3	Coagulation and sedimentation	Water treatment technology that consumes less energy
4	Physico-chemical treatment processes like ultrafiltration process	Oilfield-produced water (OPW) treatment
5	Nature-based solutions (NBS)	For greywater treatment

Table 11.2 The traditional water practices.

Name of the process		Removal of protozoa E.g. *Cryptosporidium*, *Giardia*	Removal of bacteria E.g. *Campylobacter*, *Salmonella*, Shigella	Removal of virus E.g. Enteric, Hepatitis A, Norovirus, Rotavirus	Chemical removal
1. Filtration	A) Microfiltration	Very highly effective	Moderately effective	Not effective	Not effective
	B) Ultrafiltration	Very highly effective	Very highly effective	moderately effective	Low effectiveness
	C) Nanofiltration	Very highly effective	Very highly effective	Very highly effective	Moderately effective
2. Reverse osmosis systems		Very highly effective	Very highly effective	Very highly effective	Effective in removal of some common chemical contaminants like sodium, lead, copper, chloride, chromium; may reduce arsenic, potassium fluoride, radiumcalcium, magnesium, sulphate, nitrate, and phosphorous.
3. Distillation systems		Very highly effective	Very highly effective	Very highly effective	Effective in removal of some common chemical contaminants including arsenic, barium, cadmium, nitrate , lead, sodium, chromium, sulphate, and many organic chemicals.
4. Ultraviolet treatment systems (with pre-filtration)		Very highly effective	Very highly effective	Very highly effective	Not effective
5. Water softeners		Effective in ion or chemical removal using ion exchange technology; ineffective in removal of bacteria, virus, or protozoa			

Table 11.3 The advanced scientific technologies for water purification.

S. no.	Technology name	Type/remark
1	Nanotechnology	Mexico Nanosciences and nanotechnology (N&N) Removal of chemical, biological, and physical contaminants
2	Membrane bioreactor (MBR) technology	North America Nitrate removal in drinking water Endocrine disrupting compounds removal from water and wastewater streams
3	Renewable energy-powered membrane technology (RE – Membrane Technology)	Cost-effective water purification technology
4	Household-level water purification technologies	US–Mexico border To improve community water systems
5	Photovoltaic-powered reverse osmosis (PVRO)	Yucatan Peninsula of Mexico Remote communities Solar energy-powered water purification technology
6	Advanced primary treatment – Coagulation-flocculation processes with high-rate sedimentation	Mexico Purification of municipal wastewater for agricultural reuse
7	Hybrid ozonation-ultrasonication treatment	Removal of pesticides, beta-blockers, and pharmaceuticals
8	Electrocoagulation process	Wastewater treatment
9	Electroflotation	Removal of colloidal particles, oil, and grease, as well as organic pollutants
10	Electrochemically activated persulphate oxidation process	Treatment of arsenite contaminated water

there is a need to protect water resources. A combination of biological, physical, and chemical processes is involved in treating the wastewater and also removing soluble contaminants and insoluble particles from it (Crini and Lichtfouse 2019).

11.3 Extensive Purification Traditional Technologies

Water and energy are essential medium for comfortable living and overall development and there is also a problem of water scarcity and energy crisis. Therefore, it is necessary to use water treatment technologies that consume less energy. Traditional technologies like coagulation and sedimentation consume less energy when compared to other advanced techniques (Grzegorzek et al. 2023). Extensive purification technologies like the "green filter" or the "land application system" were used as an alternative to purify wastewater

Table 11.4 The history of water contamination and mitigation strategies.

Challenges in the 21st century	How water get contaminated?	What CDC is doing?
Pathogens that grow in pipe slime contaminate the tap water. In 2014, 94% of deaths from waterborne pathogens were caused due to biofilm – associated pathogens while the remaining 6% due to other pathogens.	Biofilm-associated pathogens usually grow in pipes and other equipment inside large buildings and water facilities.	Use of new and innovative technologies to detect and analyse pathogens.
Harmful algae blooms (HABs) threaten oceans, rivers, drinking water, etc.	Broken water mains can be hazardous to health.	Actively responding to emergencies related to waterborne diseases.
Over 1 million illness each year is caused due to the pathogens present in pool water	Pools and other recreational water facilities can also spread pathogens if not maintained well.	Supporting water management programs that mainly focus on biofilm-associated pathogens.
	Harmful pathogens can grow in warm water and due to presence of excessive nutrients in water.	Creating public awareness related to outbreak and prevention of waterborne diseases.

Table 11.5 Developments and further scope of research and developments in water purification technologies.

S. no.	Technology name	Type/remark
1	Oxidation process to advanced oxidation process (AOP)	• Removal of organic contaminants from produced water and wastewater
2	Conventional membrane-based technologies to advanced MBRs	• Reverse osmosis (RO), electrodialysis, microfiltration
3	Advancements in cellulose-based nanofiltration membranes	• Use of nano-cellulose for water purification

using the ground and plot with cultivated trees. This was done by calculating the wastewater production, precipitation, and evapotranspiration of the area used in the state of Chiapas, Mexico, using the water balance technology. The precipitation occurring throughout the year was measured through a "multi-stage green filter" (Navarro and de Bustamante Gutiérrez 2013). Tables 11.1 to 11.5 depicts the traditional and advanced techniques of water purification.

Emerging contaminants (ECs) have to be removed in the wastewater treatment plant (WWTP) by various chemical, biological, and hybrid technologies. Membrane bioreactor (MBR) proved to be a very effective biological process for the removal of endocrine disruption chemicals (EDCs). Almost all type of ECs can be removed from the wastewater through microalgae treatment process compared to any other biological process. Fenton process can effectively remove pesticides and pharmaceutical contaminants from the wastewater, while ozonation can better remove EDCs. Therefore, a combination of them is used for almost 100% removal of these contaminants (Ahmed et al. 2017).

Nanotechnologies are part of a new industrial platform with the capability to surpass existing tech-intensive systems for water treatment. Numerous nations are using nanotechnology-based processes to solve problems related to water. Nanotechnology is one of the most emerging, advanced, and effective ways to remove contaminants from water. The advances in nanotechnology and nanoscience, implications and the challenges, and the potential risks of nanotechnology in Mexico even cover components of patents where extensive research is in progress (Olvera et al. 2017).

One of the by-products of oil and gas extraction is oilfield produced water (OPW). Some organic and inorganic contaminants like heavy metals, grease, etc. are present in OPW which can be hazardous for human health and can also deteriorate the environment. Traditional membrane-based technologies along with UF are used, which effectively removes oily wastewater. It must be ensured that the technique used for OPW treatment is cost-effective and also maintains the quality standards for water treatment (Samuel et al. 2022). Then another Indigenous and traditional method is photovoltaic reverse osmosis systems (PVRO), which is a technique that uses solar energy to purify and de-saline water in remote areas of Mexico. The Indigenous people, mainly the beekeepers and the farmers, residing in the remote village of the Yucatan Peninsula of Mexico were deployed with PVRO system to treat water and the idea was to make this technique "self-sustaining" in the long run (Elasaad et al. 2015).

11.4 Advanced Scientific Technologies for Water Purification

Nanotechnologies are part of a new industrial platform with the capability to surpass existing tech-intensive systems for water treatment (Olvera et al. 2017). One of the most advanced scientific technologies that is used for water purification and production of clean energy in Mexico is Nanotechnology. Nanosciences and Nanotechnology (N&N) is evolving and advancing across the globe. This technique can remove chemical, biological, and physical contaminants from water (Delgado-Ramos 2014). Numerous advanced electrochemical technologies like electrodeposition, electrocoagulation (EC), electrooxidation, and electroflotation are now used for wastewater treatment. Electroflotation can effectively remove all the organic pollutants, grease, oil, and colloidal particles from the wastewater and also flocculated sludge can be separated by this technique. Electrocoagulation is also one of the advanced electrochemical techniques that can be used to treat wastewater (Chen 2004).

There is a scarcity of clean drinking water in remote areas in developing countries as PVRO is an advanced scientific technology that uses solar energy for water purification. It is used in the Yucatan Peninsula of Mexico, primarily by beekeepers and subsistence farmers. Some materials and methods are developed so that the community members can themselves use the system (Elasaad et al. 2015). Another new technology and advancement is the MBR technology. MBR technology is a water purification technology that is advancing worldwide in both research and commercial applications. It is also used in North America for water purification. It is found that the industrial wastewater treatment and MBR application are higher in North America as compared to the rest of the world. North America's installations of MBR's commercial applications are about 11% of the worldwide installations. This scientific technique is used for removal of nitrate from drinking water and is also used for removing endocrine disrupting compounds from wastewater (Yang et al. 2006). One of the cost-effective water treatment techniques that is used by the low-income communities along the US–Mexico border is household-level water purification technologies (HWTs). HWT was earlier used as a temporary solution for water purification but later was widely used as a water purification by communities in the United States that face severe water insecurity. The rural and peri-urban subdivisions on the Texas-Mexico border also use HWT (Vandewalle and Jepson 2015). Even there are so many hazardous material in water which essentially need to be removed, else they are life-threatening. The underground water is highly polluted with arsenic. Traditional methods can successfully remove arsenate, but arsenite cannot be completely removed by those methods. Therefore, electrochemically activated persulphate oxidation process (EAOP) is an advanced scientific technology that is used for arsenite removal. This technique is effective for removal of both arsenate and arsenite (Syam Babu and Nidheesh 2022). Renewable energy-powered membrane technology (RE – membrane technology) is an advanced and cost-effective water purification technique that can potentially remove various chemical and microbiological contaminants from water. However, the non-availability of spare parts, lack of skilled personnel for installation and maintenance are hindrances in adoption of this technology to a large extent (Schäfer et al. 2014). There is advancement in healthcare technology and digitalisation in the medical industry which has led to lower mortality rates across the globe (Mehrotra and Joshi 2020).

11.5 Disease Management

There are huge dearth and need to develop proper technology and regulations for effective and efficient reuse of wastewater in irrigated agriculture in Mexico. The focus was on both the farmer's needs and the health of the people. Legislations were made that allowed nutrients and organic matters to be present in the wastewater in appropriate concentrations rather than being considered pollutants and also certain pathogens like helminth ova had to be removed before wastewater reuse (Jiménez 2005). The children living in a water reclamation area in Mexico City had a risk of enteric disease. There was a risk of diarrheal diseases among them, and therefore a study was conducted to measure it. Data related to the availability and supply of drinking water and sanitation facilities were collected. The rate of diarrhoea was high during the rainy season, that is, 11.8% compared to dry season, that is, 10.7%. Another factor that leads to diarrhoea was consumption of food consumed by street vendors. There is a risk of groundwater contamination (Enrique et al. 2002). There was an outbreak of typhoid fever in a small village in central Mexico due to a chloramphenicol-resistant Salmonella Typhi in mid-1972. Children from 1 to 14 years of age and people above 45 years were highly attacked. It was a waterborne disease that occurred due to a non-operational municipal water system due to a defective pump (Gonzalez-Cortes et al. 1973). Another advancement is IOMT – Internet of Medical

Things – which is augmenting the quality of life for even high risk cardiac patients and even other remote medical management of patients (Joshi and Mehrotra 2022). Such technological developments like water quality management and assessment if done at every level can ensure better and safer drinking water for all.

Cyanobacterial harmful algal blooms (cHABs) have an adverse impact on the water quality, drinking water, fisheries, agriculture, etc., and it also have ecological and socioeconomic cost. cHABs produces cyanotoxins which are hazardous for mammals, human health, and other organisms. The Great Lakes in North America supplies around 84% of the surface water all over the continent. cHABs presence in the lake water was found in the 1980s, while its presence in Lake Erie was noted in early 1990s. Microcystic, which produces microcystins, is the dominant toxigenic cyanobacterium (Carmichael and Boyer 2016). It is estimated that waterborne diseases would cause about 10% diseases around the globe but the risk perceptions related to this issue are ignored. Here, the risk related to waterborne diseases is understood in a region in Northern Mexico. The problems and solutions of these diseases are studied in eight small communities along the Sonora River Basin in Mexico, which is a highly contaminated river. A few citizens and officials of that region were aware that the waterborne diseases arise due to the contaminated water and also poor management of the wastewater. Therefore, proper management of wastewater, water treatment, and basic hygiene is essential to prevent waterborne diseases (Morua et al. 2011).

Water purification is essential, but it should also be ensured that the technologies used for water purification are cost-effective. The volume of water purified, and the cost incurred in it differs in various treatment processes. Wastewater treatment through conventional technologies is found cheaper when compared to vehicular transportation of water. The cost incurred for setting up the water purification unit, population of the region, precipitation, agricultural irrigation, government laws, and other financial, social, and environmental factors are also studied as all of them also affect the cost involved in water purification (Plappally and Lienhard 2013). Clean water scarcity is a problem that all the countries around the globe are facing, which is expected to worsen over time. Even the regions which are generally considered water-rich will face this problem. Therefore, there is a need to discover new, improved, advanced, and cost-effective methods for water purification that consume less energy. It should also be noted that these water purification technologies are environment friendly. There should be a healthy reuse of wastewater (Shannon et al. 2008). There are numerous developments happening and with artificial intelligence and blockchain taking over the storm of reforms, we expect that scientifically more new technologies advent will bring safe and drinking water accessibility across the globe. The advantage of North America being a developed nation is that its technology is more advanced than that of least developed countries (LDCs) and many developing economies. The emphasis on water quality and its supply to remote areas in developing economies or under-developed countries needs attention from governments. Hence, healthcare substantiality through technological innovations and interventions for any disease management as well as basic lifestyle sustenance is a need of hour (Mehrotra and Joshi 2021). If we look at micro- or macrolevel sustainable measures, a few case studies indicate the interlinkages operate from a microend of municipal to almost global scales, and this emphasis in developing economies is crucial issue (Alcamo 2019). Hence, even role of emerging markets in North America can contribute in achieving the United Nations UN – SDGs (Sustainable Development Goals) (Ait-Kadi 2016).

11.6 Conclusion

In conclusion, we conclude that in North America, a small country like Mexico is doing extensive research on water quality and adopting advanced technological purification methods. This emerging fastest-growing economy has been focusing largely on water quality and waterborne disease management quite nicely due to its worst past experiences. And technologies like MBR and nanotechnology have been a game changer in this. Even other technologies are helping in the supply of pure and safe drinking water across North America.

References

Ahmad, A. and Azam, T. (2019). 4 – water purification technologies. In: *Bottled and Packaged Water* (ed. A.M. Grumezescu and A.M. Holban), 83–120. Woodhead Publishing. https://doi.org/10.1016/B978-0-12-815272-0.00004-0.

Ahmed, M.B., Zhou, J.L., Ngo, H.H. et al. (2017). Progress in the biological and chemical treatment technologies for emerging contaminant removal from wastewater: a critical review. *Journal of Hazardous Materials* 323: 274–298. https://doi.org/10.1016/j.jhazmat.2016.04.045.

Ait-Kadi, M. (2016). Water for development and development for water: realizing the sustainable development goals (SDGs) vision. *Aquatic Procedia* 6: 106–110. https://doi.org/10.1016/j.aqpro.2016.06.013.

Alcamo, J. (2019). Water quality and its interlinkages with the sustainable development goals. *Current Opinion in Environmental Sustainability* 36: 126–140. https://doi.org/10.1016/j.cosust.2018.11.005.

Brix, H. (1999). How 'green' are aquaculture, constructed wetlands and conventional wastewater treatment systems? *Water Science and Technology* 40 (3): 45–50. https://doi.org/10.1016/S0273-1223(99)00418-7.

Carmichael, W.W. and Boyer, G.L. (2016). Health impacts from cyanobacteria harmful algae blooms: implications for the North American Great Lakes. *Harmful Algae* 54: 194–212. https://doi.org/10.1016/j.hal.2016.02.002.

Chen, G. (2004). Electrochemical technologies in wastewater treatment. *Separation and Purification Technology* 38 (1): 11–41. https://doi.org/10.1016/j.seppur.2003.10.006.

Crini, G. and Lichtfouse, E. (2019). Advantages and disadvantages of techniques used for wastewater treatment. *Environmental Chemistry Letters* 17 (1): 145–155. https://doi.org/10.1007/s10311-018-0785-9.

Crowley, F.W. and Packham, R.F. (1993). Water treatment in Europe and North America. *Water and Environment Journal* 7 (1): 81–87. https://doi.org/10.1111/j.1747-6593.1993.tb00814.x.

Delgado-Ramos, G.C. (2014). Nanotechnology in Mexico: Global trends and national implications for policy and regulatory issues. *Technology in Society* 37: 4–15. https://doi.org/10.1016/j.techsoc.2013.09.005.

Elasaad, H., Bilton, A., Kelley, L. et al. (2015). Field evaluation of a community scale solar powered water purification technology: a case study of a remote Mexican community application. *Desalination* 375: 71–80. https://doi.org/10.1016/j.desal.2015.08.001.

Enrique, C., Leticia, S., Maritsa, S., and René, S. (2002). Diarrheal diseases in children from a water reclamation site in Mexico city. *Environmental Health Perspectives* 110 (10): A619–A624. https://doi.org/10.1289/ehp.021100619.

Gonzalez-Cortes, A., Sanchez-Leyva, R., Hinojosa, M. et al. (1973). Water-borne transmission of chloramphenicol-resistant Salmonella Typhi in Mexico. *The Lancet* 302 (7829): 605–607. https://doi.org/10.1016/S0140-6736(73)92427-6.

Grzegorzek, M., Wartalska, K., and Kaźmierczak, B. (2023). Review of water treatment methods with a focus on energy consumption. *International Communications in Heat and Mass Transfer* 143: 106674. https://doi.org/10.1016/j.icheatmasstransfer.2023.106674.

Jiménez, B. (2005). Treatment technology and standards for agricultural wastewater reuse: a case study in Mexico. *Irrigation and Drainage* 54 (S1): S23–S33. https://doi.org/10.1002/ird.183.

Jiménez, S., Micó, M.M., Arnaldos, M. et al. (2018). State of the art of produced water treatment. *Chemosphere* 192: 186–208. https://doi.org/10.1016/j.chemosphere.2017.10.139.

Joshi, K., and Mehrotra, R. (2022). Internet of Medical Things-State-of-the-Art. In Tanwar, R., Balamurugan, S., Saini, R. K., Bharti, V., & Chithaluru, P. (Eds.), *Advanced Healthcare Systems: Empowering Physicians with IoT-Enabled Technologies* (1st ed, pp. 1–20). John Wiley & Sons.

Mehrotra, R. and Joshi, K. (2020). Digitalization in the healthcare industry. Sharma, R., Saini, R., Prakash, C., and Prasad, V. (Eds.). In: *Internet of Things and Businesses in a Disruptive Economy*, 27–48. https://www.scopus.com/inward/record.uri?eid=2-s2.0-85109248320&partnerID=40&md5=76a99eca58749a5b3168a58ecdd370ec. Nova Science Publishers.

Mehrotra, R. and Joshi, K. (2021). Healthcare sustainability through technological innovations. In: *Advances in Environment Engineering and Management* (ed. N.A. Siddiqui, K.D. Bahukhandi, S.M. Tauseef, and N. Koranga), 341–349. Springer International Publishing.

Morua, A.R., Halvorsen, K.E., and Mayer, A.S. (2011). Waterborne disease-related risk perceptions in the Sonora River Basin, Mexico. *Risk Analysis* 31 (5): 866–878. https://doi.org/10.1111/j.1539-6924.2010.01570.x.

Navarro, M., and de Bustamante Gutiérrez, I. (2013). Systems Design of Wastewater Treatment by Extensive Purification Technologies: Application in Chiapas, Mexico. In: Farfán González, H., Corvea Porras, J., de Bustamente Gutiérrez, I., and LaMoreaux, J. (eds) Management of Water Resources in Protected Areas. Environmental Earth Sciences. Springer, Berlin, Heidelberg. https://doi.org/10.1007/978-3-642-16330-2_3

Olvera, R.C., Silva, S.L., Robles-Belmont, E., and Lau, E.Z. (2017). Review of nanotechnology value chain for water treatment applications in Mexico. *Resource-Efficient Technologies* 3 (1): 1–11. https://doi.org/10.1016/j.reffit.2017.01.008.

Plappally, A.K. and Lienhard, J.H. (2013). Costs for water supply, treatment, end-use and reclamation. *Desalination and Water Treatment* 51 (1–3): 200–232. https://doi.org/10.1080/19443994.2012.708996.

Rubio, J., Souza, M.L., and Smith, R.W. (2002). Overview of flotation as a wastewater treatment technique. *Minerals Engineering* 15 (3): 139–155. https://doi.org/10.1016/S0892-6875(01)00216-3.

Samuel, O., Othman, M.H.D., Kamaludin, R. et al. (2022). Oilfield-produced water treatment using conventional and membrane-based technologies for beneficial reuse: a critical review. *Journal of Environmental Management* 308: 114556. https://doi.org/10.1016/j.jenvman.2022.114556.

Schäfer, A.I., Hughes, G., and Richards, B.S. (2014). Renewable energy powered membrane technology: a leapfrog approach to rural water treatment in developing countries? *Renewable and Sustainable Energy Reviews* 40: 542–556. https://doi.org/10.1016/j.rser.2014.07.164.

Shannon, M.A., Bohn, P.W., Elimelech, M. et al. (2008). Science and technology for water purification in the coming decades. *Nature* 452 (7185): 301–310. https://doi.org/10.1038/nature06599.

Syam Babu, D. and Nidheesh, P.V. (2022). Treatment of arsenite contaminated water by electrochemically activated persulfate oxidation process. *Separation and Purification Technology* 282: 119999. https://doi.org/10.1016/j.seppur.2021.119999.

Vandewalle, E. and Jepson, W. (2015). Mediating water governance: point-of-use water filtration devices for low-income communities along the US–Mexico border. *Geo: Geography and Environment* 2 (2): 107–121. https://doi.org/10.1002/geo2.9.

Yang, W., Cicek, N., and Ilg, J. (2006). State-of-the-art of membrane bioreactors: worldwide research and commercial applications in North America. *Journal of Membrane Science* 270 (1): 201–211. https://doi.org/10.1016/j.memsci.2005.07.010.

12

Water-Related Traditional and Indigenous Practices

A Global Review and Case Study from Tank-Cascade Systems in Sri Lanka

S.S.K. Chandrasekara[1], M.P. Perera[2], and M. Vithanage[3,4]

[1] *Department of Agricultural Engineering, Faculty of Agriculture, University of Peradeniya, Peradeniya, Sri Lanka*
[2] *Department of Geography, Faculty of Arts, University of Peradeniya, Peradeniya, Sri Lanka*
[3] *Ecosphere Resilience Research Center, Faculty of Applied Sciences, University of Sri Jayewardenepura, Nugegoda, Sri Lanka*
[4] *National Institute of Fundamental Studies, Kandy, Sri Lanka*

12.1 Introduction

Water is a basic need for the human being and has both physical and spiritual relationships with humans and living beings. Religious teaching and traditional organizations/societies have direct or indirect impacts on sustainable water management in a country considering harnessing, conserving, developing the water resources, and sharing the water resources effectively and efficiently among users (i.e., intra-boundary and transboundary). The monarchs, religious institutions, governments, and nongovernment institutions enacted formal and informal practices and continue as traditional and Indigenous teaching/knowledge and practices for sustainable water resources development and management. The informal rules were implemented by the communities based on day-to-day activities.

Traditional Indigenous knowledge is defined as a collective body of knowledge, beliefs, and traditions that aim to preserve, communicate, and facilitate interactions between ancestors and the natural environment (Bruchac 2014, pp. 3814–3824). The terms "knowledge," "belief," and "tradition" can be recognized as factual data, religious concepts, and practice, respectively, but occasionally the terms are misinterpreted to different definitions while describing them for Indigenous studies. The Indigenous knowledge and traditions/practices are passed to the next generations formally and informally through relatives, community elders/groups via oral, traditions, ritual practices, and other written historic scripts.

It is important to understand the traditional and Indigenous practices in water resources development and management and they are the foundation for most of the modern large-scale water resources-related strategies. Nevertheless, it is challenging to introduce a new paradigm in water resources because new strategies may get resistance from the communities who are practicing Indigenous/traditional knowledge. Therefore, gradual and systematic approaches are crucial to introduce novel water development and management strategies that would balance loyalty to traditional customs and norms (Behailu et al. 2016).

Water Scarcity Management: Enabling Technologies, First Edition. Edited by Kanchan D. Bahukhandi, Manish Kumar, Durga P. Panday, and Tushara G.G. Chaminda.

Despite that, this study is focused on identifying the traditional and Indigenous practices that existed in the water resources sector internationally and with a special focus on Sri Lanka for identifying the relevant practices that can be used for efficient and effective water resources development and management.

12.2 Ancient Hydraulic Civilization – The Backbone for Irrigated Agriculture

The earliest hydraulic civilizations were settled in valleys/banks of large rivers such as the Nile, Euphrates, Tigris, Indus, Yellow, and Yangtze rivers, etc. to satisfy primarily the domestic and irrigation water demand. Secondarily the other waterborne requirements were assured. Border irrigation, strip irrigation, and check basin irrigation were practiced by ancient civilians to lessen soil erosion in the irrigating fields, and check basin irrigation is practiced still in India. Most of these techniques modernized into modern irrigation techniques like furrow irrigation. Water harvesting from rain and glaciers was adopted in India too (Food and Agriculture Organization of the United Nations, Italy 2018). Table 12.1 shows the different types of irrigated agricultural systems practiced in the past.

12.3 Historical Knowledge of Hydrological Cycle and Measurement of Its Processes

The historical Indian scriptures mentioned the civilians who lived in the ancient Harappan Civilization (3000 BC–1500 BC) and the Vedic period (1500–500 BC) knew about the hydrological cycle and its processes (Singh et al. 2020). Malik (2016, pp. 54–59) compared the principles and processes of the modern hydrologic cycle with the statements in the ancient script – "Ramayana" and identified there were similarities between the concepts. Ancient Indian civilization was aware of the spatial and temporal variability in rainfall across India in terms of quantity and quality. The rain gauges measured rainfall as a dimension of weight and not as a volume as today (Singh et al. 2020).

The desalinization of seawater was originally practiced by the ancient Minoan civilization (2000–1400 BC). Minoan sailors' excessive sea navigation for the trade made them drink desalinized seawater. The adsorption and ion exchange for adsorbing salt was practiced in China during 475–221 BC using the thin layer extracted from bamboo mats, which were used for steaming rice over long periods. A long-term submerging of wax vessels in seawater filtered the salt and provided drinking water in Greece (384–322 BC). There were examples of experimentation with desalination technologies described by Angelakis et al. (2022). Furthermore, the "Nilometer" was used in Egypt to monitor the water level in the Nile river and historical evidence showed that there were about 20 Nilometers in the Nile river in ancient Egypt in 600–7000 BC (Chathuranika et al. 2022).

The spatial and temporal water availability encouraged ancient civilizations to adopt water treatment techniques to purify the available water. Harappan civilization in India purified water by means of (i) adding natural herbal or minerals to the water as a coagulant

Table 12.1 Different types of irrigated agriculture in the past and their descriptions.

Type of irrigation	Country	Description	Source
Flood plain-based irrigated agriculture	Egypt (5000 BC) for winter crops	• Fluctuation of the Nile river facilitated basin irrigation • Water was retained for 40–60 days. • The network of earthen banks was constructed parallel and perpendicular to the river and regulated the water via sluices. • The water was stored until the soil became saturated and excess water drained to nearby canals.	Bjornlund and Bjornlund (2010, pp. 1–12)
	South Central Iraq – In southern Mesopotamia (Tigris and the Euphrates) for Summer	• The bed of the Euphrates was located higher than that of the Tigris and therefore the water in the Euphrates was used as a supply of water, whereas the Tigris behaves as a drain. • The in-between lands were flooded and used for agriculture. • The threat of lowering the groundwater level and increasing the salinity level in the fields was abated in these areas by the farming concept of "alternate year."	Bjornlund and Bjornlund (2010, pp. 1–12)
Raised irrigation using barrages	South Central Iraq – In southern Mesopotamia (Tigris and the Euphrates) for Summer	• The lowered water levels in Euphrates and Tigris during May to September/October which impacted both river navigation and irrigation and therefore, the barrages were used to raise the water level for irrigation in dry periods in summer (i.e., October/November). • The barrages were made from reeds, brushwood, and mud to raise the water level to the inlet of the primary irrigation canal. • Barrages reduce the water velocity which negatively impacted river navigation, reduce water use in downstream water users, and increase sedimentation in dry periods. In addition, barrages can cause flooding during wet periods.	Bjornlund and Bjornlund (2010, pp. 1–12), Rost (2019)
Terrace irrigation	"Djessure" irrigation technique was used in Tunisia. Parallel techniques used in China, India, America, Peru, and North Africa	• Stonewalls and soil were made into a barrier to store runoff and rainfall water. • Used to irrigate Olives and other trees in Tunisia.	Chathuranika et al. (2022).

(*Continued*)

Table 12.1 (Continued)

Type of irrigation	Country	Description	Source
Gravity-fed irrigation	Dujiangyan Irrigation System (DIS) constructed in 211–207 BC is in Chengdu flood plain in Sichuan Province, China.	• The ancient systems which still control flood and sand discharge, and distribution of water for gravity-fed irrigation and domestic requirements.	Cao et al. (2010)
	Qanat (known as karez, kanerjing, falaji, fogorra, khettara, and galleria) in the 7th century BC for arid/ semiarid regions of the world/Iranian Plateau.	• Water distributed under gravity in an artificial underground horizontal wells/tunnel. These tunnels were dug up to alluvial aquifers to harness groundwater to a storage pond for irrigation. • The system reduced the water loss from evaporation. • Nevertheless, qanat is now in danger due to climate change, desertification, overuse of freshwater, pollution, and inadequate policies. • The replacement of the qanat by the pump wells dries up the qanat due to the overexploitation of groundwater.	Chathuranika et al. (2022) Esmaeili et al. (2022, pp. 131–144)
	Drier regions in India and Sri Lanka	• Tank cascade systems in Sri Lanka (discussed in the later part of the manuscript).	
Pitcher irrigation	Northern Kerala for irrigation of summer crops Tafilalet oasis, Morocco	• Buried porous earthen pots provide water using a technique similar to drip irrigation.	Chathuranika et al. (2022)
Drip irrigation	Northeast India	• Bamboo drip irrigation is used to deliver water directly to the roots of plants through bamboo pipes, which helps conserve water in tribal villages.	Chathuranika et al. (2022)

to sediment the unpleasant materials in the water, (ii) boiling of water and stored in iron, copper, and brass containers for disinfection from bacteria, and (iii) graded filtering using sand and gravel. The painting evidence in Egypt (1500 BC) showed the water filtering technique using wick siphons. Furthermore, suspended soils were extracted using aluminum sulfate and iron sulfate or a mix of both, and these methods still exist at present. In ancient Greece, Hippocrates Sleeve – a fabricated bag was used to drain water before boiling. Detailed descriptions of historic water treatment techniques are available in Bansal and Asthana (2019, pp. 16–18).

Understanding of groundwater such as water-bearing structures/aquifers, quality and quantity of the aquifers, and water-utilizing methods was impressive during the early civilization in India. The public and private baths, well-managed sewerage systems, water storing reservoirs, water harvesting systems, water distributary canals, and the cluster of reservoirs for flood water utilization for Rabi crops and renowned stepwells (i.e., which harness the seepage from upstream reservoirs) were prominent during ancient India. Higher water efficiency was achieved via (i) field embankment to increase water holding capacity, (ii) sluice gates for regulation of water, and (iii) conduit irrigation. Not only did the inland waters, but the Harappan civilization also developed maritime resources including dockyards and protective walls for oceanic disasters such as tsunamis. The Mauryan Empire (322–185 BC) implemented water pricing as a practice for sustainable water management during his reign, but descriptive written evidence is lacking.

In the Dujiang Weir project, in China, the water level of the canal head and the inflow discharge at the canal were measured using a depth gauge of 2.9 m stone-built man-like structure which gradually illustrates the water level from foot to shoulder. There were three similar structures identified in the systems. Majapahit civilization in the 14th century presently locates in Trowulan suburban, east Java which experiences prolonged drought and torrential rainy season. Therefore, the Majapahit gravity-fed irrigation system was built to control and collect the water for irrigation. The system consists of man-made dams connected to natural streams to collect water, underground water canals built in non-stream areas to provide water for paddy cultivation and to central reservoirs to collect water, water distribution networks, and irrigation structures decorated as a temple to aspire and bond the spirituality of living being with water. One of the important aspects of the system is divided irrigation which mitigated the impacts of unpredictable tropical climates in dryer periods by collecting water in the end reservoirs. Detailed descriptions are available at Wilbowo (2020).

12.4 Ancient Water Governance

Communal governance is a mostly practiced system in ancient water management systems. The Babylonian monarchs (1792–1750 BC) enacted the first-ever legislation for irrigation systems ensuring the communal participation of farmers in the maintenance of irrigation systems (Krasilnikoff and Angelakis 2019). In Greek history, the Draconian Constitution included the principle of sharing water in Attica to benefit farmers. Krasilnikoff and Angelakis (2019) described the legislation that evolved in ancient Greece to optimize the use of both groundwater and surface water, as well as to punish those who misused water. The community forests in Africa were managed by the Indigenous rules where the customary rights of the natives were preserved. The forests and water were shared among the community (Food and Agriculture Organization of the United Nations, Italy 2018). The cultural, religious, and spiritual taboos and beliefs protected the natural resources – the Tano river from overextraction in Ghana and secured water sources from misusing and pollution in Konso, Ethiopia (Behailu et al. 2016). In Borana, Ethiopia, the traditional Gedda administration system managed water by compulsory labor system where taxation/pricing of water was not involved but misuse or evasion of water resulted in revocation of access to communal water resources.

12.5 Sri Lanka: As a Pioneering Case Study for Tank Irrigation Practices

The dry zone of Sri Lanka is a unique place to study traditional water-related practices due to its historical hydraulic civilization which contains diverse strategies for irrigation water management, since the 3rd century B.C. These Indigenous techniques of building thousands of tanks and management practices, which prevailed throughout the dry zone of the island, could be considered one of the oldest and most successful irrigation management practices in the world (Perera 2017, pp. 597–610).

12.5.1 The Ancient Craftsmanship of Constructing Irrigation Tanks as Tank-Cascade Systems

With the absence of rainfall in the dry season of the dry zone of Sri Lanka, there was a requirement to collect rainwater in the terrains of mini basins and release the collected water to the downstream irrigated agricultural lands. The mechanism was constructing a number of irrigation tanks along the mini basins by reducing water discharge to the sea (Figure 12.1). This pattern has been defined as "Tank-Cascade Systems" (Madduma Bandara 1985, pp. 99–114).

This ancient tank constructing craftsmanship has been diversified considering the regional requirements of the dry zone landscape of Sri Lanka. While the main irrigation tanks have been focused on collecting rainwater and distribution, there were sediment trapping tanks – *Kulu wewa* in the immediate catchments, focusing on protecting the irrigation tanks from sedimentation. These tanks have been established above the major tanks without irrigation devices such as sluices, irrigation canals, and command areas. In irrigation water management, the main sluice gate is not open during the spill because of the pressure caused by the water towards the sluice gate and dam. The "Sluice Front Ridge" at *Peravetiya* was developed during the 3rd century CE, in the period of King *Mahasen*. It is very similar to a "Bracket" in front of the sluice gate and can be used to open the sluice gate when necessary, such as during overflow. Due to the front ridge, it will not cause damage or impact the sluice or dam (Perera 2013, pp. 1–9).

12.5.2 Forest-Based Tank Reservations

There were three main forest-based tank reservations around the tanks, especially for tank water conservation and purification. The first one is the "*Gasgommana*," and it is a strip of large trees located in the tank's immediate catchment/around the tank, that is roughly near the full supply level, which serves as a windbreak, to reduce the evaporation losses in the tank. The second one is the "*Perahana*," which is a strip of bushes that acts as a filter by preventing the flow of eroded soil and litter from agricultural lands in the catchment. These silt traps are one of the secrets behind the long period of sustainability of the tanks without serious siltation. The third feature is the "*Kattakaduwa*," which is a reserved land strip with well-rooted trees located between the tank bund and the paddy field on the downstream side (Figure 12.2). This plot of land helps to prevent salinization of the downstream paddy fields and minimizes water seepage going through

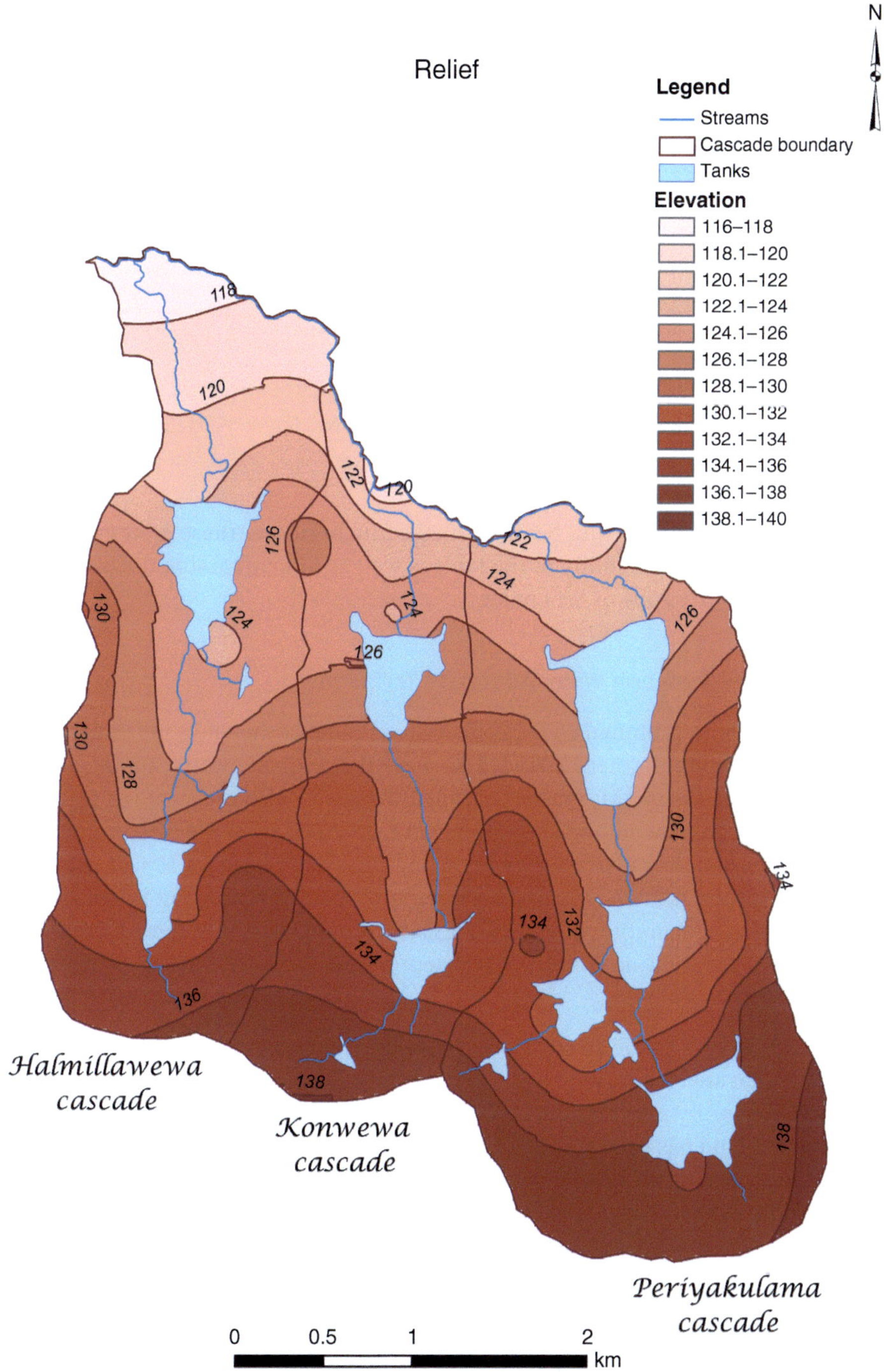

Figure 12.1 A few examples of tank cascades in the dry zone of Sri Lanka.

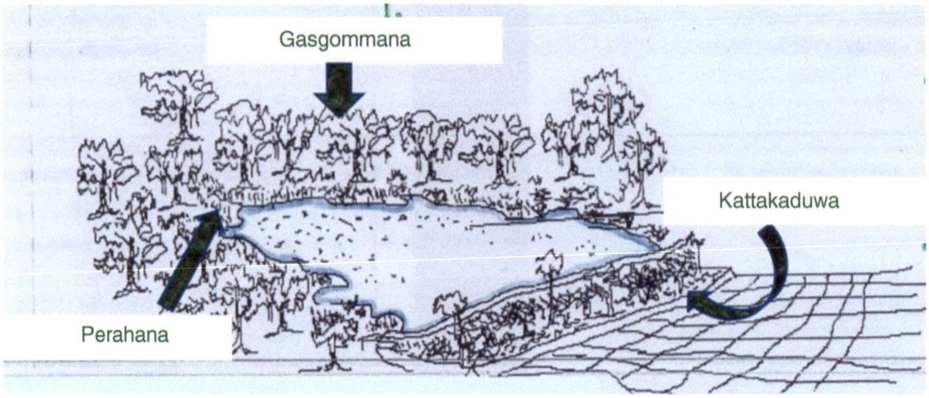

Figure 12.2 Ancient tank reservations in the dry zone of Sri Lanka. *Source:* Perera and Rekha Nianthi (2020)/IGI Global.

the tank bund (Perera and Rekha Nianthi 2020). Furthermore, these systems play a significant role in the sustainability of the tank as well as in maintaining the biological diversity spots in the dry zone of Sri Lanka.

12.5.3 Spatial and Temporal Water Allocation Technique

In the dry season, there is limited water availability in most of the tanks in the dry zone of Sri Lanka, and ancient farmers intended to share the available water, among all the farmers by sharing of water. This practice is called "*Bethma*" with the meaning of dividing the tank irrigation area among all farmers who are cultivating the same command area. Roughly 50% of the available water in the tank has been used for 50% of the land by sharing with all the farmers. It was an attractive strategy for land and water allocation by maintaining cooperation among all farmers. For this purpose, there were three criteria for land allocation and farmers selected one of these, according to regional needs (Figure 12.3).

1) Selected one major irrigation canal on one side of the command area and all the farmers cultivated in that area on a shared basis.
2) Tank water is used by selecting the closest part of the command area to the tank and that amount of land has been shared by all the farmers.
3) Selected the most suitable alluvial parts with relevant dividing canals of the command area and the land has been shared by all the farmers. Especially, this practice is linked to minimizing water usage in large and medium-scale tanks.

12.5.4 Inter-basin Water Diversions

One of the most important water-related practices was inter-basin water diversions from water-surplus river basins to water-deficit river basins. Although this strategy is used in modern watershed management concepts, earlier most historical evidence has been

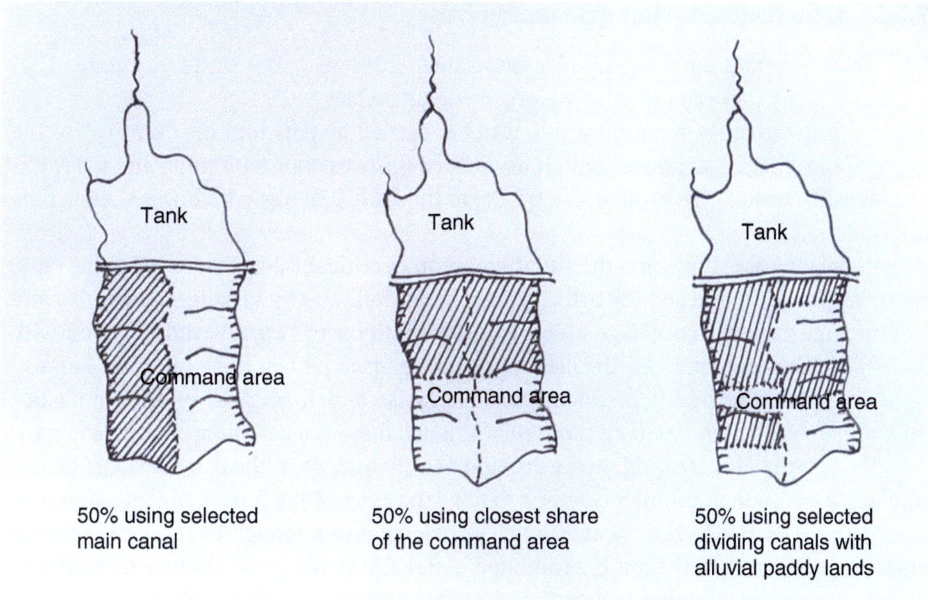

Figure 12.3 Three types of Bethma Systems in Sri Lanka. *Source:* Perera and Rekha Nianthi (2020).

reviled from Sri Lanka in the 1st century CE. It is the "*Elahera* Canal" that diverted water from the Mahaweli river towards the Minneriya basin by king *Wasabha* (66–110 CE). In the 3rd CE century, it extended to *Kantale* Ganga basin. Later in the 5th century CE king *Dhathusena* (453–477 CE) constructed *Yodha Ela* with the aim of water diverting from *Kalawewa* basin to *Malwathu Oya* Basin. In addition to that the same period "*Thekam* Anicut" was constructed to divert the water from the *Malwathu* Oya Basin to the "*Jayant* Tank" basin. This evidence has proved that the craftsmanship of "Inter-Basin Water Diversion" was practiced in Sri Lanka in the 1st century CE as the pioneering water balance strategy at an inter-basin level.

12.5.5 Indigenous Practices on Conflict Resolution

More than 95% of tank-cascade villages have farmer organizations with the aim of implementing collective decisions. It was a conflict resolution method at the bottom level. In addition to that, all farmers are responsible for maintaining the tank bund and canal system, based on land ownership of the paddy fields. At the beginning of the rainy season, all the farmers of the village tank would gather at a common place to plan the cultivation season. It has happened under the supervision of the chief of the farmer organization.[1] The seasonal meetings, sharing the seeds, and rotational water sharing among all the sublevel irrigation canals, were also practiced.

1 Local Term is *Well Vidane.*

12.5.6 Water-Related Traditional Ceremonies

There are a few traditional ceremonies associated with tank water in the dry zone of Sri Lanka and rarely these ceremonies are organized up to date.

After filling the tanks with rainwater, water is carried by pots into the "Bo tree" of the temple of the village and poured with it. In history, this was done with milk, and today it is symbolized by water. This is done as a tribute to the sacred *Bo* tree which was sheltered by Lord Buddha.

When the tank overflows and the situation becomes critical, village elders go to the tank. They offer a salute to the gods by firing two gunshots, followed by a traditional *Vathika* and a ceremonial wow by the village chief. It is the tradition of "*Puja Wedilla*" linked with *Panduru bendeema*, to pray for the blessings of the gods to protect the tank. After harvesting, there is another tradition called *Mutti mangalle*, which involves offering milk rice while remembering the *Panduru bendeema*. Finally, there is a ceremony for the god most probably near the tank. During severe drought periods and when the fish are under threat, there was a tradition to prohibit catching fish, by the chief of the farmer organization. It is called "*Ana Bola Bandeema*," which involves placing a tree branch in the tank water to demarcate areas where fishing is prohibited. After filling the tank again with sufficient rainfall, remove the branches and declare that the prohibition ended.

12.6 Conclusion

Ancient civilizations experienced some strategies to acquire optimum use of water for domestic and agricultural purposes and it is important to understand the historical traditions and practices related to sustainable water resources management. It would be an advantage for practicing successful water resources management techniques in the current world too. Irrigated agriculture is one of the main uses of water, as demonstrated by ancient civilizations such as those in southern Mesopotamia, Egypt, China, the Iranian plateau, Tunisia, India, and Sri Lanka. The irrigation systems such as (i) gravity-fed, (ii) basin, (iii) terrace, (iv) border, (v) strip irrigation, and check basin irrigation were practiced in many countries where (i) advanced techniques were not required, and (ii) existing natural topography of the region was easily utilized. Furthermore, man-made lift irrigation, pitcher irrigation, and bamboo drip irrigation were constructed for the areas where the topography was not favored for basin irrigation. These techniques have increased water productivity by minimizing the wastage of available water. Qanat system is an example of groundwater harnessing techniques in past arid/semiarid regions where (i) surface irrigation was not possible and (ii) located in alluvial aquifers. Furthermore, flood control and river navigation were successfully practiced in ancient times. Water harvesting from rain and glaciers was practiced in India to implement optimum water use. These techniques used for irrigation are still practiced in agricultural countries and founders for new technologies like sprinkler and drip irrigation systems. Knowledge of the hydrological cycle and measurement of available water is crucial for the development and management of water resources. The ancient civilization knew the hydrological process and developed man-made dams and water distribution canals, rain gauges, river gauges, apparatus for desalination and treatment of water, and mitigating coastal disasters. The lessons learned from past technologies facilitated modern techniques for water measurements, desalination, water treatment, bulk-water allocation, coastal navigation, and disaster mitigation. In addition to that, governance is the most important aspect in water resources management and communal governance adapted historically for effective and efficient (i) water distribution among users, (ii) harnessing both surface and groundwater, and (iii) water conservation. Especially in Sri Lanka, tank-cascade systems in the dry zone are one of the currently functioning traditional water management practices. In this system, the major irrigation tank/reservoir is focused on collecting runoff water and diverting it to the command area through a canal system. Some minor tanks were established in the immediate catchment to trap the sediment. Furthermore, forest-based tank reservations were maintained to minimize the evaporation and purification of water. Some structures such as Sluice Front Ridge help control the dam pressure during overflow. The water distribution and labor allocation for different periods of dry and wet were managed in a systematic way confirming effective water allocation for users. Collective and communal governance was practiced in these systems due to their continuous use and efficiency towards the sustainability of the ecosystems and farming communities. The Government of Sri Lanka focused on the rehabilitation of these tank-cascade systems through institutional interventions and the Climate Smart Irrigated Agriculture Project recently.

References

Angelakis, A.N., Valipour, M., Choo, K.-H. et al. (2022). Desalination: from ancient to present and future. *Water* 2021: https://doi.org/10.3390/w13162222.

Bansal, S.L. and Asthana, S. (2019). Water treatment techniques system in ancient Indian culture. *Indian Journal of Applied Research* 9 (5): 16–18.

Behailu, B.M., Pietila, P.E., and Katko, T.S. (2016). Indigenous practices of water management for sustainable services: case of Borana and Konso, Ethiopia. *Traditional Wisdom* 1–11. https://doi.org/10.1177/2158244016682292.

Bjornlund, V. and Bjornlund, H. (2010). Chapter 2. Sustainable irrigation: a historical perspective. In: *WIT Transactions on State of the Art in Science and Engineering*, vol. 37. UK: WIT Press.

Bruchac, M. (2014). Indigenous knowledge and traditional knowledge. In: *Encyclopedia of Global Archaeology* (ed. C. Smith), 3814–3824. New York: Springer.

Cao, S., Liu, X., and Er, H. (2010). Dujiangyan irrigation system – a world cultural heritage corresponding to concept of modern hydraulic science. *Journal of Hydro-Environment Research*. https://doi.org/10.1016/j.jher.2009.09.003.

Chathuranika, I., Khaniya, B., Neupane, K. et al. (2022). Implementation of water-saving agro-technologies and irrigation methods in agriculture of Uzbekistan on a large scale as an urgent issue. *Sustainable Water Resources Management*. https://doi.org/10.1007/s40899-022-00746-6.

Esmaeili, G., Habibi, A., and Esmaeili, H.R. (2022). Qanat system, an ancient water management system in Iran: history, architectural design and fish diversity. *International Journal of Aquatic Biology* 10 (2): 131–144.

Food and Agriculture Organization of the United Nations (2018). Addressing Water Scarcity in Agriculture: How Can Indigenous or Traditional Practices Help? Global Forum on Food Security and Nutrition. Report of Activity No. 151. Food and Agriculture Organization of the United Nations.

Krasilnikoff, J. and Angelakis, A.N. (2019). Water management and its judicial contexts in ancient Greece: a review from the earliest times to the Roman period. *Water Policy*. https://doi.org/10.2166/wp.2019.176.

Madduma Bandara, C.M. (1985). Catchment eco-systems and village tank cascades in the dry zone of Sri Lanka. *Strategies for River Basin Management. The GeoJournal Library* (ed. J. Lundqvist, U. Lohm, and M. Falkenmark), vol 6. Dordrecht:. Springer. https://doi.org/10.1007/978-94-009-5458-8_11.

Malik, S. (2016). Conceptual aspect of hydrological cycle in Indian mythology of Kishkindha Kanda, Ramayana. *Journal of Environment and Earth Science* 6 (4): 54–59.

Perera, M.P. (2013). The destruction of an ancient tank irrigation vestige: a case study on the breach of the historical Kivulekada Tank of Sri Lanka. *International Journal of Arts and Commerce* 2 (9): 1–9.

Perera, M.P. (2017). Evolution of tank cascade studies of Sri Lanka. *Saudi Journal of Humanities and Social Sciences (SJHSS)*, Scholars Middle East Publishers, Dubai, United Arab Emirates 2 (7): 597–610.

Perera, M.P. and Rekha Nianthi, K.W.G. (2020). Land use planning and management in tank cascade environment of Sri Lanka. In: *Examining International Land Use Policies, Changes, and Conflicts Hasnat* (ed. G.N. Tanjina Hasnat and M.K. Hossain). Pennsylvania: IGI Global Publishers 310–329.

Rost, S. (2019). Navigating the ancient Tigris – insights into water management in an early state. *Journal of Anthropological Archaeology*. https://doi.org/10.1016/j.jaa.2019.01.005.

Singh, P.K., Dey, P., Jain, S.K., and Mujumdar, P. (2020). Hydrology and water resources management in Ancient India. *Hydrology and Earth System Sciences* 2020: https://doi.org/10.5194/hess-2020-213.

Wilbowo, A. (2020). Ancient gravity-fed hydrology and irrigation system of Majapahit Civilization 14th century in Java Island and its impact on vegetation health during 2019. *El Nino Using NDVI Analysis* 2020: https://doi.org/10.20944/preprints202012.0216.v1.

13

Water-Related Traditional and Indigenous Practices

Nirlipta P. Nayak

Energy Cluster, School of Advanced Engineering, UPES, Dehradun, Uttarakhand, India

13.1 Introduction

The elixir of life, water, is essential to maintaining all life forms on Earth. Our planet's water system is a complex and linked web that maintains ecosystems, regulates weather patterns, and sustains life. It includes the massive seas as well as the complicated network of rivers, lakes, and groundwater reservoirs (Wright 1985). The components of Earth's water system include the hydrologic cycle, oceans, freshwater system, glaciers and polar ice, ecosystems, and biodiversity. The hydrologic cycle, a continuous and dynamic process that circulates water across the world, is at the center of the water system on Earth (Vijayalakshmi 1982). Evaporation, condensation, precipitation, and runoff are some of the steps that are involved. As a result of solar radiation, water from rivers, lakes, and seas evaporates into vapor. Rising into the atmosphere, the water vapor cools and condenses to create clouds. These clouds eventually release water droplets, which cause precipitation to fall to the ground, ending the cycle (Pacey and Cullis 1986; Reij 1988).

The oceans, which make up over 71% of the Earth's surface, are a crucial component of the water system. They serve as heat reservoirs and store enormous amounts of water, regulating world temperatures. In addition to maintaining complex food chains and controlling carbon dioxide levels through absorption, oceans are essential habitats for a variety of marine life (Willmott and Rowe 1985). Additionally, they have an impact on climate patterns like the El Nio and La Nia phenomena. The Earth's water system also consists of freshwater bodies, such as rivers, lakes, and groundwater, in addition to the oceans. Rivers, which have their origins in a variety of places, convey water over great distances, feeding terrestrial ecosystems and aiding the transfer of nutrients and debris (Gohari et al. 2013). Lakes serve as important water reservoirs, holding water and sustaining distinctive wildlife. In subsurface aquifers, groundwater is a critical resource for agriculture, human use, and supporting ecosystems during dry spells (Swanson and Baldwin 1965).

Polar ice caps and glaciers are a crucial part of the water cycle on Earth. These enormous reservoirs hold frozen freshwater. During the warmer months, they release meltwater into

Water Scarcity Management: Enabling Technologies, First Edition. Edited by Kanchan D. Bahukhandi, Manish Kumar, Durga P. Panday, and Tushara G.G. Chaminda.

rivers and oceans, providing a reliable water source (Rango 1992). Additionally, they help control sea levels and have an impact on patterns of the earth's climate. An amazing variety of ecosystems, each specifically adapted to its aquatic environment, are supported by the water system of Earth (Oerlemans 1979). For instance, wetlands act as natural filters that clean water and serve as habitats for a wide variety of species. With their careful saltwater and temperature balancing, coral reefs support unrivaled biodiversity (Frei et al. 2003). These ecosystems are crucial to maintaining life on our planet since they depend on the interaction of water, nutrients, and sunlight to grow.

The Earth's water system is a priceless resource that supports all facets of life. It is essential for human consumption, industry, and agriculture. Additionally, water ensures the existence of innumerable species by maintaining the complex balance of our planet's ecosystems. This important resource does, however, encounter many difficulties (Mekonnen and Hoekstra 2016). The integrity of the water system on Earth is in danger due to home and industrial pollution, unsustainable water management techniques, and droughts and floods brought on by climate change. To ensure the continued availability of this priceless resource, it is essential for individuals, communities, and governments to prioritize water conservation, advance sustainable practices, and safeguard freshwater ecosystems (Shen et al. 2008). Life in all its forms is supported by the Earth's intricate and linked water system. Water is a crucial component in determining how our planet's natural processes function, influencing everything from the hydrologic cycle to the oceans, freshwater bodies, glaciers, and different ecosystems. Given its importance, we must take responsible action as soon as possible to safeguard and conserve this priceless resource for future generations. All life forms depend on water, which is becoming increasingly scarce owing to population increase, industrialization, and climate change (Frederiksen 2009). It is essential that we give conservation of water top priority if we want to ensure a sustainable future. Water harvesting, a technique that involves collecting and using rainwater and other sources, is one efficient strategy (Jiang 2009).

Numerous advantages of water harvesting support sustainability and water conservation. First of all, it helps to lessen reliance on freshwater sources, relieving pressure on surface and groundwater supplies. By catching rainwater and reducing runoff, it also reduces the risk of soil erosion and flooding (Khoury-Nolde 2008). Water collection also encourages resilience and self-sufficiency, especially in rural areas and those with erratic water sources. Rainwater harvesting is one way to collect water. It entails gathering rainwater from rooftops and directing it into subterranean reservoirs or storage tanks (Prinz 2002). With this easy and affordable technique, rainwater can be used for irrigation, household needs, or groundwater recharge. Building check dams and contour bunds across rivers or streams is another tactic. These structures reduce soil erosion by slowing the flow of water, which encourages infiltration and groundwater recharge (Ray 2005).

A technique called "floodwater harvesting" includes catching and using extra water during periods of heavy rain or flooding. This strategy focuses on harnessing the floodwater's potential by putting it to useful use, as opposed to allowing it to cause harm and loss. Floodwater harvesting has a number of advantages in addition to aiding in the prevention of flood-related calamities (Konig 2001). It can refill depleted aquifers, recharge groundwater reserves, and increase water availability during dry spells by diverting and storing floodwater. Additionally, by providing water for irrigation and minimizing reliance on unpredictable rainfall patterns, floodwater collection can help agricultural activities (Kim 2000). Additionally, the gathered floodwater might be utilized to reforest land, replenish wetlands, or aid in ecological restoration projects. Overall, floodwater harvesting is an effective method for minimizing the negative consequences of floods and making the best use of surplus water supplies for the benefit of ecosystems and communities (Coombes et al. 1999; Wagner et al. 2002; Weigert and Steinberg 2002).

Groundwater harvesting is a method for obtaining and using groundwater for different uses. Groundwater is a critical supply of freshwater for human consumption, agriculture, and industrial requirements (CISRO 2006). Groundwater is stored beneath the Earth's surface. Water from subsurface aquifers is extracted via wells or boreholes as part of groundwater harvesting. This method is essential for water sustainability and conservation, especially if surface water supplies are scarce or unreliable (Furumai 2008). Communities can ensure a consistent water supply, even during dry seasons or droughts, by drawing on the large groundwater reserves. Additionally, groundwater gathering encourages independence, lessens reliance on surface water supplies, and decreases the dangers of water scarcity. To avoid over-extraction and the depletion of aquifers, it is crucial to implement sustainable and responsible groundwater management practices (Herrmann and Hasse 1998; Kim et al. 2003). We may use groundwater harvesting methods to meet the water needs of expanding populations, support agricultural output, and preserve the long-term sustainability of water supplies. In acknowledging the vital role of water-related traditional and Indigenous practices, we recognize that integrating these age-old, sustainable techniques with modern groundwater harvesting methods can offer holistic solutions to address the escalating water demands of growing populations and ensure the enduring resilience of our water resources.

Water-related traditional and Indigenous practices span a rich tapestry of ingenious solutions crafted by diverse cultures worldwide. From the qanat systems of the Middle East, with their subterranean channels harnessing gravity to transport water over vast distances, to the Apu system in the Andes, where mountain spirits are revered in the construction of intricate canals for irrigation, these practices showcase a deep understanding of local ecosystems. In Oman, the Aflaj system epitomizes communal water management through networks of underground and surface channels, reflecting a harmonious integration of human ingenuity with arid landscapes. Meanwhile, the Subak system in Bali embraces cooperative water governance, ensuring equitable distribution for rice cultivation while embodying spiritual values.

As a critical tactic to address water scarcity and maintain sustainable water management, water harvesting has received a lot of attention in India. Drawing inspiration from water-centric traditions such as the community-managed step wells, these contemporary initiatives seek to integrate Indigenous wisdom with modern practices, fostering a harmonious relationship between communities and their water resources. Similarly, reflections of historical reed-based agriculture, as seen in the Aztec Chinampas, resonate with the evolving strategies in India, emphasizing the importance of embracing traditional knowledge for resilient and culturally rooted solutions to water challenges to collect and use rainwater, the nation has established several waters harvesting strategies, including rooftop rainwater collecting, percolation ponds, check dams, and groundwater recharge (Thomas and Martinson 2007). Rainwater from rooftops is collected and stored in tanks or underground reservoirs through the process of rooftop rainwater harvesting. According to estimates, 40% of rural families and 80% of urban households, respectively, have rooftop rainwater collection systems installed. Rainwater is collected and kept in shallow excavated depressions called percolation ponds (UNEP 2002). These ponds support groundwater recharge and help replenish groundwater reserves. To reduce soil erosion and promote groundwater recharge, check dams have also been built across rivers and streams to restrict the flow of water and encourage infiltration. Aquifers that have been drained have also been refilled artificially using methods like redirecting precipitation or surface runoff into wells or boreholes (Musayev et al. 2018; Tamagnone et al. 2020). The Indian government has offered financial incentives and subsidies for water harvesting structures to promote their broad use. To guarantee maximum effectiveness and lifespan, the technical requirements for water harvesting structures in India include adequate design, sizing, and maintenance. The collection surface area, guttering systems, filtration systems, storage capacity, and overflow arrangements are just a few of the details covered by these criteria (Das et al. 2020). Even while India has achieved great strides in water harvesting, there is still much space for improvement in terms of awareness and implementation. To maintain sustainable water management practices and solve the country's water concerns, ongoing efforts are required, including education, awareness campaigns, and policy assistance (Sepehri et al. 2018; Betasolo and Smith 2020).

13.2 Common Practices of Water Harvesting Across the Globe

Water practices vary greatly between cultures and areas, with traditional and Indigenous people frequently creating unique and sustainable approaches to water management. Table 13.1 displays some traditional water practices across the globe.

13.3 Common Practices of Water Harvesting in India

The United Nations Educational, Scientific and Cultural Organization (UNESCO) defines Indigenous knowledge (IK) as, "Local and indigenous knowledge referring to understandings, skills and philosophies developed by societies with histories of their interaction with natural surroundings." Over a period of time, Indigenous people have a knowledge of practices on resource use and on the intricate ecological structures in their own vicinities (Gadgil et al. 1993). Water harvesting has been widely applied in different social–ecological contexts. It is a valuable approach to sustainable intensification of agriculture. Global estimates of water harvesting based on biophysical assessments whereas the socioeconomic dimension of agriculture is vividly neglected (Piemontese et al. 2020).

India has a long history of traditional water harvesting techniques that have been used to collect and store rainfall. These practices, which are firmly founded in Indigenous understanding and experience, have been vital in preserving communities and agriculture in areas with a lack of water. Dams and bunds were used for irrigation purpose and their maintenance was managed by rules as well (Tamaddun et al. 2018). Several penalties were imposed for causing damage to dams. Some rigorous punishments like the death penalty were also

Table 13.1 Various historical water harvesting techniques used across the globe.

Name	Country	Description
Qanat system	Iran	The Qanat system is an advanced and ancient subterranean water supply mechanism where a mother well, a deep vertical shaft bored into the ground till it reaches the water table, is the first step in the process. The qanat gallery, a gently sloping tube built horizontally along the contour of the land from the base of this well, allows gravity to convey water from the source to the surface. Vertical access shafts are dug along the Qanat gallery regularly to allow maintenance and offer ventilation.
Apu system	Pery	The Apu system is integrally linked to these tribes' agricultural activities, tackling the difficulty of managing water supplies in steep terrains. The Apu system entails the development of canal and channel networks that transport water from high-altitude sources like as mountain springs and glacier meltwater to lower-lying agricultural areas. These canals are meticulously designed to follow the curves of the landscape, allowing for efficient water distribution to fields and terraces at various elevations.
Roman aqueducts	Italy	Roman aqueducts are magnificent examples of ancient engineering that were built to carry water over great distances for the cities and towns of the Roman Empire. The water in these aqueducts was guided along precisely designed slopes by gravity. The aqueducts' stable and effective passage through a variety of topographies, including hills and valleys, was made possible by arches and tunnels. Water from springs, rivers, and lakes was funneled to urban areas via Roman aqueducts, which channeled this valuable resource. Water was distributed across towns using lead pipes, and settling tanks made sure silt settled before it got to where it was going.
Aflaj system	Oman	"Aflaj" refers to a system of surface and subterranean channels that have been carefully designed to use gravity to propel water. These channels move water from higher sources, including mountain aquifers, to lower-lying farms and settlements by carefully following the natural contours of the land. The Aflaj system has subterranean tubes or galleries that protect the water from evaporation and contaminants while preserving its temperature.
Beresheet trenches	Israel	As "Genesis" in Hebrew, the phrase "Beresheet" refers to the beginning of water conservation methods in this dry area. These trenches, which are frequently arranged strategically across slopes, serve as collecting places for rainwater, slowing down its quick runoff and enabling it to seep into the ground. The trench designs take into account the geography of the Negev, making effective use of the occasional downpour. Water that seeps into the earth replenishes groundwater and keeps vegetation alive, making it an important resource for ecology and agriculture alike.

(*Continued*)

Table 13.1 (Continued)

Name	Country	Description
Subak system	Indonesia	The Subak system, which has its roots in the long-standing agricultural and cultural traditions of the Balinese people, is intended to manage water for rice cultivation in a sustainable and community-focused way. The water organization or cooperative that manages the allocation and utilization of water resources among farmers is referred to as "Subak." Water from rivers or springs is redirected into a system of canals, which then feeds the rice terraces. Each farmer has a designated timetable and duration for using water, which is distributed among a network of complex channels that carry the water downstream.
Ayuquila River Management	Mexico	This entails a variety of actions meant to preserve or enhance the condition of a river system, guarantee water quality, control flooding, and encourage the sustainable use of water resources. It may involve actions like infrastructure for flood control, habitat restoration, monitoring water quality, and preservation of riparian zones.
Step wells	India	Step wells, sometimes referred to as baori, bawdi, or vav, are historic constructions that were typically constructed in arid areas to give communities a dependable supply of water. These wells are distinguished by their distinctive architectural layout, which includes a set of steps that descend to the water's surface. The main purpose of step well construction was to obtain groundwater, particularly in areas with a low water table and restricted water supply.
Fog nets	Chile, Peru, and Nepal	In dry and semi-arid areas with limited access to conventional water sources, fog nets are creative constructions made to collect water from fog. These nets work especially well in places where fog and low-lying clouds are prevalent. The technology is straightforward yet effective; it looks like big screens or mesh nets placed in key spots.
Water jars	African countries	Water jars are widely used in many African nations, where they are indispensable containers for carrying and storing water. Water jars can differ greatly around the continent in terms of design, material, and cultural significance, which reflects the variety of regional customs and practices. Traditional water jars built of clay or other readily accessible materials are common in many rural areas of Africa. These jars are frequently made by hand and may have distinctive patterns or decorations that are particular to the area. They are essential to daily life because they enable people to gather and transport water to their houses from public wells, rivers, and other water sources.

imposed upon breaking a reservoir or tank full of water. Depending on the volume of water drawn, taxes were imposed and collected. The tax rate depends on the method of drawing water. In the absence of the owner, the waterbodies were maintained by the people of the village. Many examples of hydraulic engineering seen (Srinivasan et al. 2017). Of those, one is found in Sringaverapura near Allahabad. It comprises three percolation-cum-storage tanks used for floodwater harvesting from the Ganga during heavy monsoons. Well irrigation became pronounced in the Pandya, Chera, and Chola dynasties. Rural depressions in the landscape were developed into large irrigation tanks. The Grand Anicut is still functional and in use (Oweis et al. 2012). It is the fourth-oldest water-regulator structure in the world. The largest artificial lake or the Bada Talab was built by King Bhoja of Bhopal. Many such well-defined and well-maintained irrigation systems are reported in Kashmir. Examples of Traditional Water Management Techniques Still in use today are as follows:

- Traditional waterways or canals called kuls, often spelled khuls, can be found in India's Himalayan regions, mainly in Himachal Pradesh, Uttarakhand, and Jammu & Kashmir. For ages, the local residents have relied heavily on these amazing irrigation systems. The significance of kuls is found in their capacity to overcome the difficulties provided by the difficult terrain and scarce water supplies in these areas. Kuls are skillfully built by chiseling narrow channels down the sides of mountains, frequently with the use of stone or concrete walls to make a path for the water. They use the land's natural gradient to direct water from rivers or streams toward crops for irrigation. The capacity of kuls to deliver a consistent and dependable water supply to agricultural regions is one of their main advantages. Farmers can cultivate their fields despite in water-scarce locations because of kuls, which ensure that crops receive the essential moisture by capturing water from natural sources. The livelihoods of mountain populations are significantly impacted by this, allowing them to maintain agriculture and achieve food security. Kuls support biodiversity and local environment protection as well. These channels preserve the water's natural flow, reducing soil erosion and preserving the area's biological balance. They assist the overall wellbeing of the ecosystem by allowing water to soak into the earth and recharge subsurface aquifers. In addition, kuls promote a sense of communal duty and cooperation. The entire community must participate in the building and upkeep of kuls, fostering a feeling of cooperation and solidarity among the populace. To provide fair access to water supplies, the management of water distribution through kuls is frequently organized collaboratively. The mountain inhabitants' rich cultural legacy and traditional knowledge are also attested to by kuls. They demonstrate the creativity and knowledge of the people who have developed solutions to the particular problems in their area over many generations (Verma 2022).
- The Araghatta irrigation system, sometimes referred to as the Persian wheel or Rahat, has been utilized in India for many years. This innovative technology is especially common in the country's dry and semi-arid regions, where a lack of water is a serious worry. An animal or person pulling on a rope turns a big wheel that has buckets or pots attached to it to power the Araghatta. In order to provide the fields with water, the pots or buckets on the wheel scoop up water from a well or other water source and deposit it into a raised channel or distribution system. Araghatta's significance stems from its capacity to utilize water resources effectively and sustainably. This time-tested approach has a number of benefits. First off, it is a financially sensible alternative with no infrastructure needs and no reliance on power or fossil fuels. Araghatta is simple enough for small-scale farmers with limited resources to use. Additionally, Araghatta enables focused watering. Controlled water release makes sure that crops get the right amount of moisture without wasting too much. In arid areas, where every drop of water counts, this smart use of

water is essential. Araghatta also helps to increase agricultural productivity and crop production. Farmers are able to develop their fields even during dry seasons by offering a steady supply of water. For communities that depend on agriculture, this results in higher food production, improved livelihoods, and greater food security. Araghatta has advantages for irrigation as well as groundwater recharge. The water used for irrigation seeps into the ground, replenishing subsurface aquifers that can be used in the future. Araghatta also encourages interdependence and community participation. Several farms in a hamlet will frequently share a water supply and manage the infrastructure alternately, promoting a sense of communal responsibility and solidarity. The importance of Araghatta should not be disregarded in the face of contemporary irrigation systems. It stands for a time-tested, effective solution that is sustainable. Araghatta's preservation and promotion can support rural development, sustainable water management, and the resiliency of farming communities in areas with limited water supplies (Hindustan Times 2017).

- In India's desert regions, especially in the state of Rajasthan, johads, also known as traditional water storage tanks, are essential components of water conservation. These reservoirs are crucial for maintaining communities in water-stressed areas and reducing water scarcity. Johads are significant because they can collect and store rainfall, making them available all year long for a variety of uses. Johads are primarily made of dirt and built by digging small depressions in the ground and lining them with clay or concrete to stop seepage. During the monsoon season, these reservoirs are strategically placed to collect runoff from nearby catchment areas. During the dry seasons, the rainwater gathered is held in the johads and gradually released for animal needs, irrigation, and drinking water. A major advantage of johads is their contribution to increased agricultural output. Johads hold water that is used to irrigate crops, enabling farmers to work their fields even when there is not a steady supply of rain. This aids in preserving agriculture and assisting farming communities, whose livelihoods significantly rely on rain-fed agriculture. Johads also have a favorable effect on groundwater recharge. The water that has been accumulated seeps into the soil, refilling the aquifers below. This aids in preserving a sustainable groundwater supply, which is essential for supplying water for home and agricultural use. In desert areas, johads are crucial water sources for cattle and wildlife. They serve as watering holes, luring several animal species and provide a lifeline during the severe dry seasons. Additionally, by fostering favorable environments for both aquatic and terrestrial organisms, johads support biodiversity preservation. Johads also contribute significantly to stopping soil erosion. They aid in maintaining moisture in the soil, limiting the erosion of topsoil, and ensuring the fertility of agricultural land by collecting rainwater and lowering surface runoff (CPREEC (C.P.R. Environmental Education Centre) 2017).
- Temple tanks, often referred to as kovil kulam, are highly significant in South Indian culture and water management practices. These customary pools of water act as rainwater reservoirs and are frequently found close to temples. The significance of kovil kulam is found in its multifaceted function, which includes components of religion, society, and the environment. Kovil kulam is regarded as sacred and significant from a religious perspective. These tanks have a strong connection to religious rituals and practices, and they are frequently connected to temples. Visitors to these ponds engage in religious

ceremonies, take holy baths, and make prayers. Kovil kulam's cultural and heritage worth is enhanced by the spiritual atmosphere in the area. Additionally, kovil kulam is essential for managing and conserving water. In monsoon seasons, these tanks serve as natural reservoirs by catching rainwater. As the stored water slowly seeps into the soil, refilling the underlying aquifers, they aid in groundwater recharge. This enhances the region's overall water resource availability. Additionally, kovil kulam is in favor of ecological harmony and biodiversity. These bodies of water provide a favorable habitat for a variety of aquatic plants and animals. These tanks serve as essential ecosystems in urban or semi-urban settings because they draw birds and other animals. Additionally, the tranquility and vegetation that surround kovil kulam offer a change of pace from the concrete jungle and improve the aesthetics of the area. kovil kulam acts as a hub for social gatherings, such as community events and cultural gatherings. Near these tanks, celebrations, processions, and social gatherings are frequent, promoting neighborhood solidarity and upholding regional customs. Kovil kulam also plays a role in the community's overall sustainability and resiliency. The water that is held in these tanks can be used for irrigation during periods of water scarcity or drought, supporting agricultural practices and maintaining community food security maintaining local customs, and promoting togetherness (Kelke Khambete 2010).

- Surangas, or tunnel wells, are amazing subterranean water features that may be found in the southern Indian states, particularly in the Malabar region of Kerala. Communities in water-scarce places have relied on these conventional engineering wonders as a lifeline. The ability of Surangas to access groundwater sources and offer a dependable and sustainable water supply for varied applications is crucial. For the purpose of capturing subsurface water sources, Surangas are built by boring tunnels into the sides of hills or mountains. Usually, stones or bricks are used to line these tunnels in order to retain their structural integrity and prevent collapse. Depending on the location and accessibility of water sources, Surangas can range in length from a few meters to several hundred meters. The Surangas' capacity to extract groundwater even during dry seasons is one of their main advantages. Surangas offers a constant supply of water for drinking, home usage, and agriculture by tapping into subsurface aquifers. This aids communities in overcoming water scarcity and guarantees that there is water available for both agricultural demands and daily needs. Additionally, Surangas are essential for replenishing groundwater. The region's water supplies are more likely to be sustained as a result of the tunnels' ability to intercept underground water sources and assist aquifer recharge. Surangas' underground location aids in further protecting the priceless resource by reducing water evaporation. Surangas represent traditional knowledge and cultural history in addition to being a source of water. These buildings need to be built and maintained, which calls for specialized knowledge and community involvement. This distinctive method of water collection has been preserved thanks to the transmission of this expertise from one generation to the next, demonstrating the intelligence and creativity of the local communities (Geography and You 2017).
- Eri, often referred to as a tank or pond, is a traditional water storage system that is common in Tamil Nadu, a state in southern India. Eris have played a crucial role in sustaining farmland, ensuring livelihoods, and conserving water supplies for generations, making them an essential part of the region's water management. Eris are significant because they can

collect and store rainfall, ensuring a steady supply of water during dry seasons. Eris are typically artificial depressions that have been constructed to collect rainfall from catchment regions. These bodies of water range in size from little hamlet tanks to substantial reservoirs. Hydrology and geomorphology are intricately woven into the planning and building of eris to enable effective water storage and maximum rainwater utilization. The contribution that eris make to agriculture is one of their main advantages. Farmers can grow their crops all year round thanks to irrigation, which uses the water that is stored in these tanks. Eris promote both groundwater- and rain-fed agriculture, increasing crop yields and agricultural production. Particularly during times of water scarcity or drought, they provide vital lifelines for farmers. Additionally, eris aid in replenishing aquifers of groundwater. Rainwater eventually seeps into the soil as it accumulates in the eris, replenishing the subsurface water sources. This aids in preserving a sustainable groundwater supply, which is necessary to meet the region's water needs, particularly those for drinking water and domestic use. Eris are also helpful in preventing floods and lessening the effects of intense rain. These tanks serve as organic storage reservoirs, soaking up extra water during downpours and avoiding flooding downstream. Water storage and flood control can coexist in harmony with the controlled release of water from Eris. Additionally, eris function as significant ecosystems, sustaining a variety of flora and wildlife. They serve as habitat for fish, fish populations, and different bird species. Eris serve as maternity grounds and hunting areas for a variety of wildlife species, assisting in the protection of biodiversity (Agrawal 1995).

- The clever network of interconnecting channels and embankments known as Ahar Pynes, or ancient water management systems, is found in the Indian state of Bihar. For generations, the region's agriculture and water management have depended heavily on these historic water features. Ahar Pynes is significant because of their capacity to collect, store, and distribute water, which supports farming communities' livelihoods and contributes to sustainable irrigation methods. Ahar Pynes is constructed to capture river water flow during the monsoon season. They are made up of embankments that divert extra water into channels so that they can later be sent to areas used for agriculture. Systematically channeling the water enables controlled irrigation all year long, even during dry seasons. The role that Ahar Pynes plays in raising agricultural output is one of their main advantages. These systems employ the water they have conserved for irrigation, ensuring that crops get enough moisture for healthy growth. Farmers can constantly cultivate their fields thanks to the controlled release of water, increasing yields and producing better agricultural results. Ahar Pynes also helps manage and conserve water resources. These systems avoid waterlogging in fields and lower the risk of flooding by holding onto extra rainwater during the monsoon. The discharged water is released gradually to reduce waste and maximize water use effectiveness. This approach to sustainable water management provides a fair allocation of water resources among farmers and maintains their overall availability. Ahar Pynes is important from an economic and social perspective. They encourage group responsibility and collaboration within the community. These systems are built, maintained, and run with the active involvement of the neighborhood, generating a sense of ownership and cohesion. Ahar Pynes facilitates various activities including fish aquaculture, animal husbandry, and domestic water supply in addition to providing water for agriculture. Ahar Pynes also helps preserve biodiversity and local ecosystems. These water features help conserve biodiversity and improve ecological harmony by providing a habitat for a variety of aquatic creatures (Rainwaterharvesting.org 2017).

- Traditional water storage structures called kunds, or kundis, can be found throughout India. For ages, these buildings have been an essential part of rural societies' water management systems. A reliable supply of water for drinking, agriculture, and domestic use, kunds are significant because of their capacity to catch and store rainwater. Kunds are normally built by digging a sizable hole in the earth, enclosing it with impermeable materials like clay or stone and then filling it with water. Kunds come in a variety of sizes and shapes based on their location and use. They are placed in catchment areas in order to strategically catch rainwater runoff during the monsoon season. The role that kunds play in groundwater recharging is one of their main advantages. Rainwater eventually seeps into the ground and replenishes the subsurface aquifers as it collects in the kund. This aids in preserving a sustainable groundwater supply, which is essential for supplying water for home and agricultural use. Kunds are essential to the support of agriculture. Farmers are able to develop their crops even during dry spells because of the storage of water that may be used for irrigation. This aids in raising agricultural output and ensuring the local inhabitants' access to food. Additionally, kunds aid in managing and conserving water. They aid in water resource conservation and stop soil erosion by collecting rainwater and lowering surface runoff. By storing water during the rainy season, kunds act as a buffer against droughts by ensuring that it is available when it is scarce. Kunds also have cultural and social importance. They frequently function as community hubs where residents congregate to collect water, have conversations, and partake in cultural events. Kunds are also connected to spiritual and religious practices; many are revered and related to temples or other holy places (Moudgil 2014).
- Innovative and efficient water irrigation systems called bamboo drip systems have grown in popularity throughout the world, especially in areas with limited access to water and agricultural resources. These systems, which were built with bamboo and other locally accessible materials, are very important for raising agricultural productivity, conserving water, and giving farmers more power. Bamboo drip irrigation systems are significant because they can distribute water precisely and effectively to plant roots. Bamboo pipes with tiny perforations make up the system, which drips water onto the soil gradually and steadily. This kind of targeted irrigation reduces water waste and guarantees that plants receive the appropriate amount of water for growth. The capacity of bamboo drip systems to maximize water efficiency is one of its main advantages. In comparison to conventional irrigation techniques, these systems significantly reduce evaporation and minimize runoff by supplying water directly to the roots. This is especially useful in areas that are dry or semi-dry, where a pressing issue is a lack of water. Bamboo drip systems are also economical and environmentally responsible. Because bamboo is a renewable and easily accessible resource, farmers in rural areas can afford and use these systems. Farmers can set up and operate the systems themselves thanks to the ease of building and maintenance, which lowers their dependency on outside help and fosters self-sufficiency. The adaptability of bamboo drip systems to different crop varieties and agricultural techniques is another crucial feature. They can be utilized for a variety of crops, including cash crops, fruits, and vegetables. Farmers can customize the system to meet their unique needs and crop requirements thanks to its

flexible design and setup, which maximizes agricultural outputs and revenue production. Additionally, bamboo drip systems encourage the use of sustainable agricultural methods by minimizing the use of synthetic fertilizers, eliminating the need for excessive water use, and halting soil erosion. These methods aid in soil health preservation, water resource conservation, and environmental sustainability in general (Zakaria et al. 2012).

- A traditional stepwell structure called a jhalara, often referred to as a baori, can be found throughout India. In the areas where they are found, these spectacular architectural wonders are of utmost historical, cultural, and utilitarian significance. The name Jhalara, which comes from the Hindi terms "jhal" (which means to quench) and "ara" (which means a platform for ornamentation), precisely captures what it does. In particular during times of water scarcity, these stepwells were created primarily to offer access to groundwater for drinking, irrigation, and domestic usage. Jhalara also served as significant gathering spaces, promoting interpersonal communication and neighborhood togetherness. Around these wells, people from all walks of life would congregate and engage in talks, knowledge exchanges, and relationship building. These stepwells developed into thriving centers of cultural activity, holding religious ceremonies, festivals, and entertainment. They promoted intellectual development and creativity by serving as catalysts for the exchange of ideas. Jhalara also serve as ecologically important miniature oasis in desert areas. With their deep wells and well-considered architecture, stepwell structures enable the gathering and preservation of rainwater, recharging groundwater and reducing water scarcity. These wells also support and shelter numerous plant and animal species, enhancing the biodiversity of the area. Jhalara has recently attracted new interest and admiration as a historical and architectural treasure. These buildings have undergone preservation and restoration efforts to protect their cultural legacy for future generations. Visitors are still in wonder and fascinated by today's jhalara, which also serves as reminders of the past's rich history, cultural legacy, and ecological practices (Stavros Yannopoulos 2019).
- In India's arid and semi-arid regions, especially in Rajasthan, khadin is a traditional method of collecting water. It is a clever technology that involves building tiny earthen embankments to collect and store rainwater for various uses. The importance of khadin is found in its capacity to support sustainable agriculture in dry places and address the problems associated with water scarcity. For Khadin to function, precipitation from nearby catchment areas is directed into small, rectangular reservoirs. The purpose of the embankments is to deliberately slow down the water flow and promote sedimentation and percolation. As a result, the groundwater table is replenished, soil erosion is reduced, and valuable rainwater is preserved. Agriculture can be supported even during dry seasons by using the collected water for irrigation. Khadin is significant for reasons other than water conservation. It is a comprehensive strategy that incorporates agriculture, soil protection, and water management. Additionally, the embankments serve as a deterrent to wind erosion, preventing the valuable topsoil from being swept away. This improves the land's general condition and fertility, making it suited for long-term agricultural production. Additionally, khadin encourages community members to be resilient and self-reliant. Effective rainwater harvesting and use allows farmers to lessen their reliance on erratic monsoons and their susceptibility to droughts. It offers a dependable source of water for

irrigation, enabling farmers to cultivate crops all year long and increase their standard of living. Khadin has social and cultural significance in addition to its functional advantages. It is well established in the customs and knowledge of the neighborhood communities. Collective efforts are used in the building and upkeep of Khadin, generating a sense of community involvement and cooperation (Saravanan and Kalpana 2016).

13.4 Current Methods of Water Harvesting

India, a nation with a variety of climates and fluctuating water supply, has put into practice a number of cutting-edge water harvesting techniques to fight water scarcity and ensure sustainable water management. Some current and advanced water harvesting methods include rainwater harvesting, check dams, percolation tanks, farm ponds, an artificial recharge well, watershed management, etc. (Kim et al. 2005).

1) India uses a lot of rainwater-gathering techniques. It entails gathering and preserving rainwater for later use. Many methods are employed, such as rooftop rainwater harvesting, in which rainwater is gathered from rooftops and routed to storage tanks or recharging structures (Bruijnzeel 2004; Dinesh Kumar et al. 2006; Central Ground Water Board, Central Region, Ministry of Water Resources Govt. of India 2013; Julius et al. 2013; Ilstedt 2016).
2) Small, low-height check dams are built over intermittent streams or small rivers to catch rainwater and stop soil erosion. These dams aid in producing small reservoirs and groundwater recharge (Das et al. 2022; Kumar et al. 2023).
3) Large, dug reservoirs called percolation tanks are used to hold extra rainwater and allow it to seep back into the soil, replenishing groundwater aquifers. They are often constructed in locations with abundant groundwater resources (Panday et al. 2021; Kumar et al. 2023).
4) Small earthen reservoirs called farm ponds are built on agricultural land to catch and store rainwater. They assist in groundwater recharge and provide a source of irrigation during dry spells (Panday et al. 2021; Panday and Kumar 2022).
5) By putting captured rainwater or surface water directly into the ground, artificial recharge wells can replenish groundwater. These wells increase the availability of water and refill depleted aquifers (Panday and Kumar 2023).
6) Conservation, development, and wise use of water resources within a specific watershed area are all elements of watershed management. It comprises methods that aid in retaining rainwater and minimizing runoff, such as contour trenching, bunding, and reforestation (Panday et al. 2023).
7) Vertical constructions known as recharge shafts are built in suitable areas to recharge groundwater. By allowing precipitation to immediately soak into the underlying aquifers, these shafts restore groundwater supplies (Asok Raja et al. 2023).
8) Traditional Indian rainwater collection devices known as tanka systems are widely employed in Rajasthan and other dry regions of the country. When it dries out, these subterranean tanks act as a reliable source of water by collecting and storing rainwater from rooftops (Asok Raja et al. 2023).

13.5 Conclusion

Water conservation is an urgent and crucial task that calls for global cooperation from people, communities, and governments. We can successfully tackle water scarcity and advance sustainable water management by putting diverse water harvesting techniques into practice. Traditional water harvesting techniques, which have their roots in ancient knowledge, provide important insights into making prudent use of natural resources. Water conservation methods that have stood the test of time and are still effective include percolation tanks, rooftop water collection systems, and rainwater harvesting. In addition, contemporary developments in water harvesting, such as desalination facilities, greywater recycling systems, and smart irrigation technologies, provide encouraging chances to satisfy the rising water needs of our expanding population while reducing waste. We can guarantee a safe water future for future generations by fusing conventional wisdom with cutting-edge strategies. Traditional methods of water harvesting include zing/Ladakhi ice stupas, khadin, jhalara/baori, zabo, bamboo drip system, kunds/kundis, virdas, ahar pynes, eri, surangas, kovil kulam, johads and araghatta. Traditional water harvesting techniques are significant because they can provide practical and long-lasting solutions for water conservation. These age-old methods are extremely valuable since they were created with a thorough grasp of the local ecosystems and water cycles. We may develop a thorough and effective water conservation strategy by combining conventional techniques with cutting-edge ideas. Traditional water-gathering techniques have shown to be useful and economical. They frequently involve straightforward construction and make use of readily available materials in the area, making them usable for communities with little resources. Through catchment surfaces like rooftops, rainfall is collected and stored through rainwater harvesting, which then directs the water into storage tanks or underground reservoirs. As a result, the groundwater is replenished and offers a consistent source of water throughout dry spells. We may take advantage of their simplicity and affordability to increase access to water conservation practices across various regions and communities by merging traditional approaches into contemporary water management systems. Traditional methods also support ecological sustainability and are friendly to the environment. They cooperate with nature to maintain soil fertility, stop erosion, and promote biodiversity. Utilizing the natural land contours to store rainfall, such as through percolation tanks, check dams, and contour bunds, promotes groundwater recharge and lowers the danger of flooding. We can take advantage of the inherent eco-friendliness of both traditional and modern water collecting techniques by combining them, resulting in a more well-rounded and long-lasting approach to water conservation. Additionally encouraging community involvement and social cohesiveness is the blending of old and modern approaches. Traditional methods frequently involve group efforts, knowledge sharing, and intergenerational cooperation. They are firmly established in local traditions. We enable communities to take control of their water resources by incorporating traditional practices into contemporary water management plans. By fostering self-sufficiency and resilience, this decentralized strategy lessens dependency on outside water supplies and fosters stronger community ties. Additionally, the effectiveness and efficiency of water conservation initiatives can be improved by combining conventional water harvesting techniques with contemporary technologies. Traditional approaches can be paired with contemporary innovations, such as intelligent irrigation systems, remote sensing technology, and water-saving appliances, to maximize water utilization. To ensure precise and effective water distribution in agricultural practices, sensor-based irrigation systems can be combined with conventional rainwater gathering methods. Similar to how traditional techniques can be combined with greywater recycling systems to reduce waste and increase usage. We can develop a thorough and all-encompassing strategy for water conservation by combining conventional water-gathering practices with contemporary ones. By integrating the two methodologies, we are able to address the intricate problems of water scarcity and sustainable water management. It permits us to take advantage of the knowledge of our forefathers while utilizing contemporary developments. Additionally, by recognizing the significance of traditional knowledge and practices in establishing sustainable water futures, this integration fosters a feeling of cultural preservation. In conclusion, combining conventional water harvesting techniques with contemporary strategies is essential for successful water conservation. Traditional techniques provide sensible, affordable, and eco-friendly solutions that are based on local knowledge. We can maximize water use, include communities, and build a sustainable water future by fusing conventional methods with contemporary technologies. Combining these strategies improves our capacity to address water scarcity while also paying homage to our predecessors' rich cultural heritage. We can create a healthy balance between human needs, environmental sustainability, and the retention of traditional knowledge through this combination.

References

Agrawal, R. (1995). Water of life. http://www.downtoearth.org.in/coverage/water-of-life-28919 (accessed 6 April 2017).

Asok Raja, S.K., Panday, D.P., and Kumar, M. (2023). Decoding the enigma of 100-year record-breaking rainfall over Tamil Nadu using wavelet analysis. *Groundwater for Sustainable Development* 101007. https://doi.org/10.1016/j.gsd.2023.101007.

M. Betasolo and C. Smith, "Rainwater harvesting infrastructure management," in *Environmental Health - Management and Prevention Practices, IntechOpen*, 2020. http://doi.org/10.5772/intechopen.90342.

Bruijnzeel, L.A. (2004). Hydrological functions of tropical forests, not seeing the soil for the trees? *Agriculture, Ecosystems and Environment* 104: https://doi.org/10.1016/j.agee.2004.01.015.

Central Ground Water Board, Central Region, Ministry of Water Resources Govt. of India (2013). *Pilot Project on Water Conservation and Artificial Recharge at Raj Bhawan Premises*. Nagpur: Central Ground Water Board, Central Region, Ministry of Water Resources Govt. of India, 2013.

CISRO (2006). *Urban Stormwater: Best Practice Environmental Management Guidelines*. Collingwood: CISRO.

Coombes, P., Donovan, I., and Cameron, C. (1999). *Water Sensitive Urban Development: Implementation issues for the Lower Hunter and Central Coast*. Lake Macquarie City: Lake Macquarie City Council.

CPREEC (C.P.R. Environmental Education Centre) (2017). http://www.cpreec.org/pubbook-traditional.htm (accessed 5 April 2017).

Das, J., Rahman, A.T.M.S., Mandal, T., and Saha, P. (2020). Challenges of sustainable groundwater management for large scale irrigation under changing climate in Lower Ganga River basin in India. *Groundwater for Sustainable Development* 11: 100449. https://doi.org/10.1016/j.gsd.2020.100449.

Das, A., Kumar, M., Jha, P.K. et al. (2022). Isotopic and hydrogeochemical tracking of dissolved nutrient dynamics in the Brahmaputra River System: a source delineation perspective. *Chemosphere* 307: 135757. https://doi.org/10.1016/j.chemosphere.2022.135757.

Dinesh Kumar, M., Ghosh, S., Patel, A. et al. (2006). Rainwater harvesting in India: some critical issues from basin planning and research. *Land Use and Water Resources Research* 6: 1–17.

Frederiksen, H.D. (2009). The world water crisis and international security. *Middle East Policy* 16: https://doi.org/10.1111/j.1475-4967.2009.00416.x.

Frei, C., Christensen, J.H., Deque, M. et al. (2003). Daily precipitation rates in regional climate models: evaluation and intercomparison for the European Alps. *Journal of Geophysical Research* 18: https://doi.org/10.1029/2002JD002287.

Furumai, H. (2008). Rainwater and reclaimed wastewater for sustainable urban water use. *Physics and Chemistry of the Earth, Parts A/B/C* 33: https://doi.org/10.1016/j.pce.2008.02.029.

Gadgil, M., Berkes, F., and Folke, C. (1993). Indigenous Knowledge for Biodiversity Conservation. *Ambio* 22.

Gohari, A., Eslamian, S., Mirchi, A. et al. (2013). Water transfer as a solution to water shortage, a fix that can backfire. *Journal of Hydrology* 491: https://doi.org/10.1016/j.jhydrol.2013.03.021.

Herrmann, T. and Hasse, K. (1998). Ways to get water: rainwater utilization or long-distance water supply? A holistic assessment. *Water Science and Technology* 36: 313–318.

Hindustan Times (2017). A look at traditional harvesting techniques across the country. http://www.hindustantimes.com/india/jammu-and-kashmir/story-4h0s8RfOTumjmgneOdv5FN.html (accessed 13 April 2017).

Ilstedt, U. (2016). Intermediate tree cover can maximize groundwater recharge in the seasonally dry tropics. *Scientific Reports* 6: https://doi.org/10.1038/srep21930.

Jiang, Y. (2009). China's water scarcity. *Journal of Environmental Management* 90: https://doi.org/10.1016/j.jenvman.2009.04.016.

Julius, J.R., Prabhavathy, R.A., and Ravikumar, G. (2013). Rainwater harvesting – a review. *International Journal of Scientific and Engineering Research* 4 (8).

Kelke Khambete, A. (2010). Surangas, the disappearing lifeline of the farmers of Kasargod, Kerala. http://www.indiawaterportal.org/articles/surangas-disappearing-lifeline-farmers-kasargod-kerala (accessed 15 April 2017).

Khoury-Nolde, N. (2008). *Rainwater Infiltration*. European Rainwater Catchment Systems Association.

Kim, R.H. (2000). Rainwater utilization and functional changes in building roof. *Constructive Technology and Review* 220: 13–19.

Kim, R.H., Lee, S., and Kim, Y.M. (2003). Development of rainwater utilization system in Korea. In: *Proceedings of the 11th IRCSA, Mexico City, Mexico*, 25–29.

Kim, R.H., Lee, S., Lee, J.H. et al. (2005). Developing technologies for rainwater utilization in urbanized area. *Environmental Technology* 26: https://doi.org/10.1080/09593332608618545.

Konig, K.W. (2001). *Rainharvesting in Building*. Dortmund: Wilo Brain.

Kumar, M., Panday, D.P., Bhagat, C. et al. (2023). Demystifying the decadal shift in the extent of groundwater in the coastal aquifers of Gujarat, India: A case of reduced extent but increased magnitude of seawater intrusion. *Science of the Total Environment* 165451. https://doi.org/10.1016/j.scitotenv.2023.165451.

Kumar, M., Mazumder, P., Silori, R. et al. (2023). Prevalence of PPCPs, microplastics and co-infecting microbes in the post-COVID-19 era and its implications on antimicrobial resistance and potential endocrine disruptive effects. *Science of the Total Environment* 166419. https://doi.org/10.1016/j.scitotenv.2023.166419.

Mekonnen, M.E.S.F.I.N.M. and Hoekstra, A.R.J.E.N.Y. (2016). Four billion people facing severe water scarcity. *Science Advances* 2: https://doi.org/10.1126/sciadv.1500323.

Moudgil, M. (2014). Kuhl, kohli and a lost tradition. http://www.indiawaterportal.org/articles/kuhl-kohli-and-lost-tradition (accessed 10 May 2017).

Musayev, S., Burgess, E., and Mellor, J. (2018). A global performance assessment of rainwater harvesting under climate change. *Resources, Conservation and Recycling* 132: 62–70. https://doi.org/10.1016/j.resconrec.2018.01.023.

Oerlemans, J. (1979). Extracting climate signal from 169 glacier records. *Science* 308: 2005. https://doi.org/10.1126/science.1107046.

Oweis, T.Y., Prinz, D., and Hachum, A.Y. (2012). *Rainwater Harvesting for Agriculture in the Dry Areas*. CRC Press. http://doi.org/10.1201/b12351.

Pacey, A. and Cullis, A. (1986). *Rainwater Harvesting. The Collection of Rainfall and Runoff in Rural Areas*. London: IT.

Panday, D.P. and Kumar, M. (2022). Climate indices and hydrological extremes: deciphering the best-fit model. *Environmental Research* 215: 114301. https://doi.org/10.1016/j.envres.2022.114301.

Panday, D.P. and Kumar, M. (2023). Application of remote sensing techniques to deal with scale aspects of GRACE data to quantify groundwater levels. *MethodsX* 102108. https://doi.org/10.1016/j.mex.2023.102108.

Panday, D.P., Khosa, R., Maheswaran, R. et al. (2021). Game-theoretic-based modelling of Krishna waters dispute: equilibrium solutions by Metagame Analysis. *Eur. Phys. J. B* 94: 101. https://doi.org/10.1140/epjb/s10051-021-00107-w.

Panday, D.P., Khosa, R., Maheswaran, R. et al. (2021). Game theoretic-based modelling of Krishna waters dispute: equilibrium solutions by hypergame analysis. *The European Physical Journal B* 94: 131. https://doi.org/10.1140/epjb/s10051-021-00135-6.

Panday, D.P., Singh, A., Gaur, S. et al. (2023). Probable maximum precipitation analysis of high rainfall regions in India. *Groundwater for Sustainable Development* 100893. https://doi.org/10.1016/j.gsd.2022.100893.

Piemontese, L., Castelli, G., Fetzer, I. et al. (2020). Estimating the global potential of water harvesting from successful case studies. *Global Environmental Change* 63: 102121. https://doi.org/10.1016/j.gloenvcha.2020.102121.

Prinz, D. (2002). The role of water harvesting in alleviating water scarcity in arid areas. In: *International Conference on Water Resources Management in Arid Regions*. Kuwait: Kuwait Institute for Scientific Research.

Rainwaterharvesting.org (2017). Rural water harvesting. http://www.rainwaterharvesting.org/Rural/Rural.htm (accessed 19 March 2017).

Rango, A. (1992). Worldwide testing of the snowmelt runoff model with applications for predicting the effects of climate change. *Hydrology Research* 23 (3): 155–172. https://doi.org/10.2166/nh.1992.0011.

Ray, K. (2005). Rainwater harvesting and utilization, Blure drop series, UN-HABITAT
Reij, C. (1988). *Water harvesting for plant production*. World Bank techn. paper, Washington DC.
Saravanan, M. and Kalpana (2016). Rainwater harvesting – a review and groundwater recharging. *Imperial Journal of Interdisciplinary Research (IJIR)* 2.
Sepehri, M., Malekinezhad, H., Ilderomi, A.R. et al. (2018). Studying the effect of rain water harvesting from roof surfaces on runoff and household consumption reduction. *Sustainable Cities and Society* 43: 317–324. https://doi.org/10.1016/j.scs.2018.09.005.
Shen, Y., Oki, T., Utsumi, N. et al. (2008). Projection of future world water resources under SRES scenarios: water withdrawal. *Hydrological Sciences Journal* 53: https://doi.org/10.1623/hysj.53.1.11.
Srinivasan, V., Konar, M., and Sivapalan, M. (2017). A dynamic framework for water security. *Water Security* 1: https://doi.org/10.1016/j.wasec.2017.03.001.
Stavros Yannopoulos, I.G.K.-S. (2019). Review: investigation of the current situation and prospects for the development of rainwater harvesting as a tool to confront water scarcity worldwide. *Water (Basel)* 11: 1–17.
Swanson, H.A. and Baldwin, H.L. (1965). *A Primer on Water Quality*. Washington, DC: U.S. Geological Survey.
Tamaddun, K., Kalra, A., and Ahmad, S. (2018). Potential of rooftop rainwater harvesting to meet outdoor water demand in arid regions. *Journal of Arid Land* 10: https://doi.org/10.1007/s40333-017-0110-7.
Tamagnone, P., Comino, E., and Rosso, M. (2020). Rainwater harvesting techniques as an adaptation strategy for flood mitigation. *Journal of Hydrology (Amst)* 586: 124880. https://doi.org/10.1016/j.jhydrol.2020.124880.
Thomas, T.H. and Martinson, D.B. (2007). *Roofwater Harvesting: A Handbook for Practitioners*. Delft: IRC.
UNEP (2002). *Rainwater Harvesting and Utilisation: An Environmentally Sound Approach for Sustainable Urban Water Management: An Introductory Guide for Decision-Makers*. Osaka: UNEP.
Verma, S. (2022). Traditional water conservation techniques in India. *Ijraset Journal for Research in Applied Science and Engineering Technology* 10 (1): 705–722. https://doi.org/10.22214/ijraset.2022.39904.
Vijayalakshmi, K. (1982). Water harvesting and re-use. In: *A Decade of Dryland Agricultural Research in India 1971–1980* (ed. ICAR), 103–119. New Delhi, India.
Wagner, W., Gawel, J., Furuma, H. et al. (2002). Sustainable watershed management: an international multi-watershed case study. *Ambio* 31 (1): 2–13. https://doi.org/10.1579/0044-7447-31.1.2.
Weigert, F.B. and Steinberg, C.E.W. (2002). Sustainable development: assessment of water resource management measures. *Water Science and Technology* 46: 55–62.
Willmott, C.J. and Rowe, C.M. (1985). Climatology of the terrestrial seasonal water cycle. *Journal of Climatology* 5: https://doi.org/10.1002/joc.3370050602.
Wright, P. (1985). *Soil and Water Conservation by Farmers*. Burkina Faso: OXFAM.
Zakaria, S., Al-Ansari, N., Knutsson, S., and Ezz-Aldeen, M. (2012). Rain water harvesting and supplemental irrigation at Northern Sinjar Mountain, Iraq. *Journal of Purity, Utility Reaction and Environment* 1: 100–108.

14

Modern Water Treatment and Technological Solutions

Aseem Saxena[1], *Durga P. Panday*[1], *and Manish Kumar*[1,2,3]

[1] *Sustainablity Cluster, School of Advanced Engineering, UPES, Dehradun, Uttarakhand, India*
[2] *Escuela de Ingeniería y Ciencias, Tecnológico de Monterrey, Monterrey, Nuevo León, Mexico*
[3] *Research and Development Initiative, Chuo University, Korakuen Campus 1-13-27, Kasuga, Bunkyo-ku, Tokyo, Japan*

14.1 Introduction

Water is a vital resource of our planet for survival and water quality is an essential aspect for all living organisms, making it a critical global issue. The demand for sustainable water treatment technologies is at an all-time high due to worsening global water scarcity (Zhang et al. 2020). The evolution of water treatment technologies has emerged as a crucial step in improving water quality worldwide. Previously, waterborne diseases did not receive much recognition, leading to limited scientific and engineering attention on water treatment technologies. Later in the late 1990s, the correlation between diseases and water quality was established (Pond 2005; Das et al. 2022; Kumar et al. 2023a, b). About 780 million people do not have access to a purified water source, and an estimated 2.5 billion people lack access to improved sanitation worldwide. It is necessary to remediate contaminated or polluted water sources before consuming them since they may include chemical or microbiological risks that might cause illness (Katsanou and Karapanagioti 2019). Groundwater, which makes up 99% of liquid freshwater on Earth, may provide nations with many social, economic, and environmental advantages. Groundwater supplies half of the world's domestic water, including drinking water for most rural residents who do not have public or private supply systems, and 25% of irrigation water. Unfortunately, this natural resource is often misunderstood, mishandled, and exploited (UN World Water Development Report 2022).

In America, 269 million people use public supply water, 107 million are supplied by groundwater, and 162 million are supplied by surface water (Panday et al. 2021a, b; Johnson et al. 2022; Panday and Kumar 2022). Unlike America, in developing countries, most of the rural population depends upon groundwater sources for all purposes. In suburban area, surface water also plays a vital role in fulfilling public water demand. In our technologically advanced world, the primary necessity involves not only the demand for water but also the crucial criterion of accessing clean water, giving rise to the need for water purification or treatment. Water treatment methods are old-fashioned ways to manage water, but they are

Water Scarcity Management: Enabling Technologies, First Edition. Edited by Kanchan D. Bahukhandi, Manish Kumar, Durga P. Panday, and Tushara G.G. Chaminda.

closely connected with the existing pollution conditions of a particular location. Water treatment has evolved throughout time from conventional techniques to cutting-edge technology, moving from simple filtration to complex procedures including membrane filtration, UV disinfection, and chlorination (Yusuf et al. 2020; Panday and Kumar 2022).

14.1.1 Importance of Water Treatment

Since water treatment addresses a wide range of issues about water quality, it is crucial for preserving both public health and environmental sustainability. Effective water treatment techniques are required to get rid of contaminants, pathogens, and microorganisms in order to guarantee that the water meets stringent quality standards for industrial and consumer use. It is hard to overestimate the significance of employing state-of-the-art coagulation, flocculation, sedimentation, filtration, and disinfection techniques to get the necessary degree of water purity. Moreover, managing residual by-products and following regulations contribute to mitigate the environmental impact of wastewater discharge and preventing waterborne diseases (Wang et al. 2014). The construction of a strong and reliable water supply system requires water treatment since it is vital to the survival of human life and ecosystems.

14.1.2 Evolution of Water Treatment

The evolution of water treatment spans centuries, marked by a relentless pursuit of ensuring safe and clean water for human consumption. Early civilizations practised rudimentary filtration methods, relying on natural materials like sand and gravel to purify water. In the late 18th century, the ideas that are the base of modern chemistry really took off. Scientists have been studying the features of carbons and charcoals for a long time, but they have not found much useful information. One thing they have found is that charcoals can change the colours and smells of gases and solutions (Karge and Weitkamp. 2008). The Industrial Revolution witnessed the advent of more sophisticated water treatment techniques, with the introduction of chlorine disinfection in the late 19th century as a watershed moment. The 20th century brought about significant advancements, including the development of coagulation, flocculation, and sedimentation processes. With the emergence of membrane filtration technologies in the latter half of the century, water treatment entered a new era, offering finer purification capabilities. Today, the evolution continues with a focus on sustainable and energy-efficient solutions, incorporating cutting-edge technologies such as ultraviolet disinfection, reverse osmosis (RO), and advanced oxidation processes (AOPs) (Aslam et al. 2023). Individual contaminant removal technologies, particularly adsorption-based ones, have emerged in recent years. This ongoing journey reflects a collective commitment to safeguarding public health and preserving water resources for future generations (Asok Raja et al. 2023; Panday et al. 2023).

14.2 Era of Water Treatment Methods

In ancient times, people generally used charcoal filtration, boiling, and alum to get clean and potable water. They were much aware of the coagulation and sedimentation through alum. Basically in ancient times, the majority of the population were concerned about visible impurities and odour. To get rid of these impurities, sediment filtrations and some chemical treatment ($KMnO_4$) were evolved (Lee and Schwartz 2007). Some Indian texts directed to store water in copper vessels and use of different herbs in wells like amla (*Phyllanthus emblica*) and khus (*Chrysopogon zizanioides*) (Mishra 2019). Sand filters were mostly used in the ancient era. In Greece, a fabric bag, called the Hippocrates Sleeve, was used to strain water before boiling it. Other methods used for water purification, based on herbs and naturally occurring products and materials, are nirmali seeds, rhizomes of algae's, roots of water lily, and different types of stones such as quartz crystals, garnet, and peals. After a long history of water purification, in the early 20th century, disinfection of water was also introduced by different scientists. Primarily calcium hypochlorite was used for disinfection, later other techniques were identified like ozone and UV light (Ellis 1991). Although waterborne diseases like diarrhoea, giardiasis, dysentery, and typhoid fever were identified in the 19th century, water treatment evolution spiked in the late 20th century. Previously it was limited to sand filtration, but then after different membrane filtrations, carbon nanotubes and bioremediation along with different disinfection methods were introduced. Further advancement in contaminant removal involves different adsorbents like biochar and zerovalent iron. At present, biochar is a subject of interest among environmentalists and researchers owing to its cost-effectiveness, high surface area, and adaptability in altering its properties (Dwibedi et al. 2022).

14.2.1 Limitations and Challenges

During the historical period of water treatment, notable obstacles surfaced, including environmental concerns, substantial energy usage, and constraints on treatment efficacy (Qu et al. 2013). Previous approaches placed significant emphasis on chemical purification, which resulted in adverse environmental consequences as a consequence of inadequate disposal of residual chemicals and sediment. A lack of emphasis on energy efficiency and inefficient technologies simultaneously contributed to the high-energy consumption of water treatment processes (Gogate and Pandit 2004). Insufficient monitoring and a restricted comprehension of contaminants further impeded the overall efficacy of the treatment process. At present, despite advancements, novel obstacles continue to endure. Ongoing hazards are presented by emerging contaminants and the environmental consequences of disinfection by-products. The ongoing concern regarding energy-intensive technologies, specifically those involved in desalination, underscores the need for the incorporation of renewable energy sources (Deshmukh et al. 2018). The issue of treatment efficacy is further complicated by the influence of climate change on water quality and the complexity of confronting diverse contaminant mixtures in water sources. It is crucial to maintain a balance between sustainable practices and technological advancements in order to surmount the historical and present challenges in water treatment.

14.3 Emerging Water Treatment Technologies

14.3.1 Membrane Filtration

Membrane filtration, a pivotal technological advancement in water treatment, operates on the intricate principles of selectively permeable barriers, commonly composed of polymers or ceramic materials. These membranes work efficiently because they can let water molecules pass through while blocking unwanted substances from going through. Ceramic membranes are a category of synthetic membranes composed of inorganic substances, such alumina, titania, zirconia oxide, or silicon carbide (Rashad et al. 2022). By using membranes with pore diameters that are precisely between nano and micrometres, this method aids in the removal of impurities from water according to their molecular size. Membrane filtration is widely used in many different applications, including RO, nanofiltration (NF), and ultrafiltration (UF) (Charcosset 2016).

It guarantees that contaminants, impurities, and microbes are removed, resulting in pure and potable water. This detailed filtering process not only reduces the environmental effects linked to traditional treatment methods but also highlights the crucial importance of membrane filtration in meeting strict water quality standards (Figure 14.1).

14.3.1.1 Reverse Osmosis

The process involves applying pressure to force water molecules through the membrane, leaving behind undesirable substances. Generally, semipermeable membrane has a pore size of 10^{-4}–10^{-3} μm. RO technology is paramount in addressing the burgeoning global demand for high-quality water, especially in regions grappling with compromised water quality (Garud et al. 2011). This method not only ensures the production of potable water but also plays a pivotal role in various industrial applications, such as desalination and wastewater treatment. The synergy of RO with advanced treatment technologies, including UV disinfection and artificial intelligence-based monitoring systems, underscores its crucial role in modern water treatment strategies. The precision and efficiency of RO make it an indispensable tool in safeguarding water resources and promoting sustainable water management practices worldwide (Simonovic 2012).

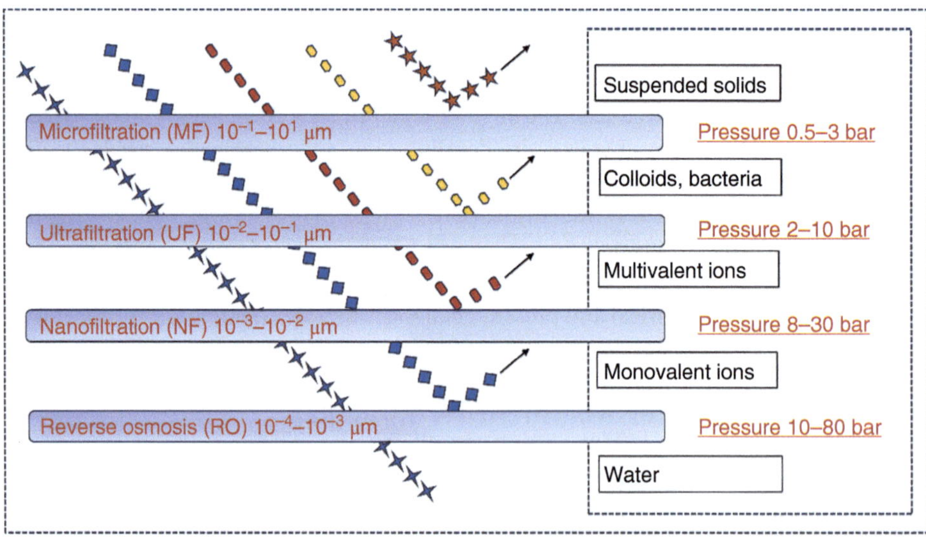

Figure 14.1 Different membrane filtration methods (Fatima et al. 2021).

14.3.1.2 Nanofiltration

NF membranes have come a long way since it was first recognized in the late 1980s. With properties in between UF and RO, NF membranes possess a pore size typically of 10^{-3}–10^{-2} μm (Mohammad et al. 2015), which corresponds to molecular weight cut-off (MWCO) of 300–500 Da. NF membranes have low rejection of monovalent ions, high rejection of divalent ions, and higher flux compared to RO membranes. Its versatility, energy efficiency, and ability to operate at lower pressures distinguish NF as a promising and environmentally sustainable solution for achieving high-quality water purification across diverse settings.

14.4 Ultrafiltration

UF, an advanced membrane-based separation process, has emerged as a pivotal technology in water treatment. Employing semi-permeable membranes with fine pores, typically in the range of 10^{-2}–10^{-1} μm, UF selectively removes suspended solids, bacteria, and macromolecules from water, ensuring a high degree of purification (Othman et al. 2020). This versatile technique is particularly effective in eliminating pathogens and contaminants while allowing essential minerals to pass through. Widely utilized in both municipal and industrial settings, UF offers a compact and efficient solution for water purification, finding applications in drinking water production, wastewater treatment, and various industrial processes. Its ability to operate at relatively low pressures, coupled with its scalability and adaptability to various water sources, positions UF as a crucial component in achieving sustainable and reliable water treatment practices (Yusuf et al. 2020).

14.4.1 Advanced Oxidation Processes (AOPs)

AOPs encompass a set of innovative water treatment techniques designed to efficiently remove a wide range of pollutants. These methods, including UV-based processes, ozonation, and electrochemical oxidation, leverage highly reactive species to break down and transform contaminants into less harmful by-products (Saravanan et al. 2022). AOPs are particularly valuable for addressing persistent organic pollutants, emerging contaminants, and pathogens, contributing to enhanced water quality and safety. The versatility of AOPs makes them vital tools in modern water treatment strategies, allowing for targeted and effective removal of diverse contaminants.

14.4.1.1 UV-Based Processes

UV-based AOPs involve the use of ultraviolet (UV) radiation to generate highly reactive radicals and species that promote the oxidation of organic and inorganic contaminants in water. UV-C light, typically at wavelengths between 100 and 280 nm, breaks down chemical bonds and initiates photochemical reactions (Luste and Sillanpaa 2020).

This process is particularly effective in degrading organic pollutants and pathogens, providing a versatile and environmentally friendly method for water treatment. However, the efficiency of UV-based AOPs can be influenced by factors such as water quality, turbidity, and the presence of certain compounds that may absorb or interfere with UV light (Zheng et al. 2019).

14.4.1.2 Ozonation

Ozonation is a potent AOP that involves the introduction of ozone (O_3), a powerful oxidizing agent, into water for the degradation of contaminants. Ozone readily reacts with a wide range of organic and inorganic substances, breaking them down into simpler, less harmful by-products (Wei et al. 2017). This process is effective in disinfection, colour removal, and the oxidation of persistent pollutants. Ozone's advantage lies in its ability to target a broad spectrum of contaminants and its rapid decomposition into oxygen without leaving harmful residues (Rekhate and Srivastava 2020). However, the cost of ozone generation and potential by-product formation are aspects that necessitate careful consideration in its application.

14.4.1.3 Electrochemical Oxidation

Electrochemical oxidation is an AOP that involves the application of an electric current to generate reactive species, such as hydroxyl radicals, for the oxidation of pollutants (Moreira et al. 2017). This method is highly controllable, allowing for precise tuning of the oxidation potential. Electrochemical oxidation can efficiently degrade a variety of contaminants, including organic pollutants and emerging micropollutants (Martínez-Huitle and Panizza 2018). The process is adaptable to various water sources and can be integrated into existing treatment systems. However, challenges include energy consumption, electrode fouling, and the need for appropriate electrode materials to enhance the overall efficiency of the electrochemical oxidation process.

14.4.2 Desalination Technologies

Desalination, the process of removing salt and other impurities from seawater or brackish water, has become increasingly crucial in addressing global water scarcity. Various desalination technologies have been developed to meet the growing demand for freshwater.

14.4.2.1 Forward Osmosis

Forward osmosis is a desalination process that leverages the osmotic pressure difference between two solutions, typically seawater and a more concentrated draw solution. Water naturally flows through a semi-permeable membrane from the seawater side to the draw solution side, leaving behind salts and impurities (Ahmed et al. 2018). The draw solution is then separated, and water is extracted, resulting in freshwater. Forward osmosis is known for its energy efficiency and has potential applications in low-energy desalination processes.

14.4.2.2 Electrodialysis

Electrodialysis employs ion-selective membranes to separate ions from water. In this process, seawater is passed through alternating cation and anion exchange membranes under the influence of an electric field (Lee et al. 2015). The charged ions are driven through the membranes, leaving desalinated water between them. Electrodialysis is recognized for its lower energy consumption compared to some traditional desalination methods, making it a promising technology for certain applications.

14.4.2.3 Pressure-Retarded Osmosis

Pressure-retarded osmosis utilizes the osmotic pressure difference between freshwater and seawater to generate power and desalinate water simultaneously. In this process, seawater and freshwater are separated by a semi-permeable membrane. As freshwater flows into the seawater side due to osmotic pressure, the increased volume and pressure can be harnessed to generate electricity. The desalination occurs as a by-product, making pressure-retarded osmosis a potential dual-benefit technology.

14.4.3 Smart Sensor Technologies

Advancements in smart sensor technologies have revolutionized the field of water treatment, offering real-time monitoring capabilities and enhancing the overall efficiency of water purification processes.

14.4.3.1 Real-Time Water Quality Monitoring

Smart sensors enable real-time monitoring of water quality parameters such as pH, turbidity, dissolved oxygen, and contaminants. This capability allows water treatment facilities to promptly identify fluctuations in water quality, ensuring a swift response to potential issues. By providing continuous, on-site data, smart sensors contribute to proactive decision-making, enabling rapid adjustments to treatment processes based on the dynamic conditions of the water source (Prasad et al. 2015). Now theses days groundwater quantity data is being used for the prediction in the water quality for the delineation of real time monitoring (Panday and Kumar 2023).

14.4.3.2 Sensor Networks and IoT in Water Treatment

Integration of sensor networks and the Internet of Things (IoT) in water treatment has ushered in a new era of connectivity and automation. Smart sensors deployed throughout the treatment infrastructure can communicate and share data seamlessly. This interconnected network facilitates a holistic understanding of the water treatment system's performance, allowing for centralized monitoring and control. IoT-enabled sensors provide real-time insights into operational parameters, enabling operators to optimize processes, minimize energy consumption, and ensure the consistent delivery of high-quality treated water (Singh et al. 2016).

In terms of water treatment, these smart sensor technologies play a pivotal role in improving the overall effectiveness and sustainability of the processes. The ability to monitor water quality in real time ensures timely detection of anomalies, preventing the release of inadequately treated water into distribution systems (Olatinwo and Joubert 2020). Moreover, the integration of sensor networks and IoT enables remote monitoring and control, reducing the need for manual intervention and enhancing the overall efficiency of the treatment plants.

14.5 Conclusion

In conclusion, the chapter provides a holistic examination of modern water treatment and technological solutions, emphasizing the historical evolution of methods and the contemporary challenges and opportunities within the field. From ancient practices using sand filtration to the advent of sophisticated membrane technologies, the commitment to innovation and resource preservation is evident. The challenges of the past, including environmental impact and high energy consumption, pave the way for current concerns such as emerging contaminants, urging a delicate balance between sustainable practices and technological advancements. The discussion on emerging technologies, spanning membrane filtration, AOPs, and desalination methods, highlights their pivotal role in addressing water scarcity and enhancing treatment efficiency. The burden on municipal entities is lessened when individual water purifier systems are used in household settings. Crucially, smart sensor technologies emerge as indispensable tools for real-time monitoring, underscoring the ongoing commitment to technological progress, sustainability, and the safeguarding of water resources for future generations. These technologies are aiding to achieve sustainability goal numbers 6 and 14.

References

Ahmed, M., Kumar, R., Al-Wazzan, Y. et al. (2018). Assessment of performance of inorganic draw solutions tested in forward osmosis process for desalinating Arabian gulf seawater. *Arabian Journal for Science and Engineering* 43 (11): 6171–6180.

Aslam, M.M., Tufail, A., Kim, K.H. et al. (2023). A comprehensive study on cyber attacks in communication networks in water purification and distribution plants: challenges, vulnerabilities, and future prospects. *Sensors* 23 (18): 7999.

Asok Raja, S.K., Panday, D.P., and Kumar, M. (2023). Decoding the enigma of 100-year record-breaking rainfall over Tamil Nadu using wavelet analysis. *Groundwater for Sustainable Development* 101007. https://doi.org/10.1016/j.gsd.2023.101007.

Charcosset, C. (2016). Ultrafiltration, microfiltration, nanofiltration and reverse osmosis in integrated membrane processes. *Integrated Membrane Systems and Processes* 1–22.

Das, A., Kumar, M., Jha, P.K. et al. (2022). Isotopic and hydrogeochemical tracking of dissolved nutrient dynamics in the Brahmaputra River System: a source delineation perspective. *Chemosphere* 307: 135757. https://doi.org/10.1016/j.chemosphere.2022.135757.

Deshmukh, A., Boo, C., Karanikola, V. et al. (2018). Membrane distillation at the water-energy nexus: limits, opportunities, and challenges. *Energy & Environmental Science* 11 (5): 1177–1196.

Dwibedi, S.K., Pandey, V.C., Divyasree, D., and Bajpai, O. (2022). Biochar-based land development. *Land Degradation & Development* 33 (8): 1139–1158.

Ellis, K.V. (1991). Water disinfection: a review with some consideration of the requirements of the third world. *Critical Reviews in Environmental Science and Technology* 20 (5–6): 341, 233–407, 234.

Fatima, F., Du, H., and Kommalapati, R.R. (2021). Treatment of poultry slaughterhouse wastewater with membrane technologies: a review. *Water 13* (14): 1905.

Garud, R.M., Kore, S.V., Kore, V.S., and Kulkarni, G.S. (2011). A short review on process and applications of reverse osmosis. *Universal Journal of Environmental Research & Technology* 1 (3).

Gogate, P.R. and Pandit, A.B. (2004). A review of imperative technologies for wastewater treatment II: hybrid methods. *Advances in Environmental Research* 8 (3–4): 553–597.

Johnson, T.D., Belitz, K., Kauffman, L.J. et al. (2022). Populations using public-supply groundwater in the conterminous US 2010; identifying the wells, hydrogeologic regions, and hydrogeologic mapping units. *Science of the Total Environment* 806: 150618.

Karge, H.G. and Weitkamp, J. (ed.) (2008). *Adsorption and Diffusion*, vol. 7. Springer Science & Business Media.

Katsanou, K. and Karapanagioti, H.K. (2019). Surface water and groundwater sources for drinking water. *Applications of Advanced Oxidation Processes (AOPs) in Drinking Water Treatment* 67: 1–19.

Kumar, M., Panday, D.P., Bhagat, C. et al. (2023a). Demystifying the decadal shift in the extent of groundwater in the coastal aquifers of Gujarat, India: a case of reduced extent but increased magnitude of seawater intrusion. *Science of the Total Environment* 165451. https://doi.org/10.1016/j.scitotenv.2023.165451.

Kumar, M., Mazumder, P., Silori, R. et al. (2023b). Prevalence of PPCPs, microplastics and co-infecting microbes in the post-COVID-19 era and its implications on antimicrobial resistance and potential endocrine disruptive effects. *Science of the Total Environment* 166419. https://doi.org/10.1016/j.scitotenv.2023.166419.

Lee, E.S. and Schwartz, F.W. (2007). Characteristics and applications of controlled–release KMnO4 for groundwater remediation. *Chemosphere* 66 (11): 2058–2066.

Lee, J.Y., Kim, J.H., Lee, J.H. et al. (2015). Morphologically aligned cation-exchange membranes by a pulsed electric field for reverse electrodialysis. *Environmental Science & Technology* 49 (14): 8872–8877.

Luste, S. and Sillanpää, M. (2020). UVC irradiation-based water treatment. In: *Advanced Water Treatment* (ed. M. Sillanpää), 95–128. Elsevier.

Martínez-Huitle, C.A. and Panizza, M. (2018). Electrochemical oxidation of organic pollutants for wastewater treatment. *Current Opinion in Electrochemistry* 11: 62–71.

Mishra, S.B. (2019). *Essentials of Herbal Drug Technology: A Guide of Standardization Quality Control*. Educreation Publishing.

Mohammad, A.W., Teow, Y.H., Ang, W.L. et al. (2015). Nanofiltration membranes review: recent advances and future prospects. *Desalination* 356: 226–254.

Moreira, F.C., Boaventura, R.A., Brillas, E., and Vilar, V.J. (2017). Electrochemical advanced oxidation processes: a review on their application to synthetic and real wastewaters. *Applied Catalysis B: Environmental* 202: 217–261.

Olatinwo, S.O. and Joubert, T.H. (2020). Energy efficiency maximization in a wireless powered IoT sensor network for water quality monitoring. *Computer Networks* 176: 107237.

Othman, M.H.D., Adam, M.R., Pauzan, M.A.B. et al. (2020). Ultrafiltration membrane for water treatment. *Self-Standing Substrates: Materials and Applications* 119–145. https://doi.org/10.1007/978-3-030-29522-6_4.

Panday, D.P. and Kumar, M. (2022). Climate indices and hydrological extremes: deciphering the best-fit model. *Environmental Research* 215: 114301. https://doi.org/10.1016/j.envres.2022.114301.

Panday, D.P. and Kumar, M. (2023). Application of remote sensing techniques to deal with scale aspects of GRACE data to quantify groundwater levels. *MethodsX* 102108. https://doi.org/10.1016/j.mex.2023.102108.

Panday, D.P., Khosa, R., Maheswaran, R. et al. (2021a). Game-theoretic-based modelling of Krishna waters dispute: equilibrium solutions by Metagame Analysis. *Eur. Phys. J. B* 94: 101. https://doi.org/10.1140/epjb/s10051-021-00107-w.

Panday, D.P., Khosa, R., Maheswaran, R. et al. (2021b). Game theoretic-based modelling of Krishna waters dispute: equilibrium solutions by hypergame analysis. *Eur. Phys. J. B* 94: 131. https://doi.org/10.1140/epjb/s10051-021-00135-6.

Panday, D.P., Singh, A., Gaur, S. et al. (2023). Probable maximum precipitation analysis of high rainfall regions in India. *Groundwater for Sustainable Development* 100893. https://doi.org/10.1016/j.gsd.2022.100893.

Pond, K. (2005). Water recreation and disease: plausibility of associated infections: acute effects, sequelae, and mortality.

Prasad, A.N., Mamun, K.A., Islam, F.R., and Haqva, H. (2015). Smart water quality monitoring system. In: *In 2015 2nd Asia-Pacific World Congress on Computer Science and Engineering (APWC on CSE)*, 1–6. IEEE.

Qu, X., Alvarez, P.J., and Li, Q. (2013). Applications of nanotechnology in water and wastewater treatment. *Water Research* 47 (12): 3931–3946.

Rashad, D., Amin, S.K., Mansour, M., and Abdallah, H.A. (2022). A systematic literature review of ceramic membranes applications in water treatment. *Egyptian Journal of Chemistry* 65 (7): 497–512.

Rekhate, C.V. and Srivastava, J.K. (2020). Recent advances in ozone-based advanced oxidation processes for treatment of wastewater – a review. *Chemical Engineering Journal Advances* 3: 100031.

Saravanan, A., Deivayanai, V.C., Kumar, P.S. et al. (2022). A detailed review on advanced oxidation process in treatment of wastewater: mechanism, challenges and future outlook. *Chemosphere* 308: 136524.

Simonovic, S.P. (2012). *Managing Water Resources: Methods and Tools for a Systems Approach.* Routledge.

Singh, S., Kumar, A., Prasad, A.B.N., and Bharadwaj, N. (2016). IoT based water quality monitoring system.

UN (2022). UN World Water Development Report. https://www.unwater.org/publications/un-world-water-development-report-2022 (accessed 20 March 2024).

Wang, H., Wang, T., Zhang, B. et al. (2014). Water and wastewater treatment in Africa–current practices and challenges. *CLEAN–Soil, Air, Water* 42 (8): 1029–1035.

Wei, C., Zhang, F., Hu, Y. et al. (2017). Ozonation in water treatment: the generation, basic properties of ozone and its practical application. *Reviews in Chemical Engineering* 33 (1): 49–89.

Yusuf, A., Sodiq, A., Giwa, A. et al. (2020). A review of emerging trends in membrane science and technology for sustainable water treatment. *Journal of Cleaner Production* 266: 121867.

Zhang, D., Sial, M.S., Ahmad, N. et al. (2020). Water scarcity and sustainability in an emerging economy: a management perspective for future. *Sustainability* 13 (1): 144.

Zheng, M., Daniels, K. D., Park, M., Nienhauser, A. B., Clevenger, E. C., Li, Y., & Snyder, S. A. (2019). Attenuation of pharmaceutically active compounds in aqueous solution by UV/CaO_2 process: influencing factors, degradation mechanism and pathways. *Water Research*, 164, 114922.Fatima, F., Du, H., & Kommalapati, R. R. (2021). Treatment of poultry slaughterhouse wastewater with membrane technologies: a review. Water, 13(14), 1905.

15

Current Challenges and Solutions to Achieving Sustainable Solutions to Water Security

Gagan Matta[1], Gaurav Pant[1], Nitin Kamboj[1], Sunil Kumar[2], Avinash Kumar[1], Pawan Kumar[1], Anjali Nayak[1], Aditi Bisht[1], Neeraj Pandey[1], Amrit Kumar[1], Vishal Bhardwaj[1], and Himanshu Saini[1]

[1] *Department of Zoology and Environmental Science, Gurukul Kangri (Deemed to be University), Haridwar, Uttarakhand, India*
[2] *Gurukula Kangri (Deemed to be University), Haridwar, Uttarakhand, India*

15.1 Introduction

Water is essential for all and the basis of life. For millions of users around the globe, it is a decreasingly scarce and degraded natural resource. One of the most significant issues recently is the insufficient water supply, which is required for many purposes by a constantly expanding population. This is a crucial issue since we need to produce more food with the same amount of water to feed the world's population in the future. The worldwide water dilemma is made worse by expanding populations and climatic change. Roughly more than 2 billion people worldwide have less access to even the most basic sanitary services, and more than 1 billion have insufficient access to clean drinking water. Economic paucity is a significant problem for water insecurity and physical shortage. Economic water shortage, which prevents people from accessing available water sources, affects around 1.5 billion people. Millions of individuals worldwide are experiencing a crisis caused by a lack of safe water, ruining their livelihoods, causing lousy health, and causing unnecessarily long periods of agony for deprived communities (Asian Water Development Outlook (AWDO) 2016).

The biggest obstacle to economic and societal growth is water scarcity. Lack of water can impact ecosystems, deteriorate health, and ruin livelihoods. In 2015, more than 1.8 million people died due to water-related diseases compared to 0.8 million people in 2012 (GBD 2015 Risk Factors Collaborators 2016). In arid and semiarid areas, which are especially sensitive to climate change and land use changes, rising population pressure reduces the capacity to provide appropriate water resources and effectively function environmental service. Water deterioration has become a worldwide issue as anthropogenic activities increase rapidly (Biswas and Tortajada 2019). Figure 15.1 shows the water use and stress on resources globally. A huge concern in the climate change set is the increase in water-related calamities,

Water Scarcity Management: Enabling Technologies, First Edition. Edited by Kanchan D. Bahukhandi, Manish Kumar, Durga P. Panday, and Tushara G.G. Chaminda.

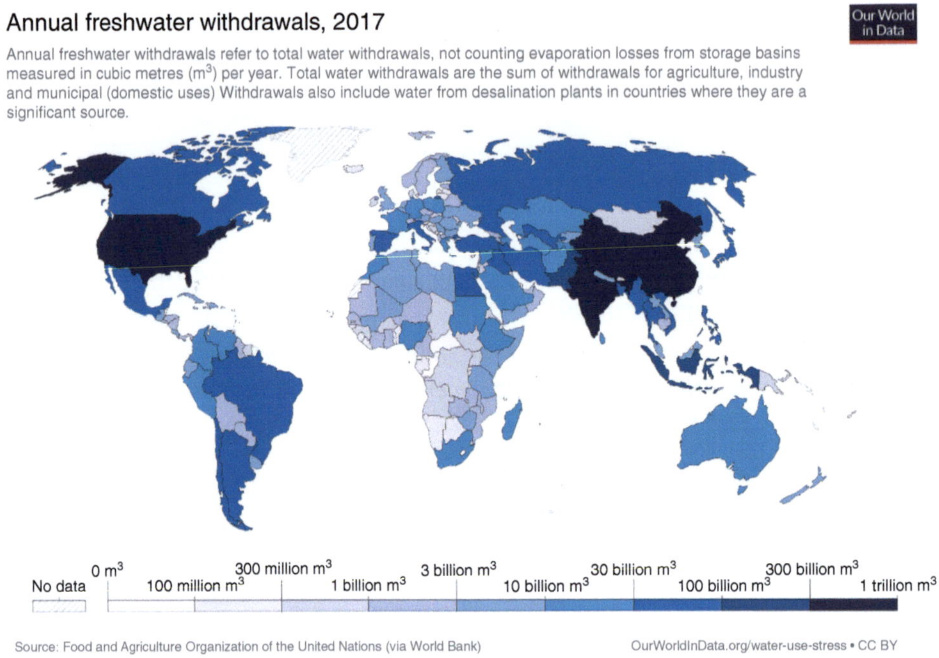

Figure 15.1 This figure is covered by the Creative Commons Attribution 4.0 International License as mentioned. *Source:* Water Use and Stress, 2015/https://ourworldindata.org/water-use-stress. Last accessed 4 June 2024/Our World in Data/CC BY 4.0.

which accounts for a significantly more significant percentage of natural and economic losses. Much more unpredictable threats can be made worse by reasons including climate change and uncertainty, shifts in land use, urban development, energy issues, and changes in food production brought on by population expansion and anthropogenic activities (UNESCO, UN-Water 2020; Matta et al. 2022a). Figure 15.1 illustrates the water use and stress on global water resources.

Because of harsh meteorological conditions, floods are frequently considered a natural event, although environmental changes due to human-induced activities alter the water cycle in nature, resulting in flooding events. Another major reason for land use change other than growing urban development is the expansion of agricultural fields. Likewise, there are numerous social and physical dimensions of drought. Depending on how resilient local towns and populations are, the effects of a lack of rainfall on freshwater resources and agricultural systems can be severe. Depletion mainly results in groundwater resources. During periods of drought, better quality and easy access to groundwater have increased the lack of water in groundwater resources over the last five decades (Matta et al. 2022b).

15.1.1 Water Conservation

The most important method for managing and planning water use shortly is becoming increasingly popular worldwide. Water conservation has somewhat succeeded in reducing demand in developed world cities' metropolitan regions, but in developing countries, due to increasing population, high water demand is there. India's water resources are running out fast, which is 4% of the world's water resources. The increasing demand for water due to the population expansion will increase by 220 billion cubic meters (bcm) from 40 bcm, and the shortage of water resources also impacts the growth and development of agriculture (Di Baldassarre et al. 2019). In essence, water conservation tries to balance supply and demand. The solutions for conserving water could be demand-, supply-, management-, or any combination of these. The approaches may change depending on the type of water use – domestic, agricultural, or industrial. The processes for water conservation are rainwater harvesting, better irrigation practices, using saline water for irrigation, contour farming, soak pit construction, tree plantation, and so on (UN 1982). Improving residential water conservation offers a possible alternative or essential complement to lower expenses and encourage environmentally friendly behavior (UN 1992; Matta et al. 2020a).

15.1.2 Sustainable Development Goals

SDGs serve as a path to reducing inequality worldwide and securing a sustainable society for all people (Matta 2020). The phrase "sustainable development" (SD) originally appeared in the World Charter for Nature, 1982, indicating that it was first used from the perspective of genuine concerns and it was further discussed and defined in Agenda 21 of Earth Summit, 1992 (United Nations 2018; Matta 2020). And after several suggestions and recommendations, a group of global SDGs with 17 goals and 169 targets was suggested at the UN in New York by the Open Working Group, which the UN General Assembly established (UN-United Nations 2018). Additionally, a proposed set of 330 indicators was released in 2015. While some SDGs expand upon earlier Millennium Development Goals, others adopt fresh concepts (Matta 2020). The management and allocation of water resources must take an integrated strategy that involves all stakeholders and considers how water resources connect many societal sectors to achieve the SDGs for water. A lack of consideration for the interactive comments between human-water systems' natural, technical, and social aspects has led to the dominance of technocratic, scenario-based approaches in water management practice. While these approaches may be effective in the short term, they may have unexpected long-term effects (United Nations University (UNU) 2013).

15.1.3 SDG 6

All freshwater natural resources worldwide are deteriorating rapidly, with population growth and increasing water withdrawal rates. It makes this issue a primary concern for society regarding water conservation (Nayak et al. 2022). One of the 17 development goals in the SDG initiative, SDG 6: "Achieve universal access to water and sanitation with effective management by 2030," was proposed to address this issue (Kumari and Singh 2016; Matta et al. 2022a). The objectives of this SDG 6 – also known as the "water goals" – are to ensure water security and to advance the achievement of the other SDGs. It has eight water aims, including offering access to clean, affordable drinking water; providing services for hygiene and sanitation; treating wastewater and reusing it; improving the efficiency of water use as well as supply; implementing integrated water resources management; protecting and restoring water-related ecosystems; receiving global assistance and support for capacity building; and exerting arduous efforts (Kumari and Singh 2016; Gjyli et al. 2018). The aims of SDG 6 have proven challenging to attain due to a finite amount of funding, rapid urbanization, and a growing population, and many regions still do not have essential water and sanitation infrastructure. Overall progress in sectors like agriculture, energy, combating climate change, protecting human and environmental health, and, eventually, economic expansion is all threatened by this paradigm. A more inclusive strategy to manage and distribute water resources is necessary to meet the targets outlined in SDG 6, along with a cost–benefit approach, collaborations, and the simplification of financial systems (Matta et al. 2022a).

15.1.4 Conceptualization of Water Security

The concept of water security is said to be the sustainable use of water, which means access to safe, clean water for the general public for a healthy life – a place where no water-related disasters or diseases will occur. A sustainable environment and social equality are fostered through water security, which also addresses the detrimental effects of ineffective water management. A variety of definitions have been defined for water security. According to the United Nations Agency on Water (UN Water), "*The capacity of a population to safeguard sustainable access to adequate quantities of acceptable quality water for sustaining livelihoods, human well-being, and socioeconomic development, for ensuring protection against water-borne pollution and water-related disasters, and for preserving ecosystems in a climate of peace and political stability*" (Matta 2020; Kumar et al. 2021).

Water shortages, clean water, ecological services, and human well-being have all been added to the emphasis on water security in the past few years. Traditional methods such as developmental studies predominantly focused on the country level, economic studies on the financial side of the issue, hydrological studies concentrated only on watershed aspects, and social studies on the local livelihood of the communities are focused on only one aim and scope. So a new perception or paradigm is necessary with a long-term commitment to water security in a sustainable way considering the environment, society, and economy (Demographia 2016).

The study of spatial and temporal analysis of the scenario of water resources can determine water security. The same place may experience droughts and floods during the same year due to hydrological extremes. As a result, average water use cannot be used to determine whether there is a water shortage or surplus. Many Asian countries experience water shortages throughout the monsoons, even though the abundance of water resources at an annualized rate seems plentiful. Water shortages and flooding can coincide, even in different sections of large nations such as India. Climate change will make the problem of the unbalanced sharing of water sources worse. The ability to comprehend the issue of water scarcity is essential because it influences opinions on the best strategies for addressing the problem (Matta et al. 2020b).

The availability of water has already started to fluctuate due to climate change. Severe hydrological events that seriously threaten urban water supplies appear to expand in size and frequency. As a result, urban flooding disasters and water shortages are becoming more severe (Perveen et al. 2012).

Climate change is anticipated to have a considerable impact on the timing and amount (depth) of extreme precipitation, which would affect the flooding that occurs in many urban places across the world and, in the absence of effective prevention, generate higher flood hazards and related losses (Jiang 2015). It is crucial to alter how management responds to social and political issues to address the new issue of water security.

15.2 Challenges to Water Security

Increasing domestic consumption because of rapidly increasing living standards, industrial demand, rising energy consumption, mining, climate change, urbanization, forest destruction, and population movement all impact water security. These factors also include agricultural irrigation and changing demography.

15.2.1 Changing Demography

Although there are now 7.7 billion people on the planet, there were only 3 billion 50 years ago (Figure 15.2). However, water usage has tripled. By 2025, there will most likely be an additional 2 billion people, and by 2050, there will be 3 billion (Figure 15.3). As a result – or much more – the water demand will increase. Diets and habits that have a large impact are

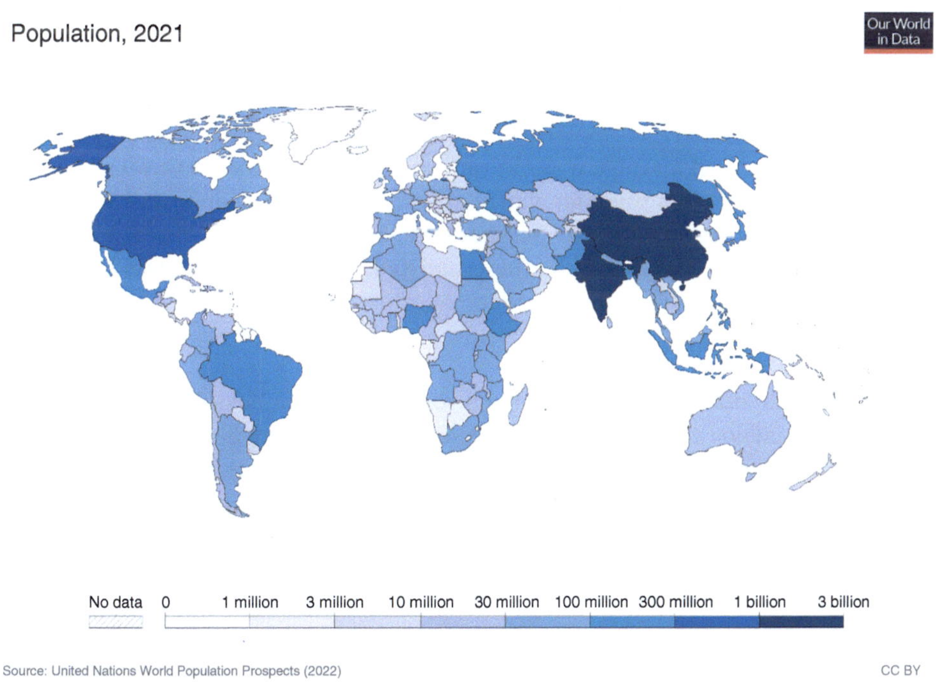

Figure 15.2 This figure is covered by the Creative Commons Attribution 4.0 International License as mentioned. *Source:* Roser and Rodés-Guirao (2013)/https://ourworldindata.org/population-growth. Last accessed 4 June 2024/Our World in Data/CC BY 4.0.

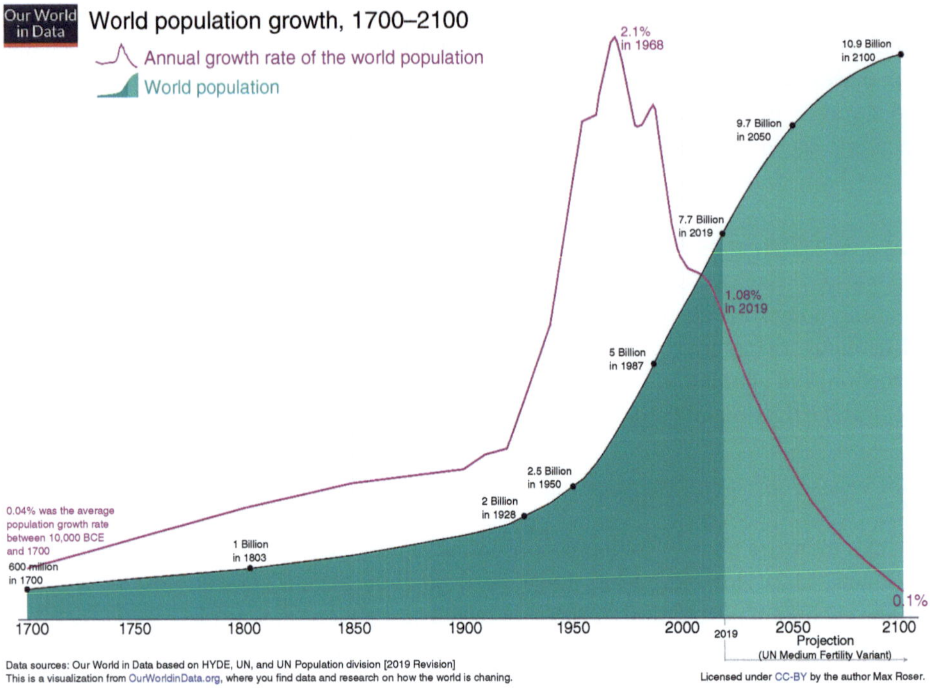

Figure 15.3 Future population growth (this figure is covered by the Creative Commons Attribution 4.0 International License as mentioned. *Source:* Roser and Rodés-Guirao (2013)/https://ourworldindata.org/population-growth#licence. Last accessed 4 June 2024/Our World in Data/CC BY 4.0.

shifting to ones that call for more water. If agricultural productivity remains unchanged, the world's requirement for water for agriculture might increase by up to 60% to feed an additional 2 billion people This area contains as much freshwater as is used worldwide outside of Asia, amounting to 1,500 km^3 (Roser and Rodés-Guirao 2013; Matta et al. 2022c).

15.2.2 Increasing Water Consumption

Urban freshwater usage for commercial and residential usage is rapidly expanding in response to population growth. Water usage in municipalities is confined to such a greater amount of consumption in a comparatively limited geographical region, despite the amount of water provided to urban inhabitants in absolute terms being much less than that consumed in farming, especially in South Asian urban cities (Matta et al. 2020c). Additionally, every water droplet has significant economic and social importance compared to agricultural water. It seems to be because families need regular access to water for necessities like drinking, food, washing, and cleanliness, in addition to industrial operations producing relatively significant economic benefits from water consumption (Rawat and Matta 2021). Sectors, especially water-intensive sectors like metal, fabrics, and paper, have a huge requirement for urban water in addition to residential consumption. Water via the municipality distribution system is used by certain urban industries. Water from nearby groundwater and surface water sources is used by other sectors (Rawat and Matta 2021). To provide an appropriate urban water supply in places experiencing water shortages, such commercial water requirement tries to compete with domestic water needs.

15.2.3 Over Exploitation of Surface and Groundwater Resources

Several river systems are said to be closed due to their full exploitation, and these are especially of the rivers in developing countries such as India and Pakistan. Those rivers use all or almost all of the yearly freshwater supply that is accessible. To satisfy the increasing pressure from towns and farmland, these areas have insufficient to no potential to boost freshwater withdrawal (Matta et al. 2021).

Freshwater is used for various purposes, but ecological flows are necessary to preserve other natural river functions and river and estuary habitats. An appropriate quantity of river water is assigned to environmental processes as part of better management systems, just as water is allocated to households, businesses, and other uses. The amount, purity, and time of the water that flows must all be considered when determining the watershed management needs. Because of South Asia's extreme hydrological spatial heterogeneity, the complexity and cost of scaling up waste treatment, disagreements over water resources between governments and territories, and the shortage of statistical data on the connections among river flows and river ecology, establishing and sustaining the necessary environmental flows completely are complex (Matta et al. 2020d).

Similarly, groundwater resources have also over-exploited over the last six decades worldwide. The inhabitants of South Asia have long used ancient technology to harvest water from wells and utilize groundwater. However after the late 1950s, groundwater consumption in South Asia significantly grew due to the Green Revolution's use of motorized equipment and borewells in farming. The quantity, frequency, and types of rainfall and the underlying geological conditions all affect groundwater replenishment. Spontaneous runoff and borehole extraction are two methods used to remove groundwater in aquifers. Groundwater levels drop as removal rates outpace recharging rates.

The much drier northern area's groundwater over-extraction is causing water tables to continuously decline, necessitating deep boreholes and more intensive pumps. Water tables in certain sections of the Punjab province of India and Pakistan, in addition to northwestern Rajasthan, north Gujarat, and Haryana provinces in India, are eroding by roughly 1 m each year as a result of this (Matta et al. 2020d).

15.3 Other Challenges

15.3.1 Inefficient Water Distribution

Urban dwellers need to receive life-sustaining water consistently for civil life to exist. Widespread distribution channels are employed to meet this demand and make water accessible to all residents. Pressure transients, corrosive activities, and pollution sources are only a few of these systems' difficulties. Developing countries in South Asia face these types of problems in water distribution (Falkenmark and Molden 2008).

15.3.2 Inadequate Drinking Water Accessibility for Poor Urban Citizens

Even if freshwater is readily accessible regionally in the environment to support human needs, hydrological drought develops when access to water is restricted due to a lack of human, organizational, and economic support. Water is plentiful in certain regions, yet some local communities struggle with water shortage because they cannot get it. Economic water shortage in South Asia has a significant urban aspect even though it has typically been seen as a rural community issue (Hairan et al. 2021).

Poor urban residents sometimes dwell in urban slums lacking access to public water, forcing them to pay higher retail costs for necessary water distribution by private suppliers or risk illness by using polluted native water sources. Cities are seeing an increase in commercial potable water marketplaces that cater to both affluent and slum dweller groups (CGWB (Central Ground Water Board) 2015).

15.3.3 Expanding Water Supplies Is a Challenge in Urban Regions

Insufficient water supply is already a problem in several South Asian areas. Some areas presently use more water than their locally sustainable resources for surface and groundwater. Expanding urbanization's constantly rising water need contributes significantly to this rapid depletion. Towns in Asia that rely on groundwater sources for essential local needs, like cities of Bangladesh and India, have dropped water tables due to rising excavation and pumping expenses. Surface water is intermittently or insufficiently available in towns that rely on it, notably Bengaluru and Hyderabad. The increased demand and an outdated and inadequate infrastructure make it difficult for many communities to maintain continuous water supplies (HPEC (High Powered Expert Committee for Estimating the Investment Requirements for Urban Infrastructure Services) 2011).

15.4 New Perspectives on Water Security

Water security encompasses a problematic, multifaceted, and interconnected range of problems. There is increased competitiveness for water consumption across various industries and authorities at local, regional, and global levels due to the mounting strain on water supplies. This includes fundamental social requirements, including irrigation, hydroelectric, fresh water supply, and industrial usage. Water quality, another crucial aspect of water utilization, is also necessary for achieving water security. The increase in human activities is placing strain on the freshwater and groundwater quality, affecting water usage for diverse applications. Rivers are used to accept, carry, and neutralize pollutants (Cohen and Ray 2018).

Water security is a prerequisite for the people and the environment for sustainable development, eradication of poverty and hunger, and accomplishing the Sustainable Development Goals. Geographically and across time, different concerns about quality, supply, and availability predominate. This implies that attaining water supply is not a dynamic aim that changes in response to shifting dynamics. Moreover, the objective of obtaining water security changes along with changes in social, cultural, political, economic, and value concerns. In light of this context, the research made a case for analyzing water security problems in light of new perspectives and looking for responses (Cohen and Ray 2018).

Four paradigms or institutional prescriptions are also described for achieving sustainable water security based on adaptive management and polycentric approaches. These approaches include polycentric governance, public involvement, experimental practices for resource management, and bioregional scale management (Gunda et al. 2019). It is common among developing countries to view integrated water resource management (IWRM) from how it is currently evolving. The idea of using nature-based solutions to solve problems is spreading worldwide, yet many academics still emphasize combining hard and soft ways to solve problems. Moving beyond system thinking techniques involves cooperation at various levels involving local and national coordination and partnerships. In new paradigms where forming relationships with the private sector and lowering non-revenue water are priorities, a change in funding the water industry is crucial (Gunda et al. 2019). This study focused on the solutions for achieving water security goals in the current challenges and growing new paradigms.

15.4.1 Consideration of Water Security and Its Predictors

The complicated idea of comprehensive water management, including conserving natural resources and their usage, is captured by the phrase "water security" (Gunda et al. 2019). Predictors for water security must be used to help quantify it. The indicators of water security were classified into four dimensions based on the UN definition of water security, as shown in Figure 15.2. These indicators or predictors have been developed by research earlier such as the economic water scarcity index, water stress indicator, and water resources vulnerability index. To encourage transparency and establish a common understanding of a topic in which there is currently a considerable amount of confusion, it is advantageous to translate water security into a quantitative form for assessing the level of water security (Huitema et al. 2009).

Recycling wastewater is an essential factor in this aspect of water security. Because if it is adequately treated, it can minimize pollution in the water and be used in various domestic and industrial uses and somehow for recharging the groundwater (Grant et al. 2012). The uncertain climate models are the indicator in which results are essential in evaluating the ambiguity in water security assessment – more frequent weather events, such as floods and droughts, result from climate change. According to IPCC, approximately 87% of climate change impact will be on water-related infrastructure (Marcal et al. 2021).

Taking action at the evaluation level is challenging in response to nationwide evaluations. Agriculture utilizes the majority of water in Asia-Pacific, although it concentrates on water productivity because it is frequently utilized to increase production (Wiek and Larson 2012).

Water security indicators must be developed and used with a management control or evaluation action in mind. Multiple impacts had to be measured via indicators, which had to be combined to create single-value indications. It is obvious that an integrated strategy is needed to ensure water security, and exchange may be necessary between various water-related industries. Politicians and water experts must concentrate on the indicators that communicate the most fantastic story to other areas if they want to support water-smart investments. Indicators are often intuitive in nature (Van-Rijswick et al. 2014).

15.5 Sustainable Solutions for Managing Risks Related to Water

Water security is a significant problem, given the consequences of climate change. A holistic approach is needed, integrating all aspects and providing scientific and technological solutions. Water security is complicated, indicating that various areas and times require different water crisis control methods. It is crucial to practice sustainable water management, which focuses on the coordinated and efficient use of diverse water resources and water allocation strategies that boost water production and have positive social and economic benefits. The social impacts of water allocation policies and equity in access to water must also be given special consideration.

This part promotes many sustainable methods to attain water security under three main themes, taking into account the dimension and multiple paradigms associated with it and its complicated and convoluted character.

15.5.1 Adaptive Water Governance

Governance may be sustainable if it can safeguard ecosystems while considering people's social and economic well-being (Saraswat and Kumar 2016). Along with national water legislation, standards, and organizations, governance also encompasses the procedures that encourage stakeholder or society involvement in developing safe water supply and sanitation. These practices also aim to educate communities and give them more participation in management decisions. The social mobilization and measures intended to promote ownership, capacity building, co-investment, readiness to pay for services, and incentives for community engagement are referred to as the restructuring of water governance techniques (Saraswat and Kumar 2016). Water governance strategies also support equal representation in decision-making (UNESCO 2019). To ensure the provision of water and sanitation and maintain surveillance of public and private participants, it specifies the applicable duties of governmental institutions. A procedure that is particularly effective at resolving market imbalances, rewards, and modifications to reasonable pricing is promoting sustainable governance as an alternative solution for water security.

Socioeconomic activities should not diminish locally accessible water bodies in an area to increase their sustainability. It is necessary to establish co-management, which includes the processes of co-design, co-implementation, and co-delivery methods throughout the water cycle.

a) By creating a more extensive network of economic and environmental linkages, the Regional Circular and Ecological Sphere (RCES) is a concept that supports and promotes local resource use and creates synergy (UNESCO 2019). It will also encourage us to use renewable energy resources rather than water.
b) Another exceptionally creative and effective strategy for more effective management of water resources is Participatory Watershed Land-use Management (PWLM) (NASAC (Network of African Science Academies) 2014). It aids in increasing the efficiency of the local changing climate and land use policy. This integrated approach uses participation strategies and computer-simulating modeling to manage water resources at a regional scale. Four significant phases comprise the entire process,

including situation assessment, impact analysis, the land-use plan's development, mitigation and adaptation strategies formulation, and their incorporation into local government policies.

c) Integrated water resources management (IWRM) is a crucial strategy for enhancing water security and sustainability by increasing water usage efficiency and conservation. Even though the word or idea IWRM has been around for a while and is often used by many stakeholders, its practical implementation is still debatable (NASAC (Network of African Science Academies) 2014). Figure 15.3 demonstrates the different stages of IWRM process. For a more profound knowledge of the consequences of action policy and research for environmentally friendly water management, it is necessary to analyze water-related hazards holistically using socio-hydrological channels (NASAC (Network of African Science Academies) 2014).

15.5.2 Environmental Approaches

Using or simulating environmental mechanisms to solve modern problems, such as water management, are known as "environmental approaches." Improving water supply through recharge of the water table and retention capacity of soil; enhancing water quality by natural and artificial wetlands; lowering the probability of water-related disasters and climate change through flood plains; and restoration strategies. The soil with vegetation in forests, meadows, moist areas, and rural and urban settings are driving forces behind natural systems that are essential for water movement, storage, and conversion (Gjyli et al. 2020).

Studies that consider whole ecosystems, such as rivers, are rare. Additionally, most river systems have undergone significant natural or human-induced change, which poses a major obstacle to sustainable water management (Matta et al. 2022d). This poses a significant impediment to the management of sustainable water resources because it prevents the inclusion of crucial "pieces" or functions that relate to the diversity of the ecosystem and are required to develop sustainable solutions. Because they are frequently highly designed, freshwater ecosystems are susceptible to missing essential components of their ecosystem variety, which is extremely common and eventually causes the freshwater systems to lose their robustness. For example, rivers are continuously changing in terms of their structure. The community that resides inside freshwater ecosystems, such as a river basin, is not exempt from the decline in resiliency, which results in indecision. To say this, for example, growing human interventions such as dams and barrage constructions from the past few decades are changing the flow and course of water, leading to a change in the whole ecosystem's pliability (Matta 2018). Hence, preserving as much diversification in freshwater environments can increase its robustness and act as a link or foundation to connect the ecological system.

The freshwater ecosystem may endure modifications by employing the pliability notion as a bridge despite losing its essential condition and functionality. Linking the existing information element in the freshwater environment by stakeholder involvement that develops values in various ecological processes for preserving the dynamism and discontinuities of the freshwater systems is a crucial sustainable approach in this respect (Matta 2018). Contrary to traditional "grey" technology, which the underdeveloped world is currently attempting to apply, particularly in urban regions, environmental solutions can function as efficient and eco-friendly technologies to boost the growth of freshwater services. These remedies are distinguished by various low-cost, more adaptable adaptation alternatives (Matta and Uniyal 2017). Besides this, PWLM is an essential tool for the long-term sustainability of watershed management in urban areas. But it can be challenging due to less interest among the people.

Approaches derived from nature quickly take the lead in resolving a wide range of intricate issues about water security. Environmental approaches for freshwater were widely emphasized in the report of UN Water Agency, 2018. Dams, irrigation channels, and facilities for water purification are not the primary tools at our fingertips for managing water (Matta et al. 2018a). We may take inspiration from nature and use it to our advantage rather than waiting for it to fix every issue.

15.5.3 Importance of Technological Interventions

Different types of sustainable or green technologies play a crucial role in achieving water security. Various measures are adopted with the use of technologies in this regard.

a) **Water conservation technologies (WCTs)**: Since traditional irrigation systems in developing nations like India only achieve 30–40% efficiency, WCTs are more crucial than ever. WCT is expected to be used more often and will significantly increase irrigation effectiveness and monsoon farming production. If properly implemented and managed, water management and water-saving technology bring advantages, including increased water production and agricultural income. It also increases differential profits following legal frameworks while reducing water consumption, such as using well-dripping watering or orchard crops (Matta et al. 2020e).
b) **Concept of 3 R's**: It is essential to think about water availability and quality from an integrated perspective when addressing the interconnected problem of water shortage in the context of accelerating worldwide transformation and regular severe climatic events. In the absence of fewer water resource areas in arid regions in developing countries like India, this problem becomes very acute. For achieving water security, using these 3 R's, that is, Reuse, Recycle, and Reclamation, is essential. To accomplish this, all water consumers and partners must understand the utilization of recycled water, its public acceptability, and its benefits and drawbacks (Matta et al. 2020e). Several cutting-edge methods are also being used to reduce water consumption, including cognitive assessment to reduce non-revenue water wastage, smart watering employing agriculture technology in the field, rainwater harvesting, financing urban water usage, water tariffs, etc. (Matta et al. 2020e).
c) **Models and future predictions**: Several hydrological simulation models are effective instruments for assessing the effects of globally dispersed activities that quickly modify the factors associated with water supplies. Numerous hydrological models with technical characteristics are provided for modeling water resource parameters such as outflow, groundwater flow, surface water-groundwater interplay, quality of water, and pollutant fate and transportation (Matta et al. 2022a). On the other hand, developing these nations' capacities is crucial for enabling them to better analyze the state of their water assets in the future and develop policies for their management that are appropriate for both acclimation and reduction.

15.5.4 Prospects of Water Security for SDG 6

The first and crucial step in water security is "how it can be done." The role of governance is critical to defining IWRM strategies (Figure 15.5). IWRM seeks to increase sustainability and safety by increasing water usage efficiency and protection. The UN Water Agency describes three efficiency levels. Nonstructural methods can increase first water efficiencies, such as innovation and improvement in technologies and structural measures, that is, water charges and community awareness. The second degree of efficiency deals with distributing and redistributing water sources to particular, better applications and equal distribution among all partners. Water exchange across basins is a component of the third degree of efficiency (Matta et al. 2022a). It is expensive to trade water in its solid form since it is such a heavy object.

Consequently, the idea of virtual water is introduced. Per the global trading theory, countries should export goods with a relative or comparative manufacturing advantage while importing goods with a comparative disadvantage. This is the economic justification for virtual water trading (Matta et al. 2020a).

According to this concept, the fourth efficiency level is for resilient economies and societies. SDG 6 is a key precursor in achieving several other SDGs by sustainable water management soon. To increase efficiency or achieve SDG 6 goals, a panel was established by the UN Secretary-General and World Bank President for effective implementation (UN- United Nation, 2018; Matta et al. 2022d) (Figure 15.4).

In a developing socioeconomic, political, and ecological context, managing water resources in various locations must fulfill all demands. Therefore, solutions must include the present water availability disparities and the requirement for equity. Global cooperation, knowledge sharing, and water for ecosystem management are also essential for efficient management (Matta et al. 2022d).

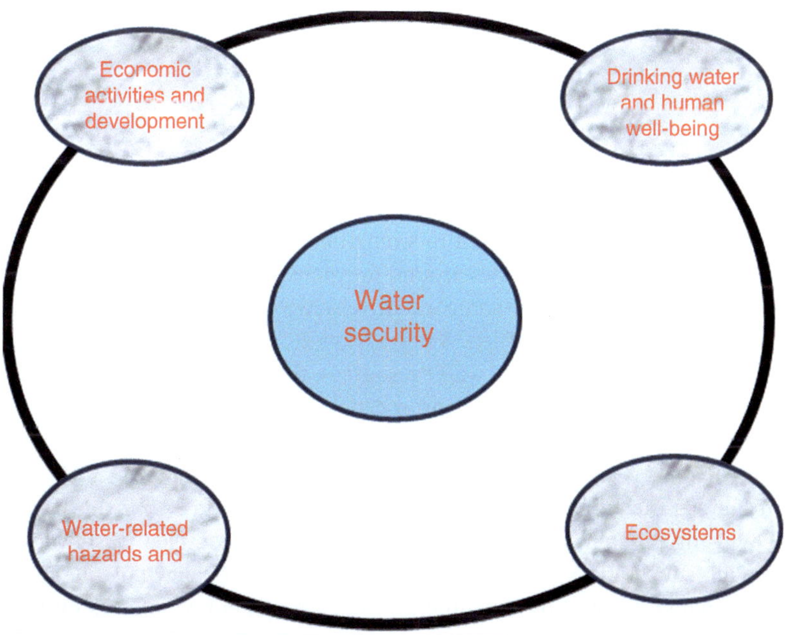

Figure 15.4 Water security dimension (UN water). *Source:* Adapted from Van-Rijswick et al. (2014).

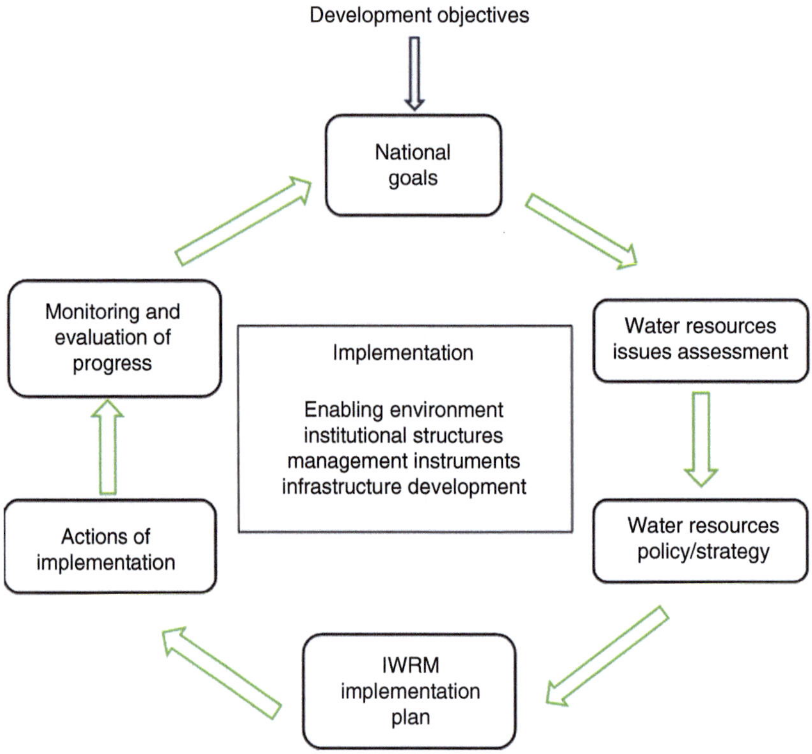

Figure 15.5 IWRM stages of planning and implementation. *Source:* Adapted from Van-Rijswick et al. (2014).

15.6 Future Prospects of Water Security for Long-Term Sustainability

Considering the current situation and difficulties, what are the future chances for water security? Can we attain sustainable water security in the future through sustainable socio-economic growth? What steps should be taken to strengthen water security as the world develops and modernizes? Different countries around the globe are focusing on the policy perspective to tackle this problem and also some initiatives have been taken by the nations (Matta 2018); Since the launch of the Africa Water Vision 2025 that was drawn by the African group of countries in the use of the water resources, the implementation projects of the vision seeks to reduce poverty levels, support social and economic development and the conservation of the natural resource (Matta and Uniyal 2017).

A systematic approach has been widely used by developing countries such as IWRM for managing water resources, including Asian and African countries (Matta et al. 2018a), and this can be done by using different green technologies, innovation, recycled wastewater, and awareness through education. The Chinese government issued a policy document of high priority in 2011, which focused on speeding up the development of reforms on the conservation of water, and also a guide on water pricing was issued by the government agencies collaboration in 2013 (Matta et al. 2020e).

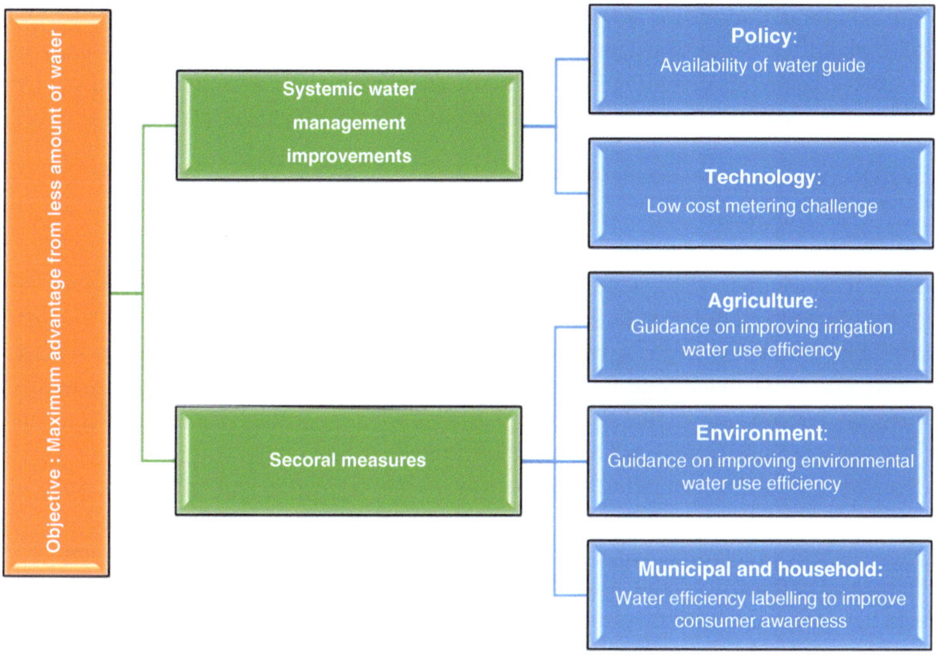

Figure 15.6 High-level panel on water (HLPW) approach for efficiency enhancement. *Source:* Adapted by Matta et al. (2020a).

Sometimes harmony in the policy is a big challenge for sustainability. Sometimes one agency, such as an agency on forestry and environmental affairs, makes policy for ecological protection, but at the same time, the agricultural agency works on fostering farming practices in fragile areas such as mountainous regions and river banks. This can obstruct the hydrology of that area and promote soil erosion events (Matta et al. 2020f). Suitable measures, organizations, and operational tools must be devised to ensure sustainable water resource management, particularly in times of scarcity.

Other smarter water conservation techniques have been used, for example, the technology and system of smart water grid (Figure 15.6). The three basic layers that build up this system are physical observation, control functions and collection handling, which includes displays, and fusion analysis. Water resources, water usage, and waste treatment plants are the three mechanisms that work here. Digital pumps, smart water meters, and smart valves have been fitted throughout the water extraction process. Data on the water's speed, pressure, and quality are collected using such intelligent devices before it reaches the use system. A pollution sensor tracks water quality while a flood sensor gathers flood data. Then, the water is divided into household and industrial water and continues to reach the end device via water and industry water channels. The equilibrium between the supply and demand for water can be controlled in real time by using intelligent sensors to monitor the water usage system's water use. After entering the sludge treatment infrastructure, home and industry wastewater can be partially reused (Matta et al. 2018b; Panday et al. 2021; Das et al. 2022; Kumar et al. 2023a, b). Only two countries, Australia and Singapore, have the smart water management system and America and South Korea are working on it (Panday et al. 2021; Panday and Kumar 2022, 2023).

Drinking water, surface water, groundwater, and wastewater must all be considered in a well-thought-out water management plan. A city's site-specific circumstances must be considered before choosing the best methods. The design should respond to the site circumstances as part of the larger planning concept (Asok Raja et al. 2023; Panday et al. 2023).

15.7 Conclusion

Several most common water issues worldwide include floods, droughts, poor accessibility to drinkable water, and hygiene. So, sustainable solutions are needed in various contexts by considering local values, synergizing with other sources, etc. We must act immediately to manage, service, and safeguard our finite freshwater resources as efficiently as possible. We also need to strengthen laws and public understanding of this issue. Along with giving remedial solutions to decrease or control the effect of the pollution, this also entails implementing water pollution prevention laws and safeguarding the water. Good water regulation and a unique perspective on managing water resources are likely to emerge from an integrated strategy. The new approach for attaining water security will provide enough water for residential, economic, and commercial entities, drinkable water to fulfill basic requirements at a decent cost with good sanitation, and wastewater treatment and collection to reduce water pollution.

We can conclude that several intricately linked aspects to water security exist and that everyone entails intricate interconnections between human culture and the environment. Finally, attaining an equilibrium among social, environmental, and economic factors is necessary for a society where everyone can access clean water. It necessitates the proper blending of soft and harsh actions, for example, in transporting and storing water and in defense of the source itself.

References

Asian Water Development Outlook (AWDO) (2016). Strengthening water security in Asia and the Pacific. https://www.adb.org/publications/asian-water-development-outlook-2016 (accessed 27 December 2022).

Asok Raja, S.K., Panday, D.P., and Kumar, M. (2023). Decoding the enigma of 100-year record-breaking rainfall over Tamil Nadu using wavelet analysis. *Groundwater for Sustainable Development* 101007. https://doi.org/10.1016/j.gsd.2023.101007.

Biswas, A.K. and Tortajada, C. (2019). Water quality management: a globally neglected issue. *International Journal of Water Resources Development* 35 (6): 913–916. https://doi.org/10.1080/07900627.2019.1670506.

CGWB (Central Ground Water Board) (2015). *Ground Water Year Book: India 2014–2015; India Ministry of Water Resources, River Development and Ganga Rejuvenation*. New Delhi: CGWB (Central Ground Water Board).

Cohen, A. and Ray, I. (2018). The global risks of increasing reliance on bottled water. *Nature Sustainability* 1: 327–329. https://doi.org/10.1038/s41893-018-0098-9.

Das, A., Kumar, M., Jha, P.K. et al. (2022). Isotopic and hydrogeochemical tracking of dissolved nutrient dynamics in the Brahmaputra River System: a source delineation perspective. *Chemosphere* 307: 135757. https://doi.org/10.1016/j.chemosphere.2022.135757.

Demographia (2016). World urban areas: 12th annual edition. http://demographia.com/db-worldua.pdf (accessed 15 January, 2023).
Di Baldassarre, G., Sivapalan, M., Rusca, M. et al. (2019). Sociohydrology: scientific challenges in addressing the sustainable development goals. *Water Resources Research* 55 (8): 6327–6355. https://doi.org/10.1029/2018WR023901.
Falkenmark, M. and Molden, D. (2008). Wake up to realities of river basin closure. *Water Resources Development* 24: 201–215. https://doi.org/10.1080/07900620701723570.
GBD 2015 Risk Factors Collaborators (2016). Global, regional, and national comparative risk assessment of 79 behavioural, environmental and occupational, and metabolic risks or clusters of risks, 1990–2015: a systematic analysis for the Global Burden of Disease Study. *Lancet* 388 (10053): 1659–1724. https://doi.org/10.1016/S0140-6736(16)31679-8.
Gjyli, L., Mucaj, O., Matta, G., and Krasniqi, M. (2018). Inorganic toxicity: barium. In: *Inorganic Toxicity: Environment and Human Health* (ed. G. Matta and L. Gjyli). Germany: Lambert Publication House.
Gjyli, L., Vlachogianni, T., Kolitari, J. et al. (2020). Marine litter on the Albanian coastline: baseline information for improved management. *Ocean and Coastal Management* 2020 187: 105108. https://doi.org/10.1016/j.ocecoaman.2020.105108.
Grant, S.B., Saphores, J.D., Feldman, D.L. et al. (2012). Taking the "waste" out of "wastewater" for human water security and ecosystem sustainability. *Science* 337 (6095): 681–686.
Gunda, T., Hess, D., Hornberger, G.M., and Worland, S. (2019). Water security in practice: the quantity-quality-society nexus. *Water Security* 6: 100022.
Hairan, M.H., Jamil, N.R., Looi, L.J., and Azmai, M.N. (2021). The assessment of environmental flow status in Southeast Asian Rivers: a review. *Journal of Cleaner Production* 295: 126411.
HPEC (High Powered Expert Committee for Estimating the Investment Requirements for Urban Infrastructure Services) (2011). *Report on Indian Urban Infrastructure and Services*. New Delhi: India Ministry of Urban Development.
Huitema, D., Mostert, E., Egas, W. et al. (2009). Adaptive water governance: assessing the institutional prescriptions of adaptive (co-) management from a governance perspective and defining a research agenda. *Ecology and Society* 14 (1). Article 26. http://www.ecologyandsociety.org/vol14/iss1/art26/.
Jiang, Y. (2015). China's water security: current status, emerging challenges and future prospects. *Environmental Science & Policy* 54: 106–125. https://doi.org/10.1016/j.gloenvcha.2011.10.011.
Kumar, A., Matta, G., and Bhatnagar, S. (2021). A coherent approach of water quality indices and multivariate statistical models to estimate the water quality and pollution source apportionment of River Ganga System in Himalayan region, Uttarakhand, India. *Environmental Science and Pollution Research* https://doi.org/10.1007/s11356-021-13711-1.
Kumar, M., Panday, D.P., Bhagat, C. et al. (2023a). Demystifying the decadal shift in the extent of groundwater in the coastal aquifers of Gujarat, India: a case of reduced extent but increased magnitude of seawater intrusion. *Science of the Total Environment* 165451. https://doi.org/10.1016/j.scitotenv.2023.165451.
Kumar, M., Mazumder, P., Silori, R. et al. (2023b). Prevalence of PPCPs, microplastics and co-infecting microbes in the post-COVID-19 era and its implications on antimicrobial resistance and potential endocrine disruptive effects. *Science of the Total Environment* 166419. https://doi.org/10.1016/j.scitotenv.2023.166419.

Kumari, M. and Singh, J. (2016). Water conservation: strategies and solutions. *International Journal of Advanced Research and Review.* 1 (4): 75–79.

Marcal, J., Antizar-Ladislao, B., and Hofman, J. (2021). Addressing Water Security: An Overview. *Sustainability* 13 (24): 13702. https://doi.org/10.3390/su132413702.

Matta, G. (2018). Applicability of WQI and scientific communication for conservation of River Ganga system in India. In: *Frontiers in Water-Energy-Nexus—Nature-Based Solutions, Advanced Technologies and Best Practices for Environmental Sustainability: Proceedings of the 2nd Water Energy NEXUS Conference*, 2020 265–267. Salerno, Italy: Springer International Publishing. https://doi.org/10.1007/978-3-030-13068-8.

Matta, G. (2020). Science communication as a preventative tool in the COVID19 pandemic. *Humanities and Social Sciences Communications* 7: 159. https://doi.org/10.1057/s41599-020-00645-1.

Matta, G. and Uniyal, D.P. (2017). Assessment of species diversity and impact of pollution on limnological conditions of River Ganga. *International Journal of Water* 11 (2): 87–102. https://doi.org/10.1504/IJW.2017.083759.

Matta, G., Kumar, A., Kumar, A. et al. (2018a). Applicability of Heavy Metal Indexing on Ganga River System assessing heavy metals toxicity and ecological impact on river water quality. *INAE Letters: An Official Journal of the Indian National Academy of Engineering* 3: 123–129. https://doi.org/10.1007/s41403-018-0041-4.

Matta, G., Naik, P., Kumar, A. et al. (2018b). Comparative study on seasonal variation in hydro-chemical parameters of Ganga River water using comprehensive pollution index (CPI) at Rishikesh (Uttarakhand) India. *Desalination and Water Treatment* 118: 87–95. https://doi.org/10.5004/dwt.2018.22487.

Matta, G., Nayak, A., Kumar, A. et al. (2020a). Water quality assessment using NSFWQI, OIP and multivariate techniques of Ganga River system, Uttarakhand, India. *Applied Water Science* 10: 206. https://doi.org/10.1007/s13201-020-01288-y.

Matta, G., Kumar, A., Nayak, A. et al. (2020b). Determination of water quality of Ganga River System in Himalayan region, referencing indexing techniques. *Arabian Journal of Geosciences* 13: 1027. https://doi.org/10.1007/s12517-020-05999-z.

Matta, G., Rajput, A., Rajput, A. et al. (2020c). Assessment of water quality of Choyyia Nadi (River) Catchment area in Bijnor District, Uttar Pradesh, India. *Ecology Environment and Conservation* 26 (November Suppl. Issue): S402–S412.

Matta, G., Gautam, R., Kumar, A., and Kumar, P. (2020d). Suitability assessment of Groundwater quality in rural areas of Haridwar District Uttarakhand. *Pollution Research* 39 (1): 91–96.

Matta, G., Kumar, A., Tiwari, A.K. et al. (2020e). HPI appraisal of concentrations of heavy metals in dynamic and static flow of Ganga River System. *Environment, Development and Sustainability* 22: 33–46. https://doi.org/10.1007/s10668-018-01182-3.

Matta G, Lu XX, Kumar A. Hydrological assessment of River Henwal using water quality indices with reference to planktonic composition In Advances in Water Pollution Monitoring and Control: Select Proceedings from HSFEA 2018 2020f (pp. 163–179). Springer Singapore. https://doi.org/10.1007/978-981-32-9956-6_18

Matta, G., Kumar, A., Kumar, P. et al. (2021). Evaluation of ground water quality by use of water quality index in the vicinity of the Rajaji National Park Haridwar, Uttarakhand, India. *Geostatistics and Geospatial Technologies for Groundwater Resources in India* 343–356. https://doi.org/10.1007/978-3-030-62397-5_17.

Matta, G., Pamposh Kumar, D., Uniyal, P., and Joshi, D.U. (2022a). Communicating water, sanitation, and hygiene under sustainable development goals 3, 4, and 6 as the panacea for epidemics and pandemics referencing the succession of COVID-19 surges. ACS ES&T. *Water* 2 (5): 667–689. https://doi.org/10.1021/acsestwater.1c00366.

Matta, G., Kumar, A., Nayak, A. et al. (2022b). Assessing heavy metal index referencing health risk in Ganga River System. *International Journal of River Basin Management* 1–11. https://doi.org/10.1080/15715124.2022.2098756.

Matta, G., Kumar, A., Nayak, A. et al. (2022c). Appraisal of spatial–temporal variation and pollution source estimation of Ganga River system through pollution indices and environmetrics in Upper Ganga basin. *Applied Water Science* 12: 33. https://doi.org/10.1007/s13201-021-01552-9.

Matta, G., Kumar, A., Nayak, A. et al. (2022d). Pollution complexity quantification using NPI and HPI of River Ganga system in Himalayan Region. *Proceedings of the Indian National Science Academy* 88 (4): 651–663. https://doi.org/10.1007/s43538-022-00111-3.

NASAC (Network of African Science Academies) (2014). *The Grand Challenge of Water Security in Africa: Recommendations to Policy Makers*. Network of African Science Academies.

Nayak, A., Matta, G., and Uniyal, D.P. (2022). Hydrochemical characterization of groundwater quality using chemometric analysis and water quality indices in the foothills of Himalayas. *Environment, Development and Sustainability* https://doi.org/10.1007/s10668.

Panday, D.P. and Kumar, M. (2022). Climate indices and hydrological extremes: deciphering the best-fit model. *Environmental Research* 215: 114301. https://doi.org/10.1016/j.envres.2022.114301.

Panday, D.P. and Kumar, M. (2023). Application of remote sensing techniques to deal with scale aspects of GRACE data to quantify groundwater levels. *MethodsX* 102108. https://doi.org/10.1016/j.mex.2023.102108.

Panday, D.P., Khosa, R., Maheswaran, R. et al. (2021). Game-theoretic-based modelling of Krishna waters dispute: equilibrium solutions by Metagame Analysis. *The European Physical Journal B* 94: 101. https://doi.org/10.1140/epjb/s10051-021-00107-w.

Panday, D.P., Khosa, R., Maheswaran, R. et al. (2021). Game theoretic-based modelling of Krishna waters dispute: equilibrium solutions by hypergame analysis. *Eur. Phys. J. B* 94: 131. https://doi.org/10.1140/epjb/s10051-021-00135-6.

Panday, D.P., Singh, A., Gaur, S. et al. (2023). Probable maximum precipitation analysis of high rainfall regions in India. *Groundwater for Sustainable Development* 100893. https://doi.org/10.1016/j.gsd.2022.100893.

Perveen, S., Sen, R., and Ghosh, M. (2012). *Water Risks for Indian Industries: A Preliminary Study of 27 Industrial Sectors*. Delhi: Federation of Indian Chambers of Commerce and Industry (FICCI) and Columbia University Water Center (CWC).

Rawat, K. and Matta, G. (2021). Ozone: risk assessment, environmental, and health hazard. In: *Hazardous Gases, Risk Assessment on the Environment and Human Health*, 1e (ed. J. Singh, R.D. Kaushik, and M. Chawla), 301–312. Elsevier Publication House. ISBN: 9780323898577.

Roser, M. and Rodés-Guirao, L. (2013). Future population growth. In: *Our World in Data*, vol. 9. https://ourworldindata.org/future-population-growth#licence.

Saraswat, C. and Kumar, P. (2016). Climate justice in lieu of climate change: a sustainable approach to respond to the climate change injustice and an awakening of the environmental movement. *Energy, Ecology and Environment* 1: 67–74.

UN (1992). *Earth Summit Agenda 21. The United Nations programme of action from Rio.* New York: United Nations Department of Public Information.

UN – United Nations (2018). *Sustainable Development Goal 6 Synthesis Report 2018 on Water and Sanitation*. New York: United Nations Publishers.

UN. (1982). *A World Charter for Nature*. New York: United Nations.

UNESCO (2019). International centre for water security and sustainable management. *Water Security and the Sustainable Development Goals* 210 p.

UNESCO, UN-Water, 2020: United Nations World Water Development Report 2020: Water and Climate Change, Paris, UNESCO. https://unesdoc.unesco.org/ark:/48223/pf0000372985.locale=en (accessed 13 January 2023).

United Nations (2018). *Sustainable Development Goal 6: Synthesis Report 2018 on Water and Sanitation*, 10017. New York: United Nations. https://sdgs.un.org/sites/default/files/documents/19901SDG6_SR2018_web_3.pdf (accessed 10 January 2023).

United Nations University (UNU) (2013). *UN-Water Analytical Brief: Water Security and the Global Water Agenda*, 47. Hamilton, ON. https://www.unwater.org/publications/water-security-and-global-water-agenda.

Van-Rijswick, M.; Edelenbos, J.; Hellegers, P.; Kok, M.; Kuks, S. Ten building blocks for sustainable water governance: An integrated method to assess the governance of water. *Water International* 2014, 39, 725–742. https://doi.org/10.1080/02508060.2014.951828

Wiek, A. and Larson, K.L. (2012). Water, people, and sustainability—A systems framework for analyzing and assessing water governance regimes. *Water Resources Management* 26: 3153–3171.

16

Enabling Solutions to Water Security

Ganesh Datt Bhatt[1] *and Deepali Rana*[2]

[1] *Department of Agro-Forestry, College of Agriculture Sciences, An ICAR Accredited College, Teerthanker Mahaveer University, Moradabad, Uttar Pradesh, India*
[2] *Department of Zoology, Dolphin (PG) Institute of Biomedical & Natural Sciences, Dehradun, Uttarakhand, India*

16.1 Introduction

Due to the complex and quickly shifting geography of water supply, demand, and use, it is challenging to predict the future sufficiency of freshwater resources on a worldwide scale. The greatest strain on the world's water supplies is due to the demand for water expanding so quickly. Water scarcity and vulnerability to water shortages are issues in many Middle Eastern, Asian, African, and other nations. When demand and supply are taken into account in the context of potential future socioeconomic and ecological changes, water scarcity becomes more acute. Population growth is the socioeconomic element that has the largest potential influence, and climate change is the natural factor that is most concerning (Salim 2006).

Water issues are primarily the result of societalinstitutions failing to manage the resource and meet the needs of current people, the economy, the environment, and future generations, even if climate change poses formidable challenges to global water systems. Water science is dominated by a single-minded concentration on climate change and hydrological responses at the expense of research into how to protect water systems in the face of unavoidable environmental and societal change. A more integrated and sustainable strategy to managing water resources is necessary for managing climate risk and uncertainty. Greater proactive adaptation and increased resilience will be made possible through improved knowledge, policy, regulation, allocation, and cooperation. A vital and extremely delicate political problem is water.

The overuse of groundwater resources, the degradation of water quality brought on by pollution, and the shortcomings of the current infrastructure all call for quick intervention. Four major pollution sources, including home wastewater, solid waste leachate, industrial effluents, and runoff from agricultural activities, are putting a strain on water resources. As a result, it is made clear how important markets, urban planning, insurance, policy,

Water Scarcity Management: Enabling Technologies, First Edition. Edited by Kanchan D. Bahukhandi, Manish Kumar, Durga P. Panday, and Tushara G.G. Chaminda.

technology, governance, cultural attitudes and values, institutions, legal frameworks, and decision-making techniques are in reducing water stress. Governmental and nongovernmental organizations, the commercial sector, and academics have all taken an interest in the term "water security" and its underlying ideas. Despite the apparent increase in usage, there is a lack of a thorough grasp of how water security is understood and used in many global contexts.

Water security is widely diffused across geographic scales and regions, conceptualized broadly, and subject to changing approaches to indicator development. Future governance narratives emphasize the development of participatory procedures, the use of both quantitative and qualitative methods, and a combination of hard- and soft-path strategies to ensure water security. Water security has strong connections to issues of power, injustice, and poverty in addition to the physical shortage of water.

Therefore, there is a significant distinction between these two ideas. One must take into account both true, physical water scarcity and the broad lack of access to water brought on by economic, political, social, and environmental reasons. In addition to the difficulties in managing resource scarcity, there are hydropolitical and international issues to take into account. Due to its intricate feedback and interactions with other elements of the climate system, water cycling has a significant influence on climate variability (Salim 2006).

16.2 Water Security Definitions

Since the 1990s, the idea of water security has been used to express concern about problems with supply consistency, quality, quantity, safe and equitable access, and environmental provisioning. From the World Wildlife Fund and the World Economic Forum to the United Nations (UN), the idea has been used more and more in policy circles (United Nations Environment Program (UNEP) 2009; World Wildlife Fund (WWF) 2009; World Economic Forum 2011; United Nations Educational, Scientific and Cultural Organization (UNESCO) 2013; UN-Water 2013).

The term "water security" has historically been defined by several authors, and meanings vary depending on the geographical areas they are researching and the goals of their studies. Due to the multiplicity of definitions, actors (actors being scholars, planners, managers, stakeholders, etc.) are employing the idea in a variety of ways and as potential benchmarks for lowering insecurity as well as normative and instrumental framing aspects. The notion is typically modified by users to meet their own unique situations (Gerlak et al. 2018). According to Michel Jarraud, a former UN-Water chair and WMO secretary general, "definitions of security have evolved over the past few decades beyond a narrow focus on military risks and conflicts."Security today refers to human security and the accomplishment of that security through growth. Water functions as a vital link between these other concerns and fits within this broader notion of security, which includes political, health, economic, personal, food, energy, and other considerations.

Water security is defined as "the capacity of a population to safeguard sustainable access to adequate quantities and acceptable quality of water for sustaining livelihoods, human wellbeing, and socioeconomic development, for ensuring protection against waterborne pollution and water-related disasters, and for preserving ecosystems in a climate that supports human life" by UN-Water, the United Nations' interagency coordination mechanism for all water-related issues (UN-Water 2013). Since a substantial share of water resources is transboundary, "fair and equitable apportionment" has not yet been established for the distribution of water among the many nations.

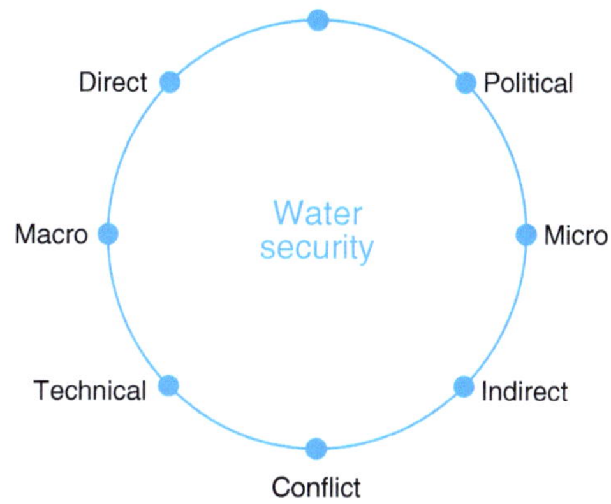

Figure 16.1 Dimensions of water security and sustainability. *Source:* Adapted from Varis et al. (2017).

Due to population expansion, rising living standards, climate change, and the requirement to expand irrigated agriculture and industrial operations, there is a possible decrease in the availability of fresh water. As a result, both water demand and the resulting gap between water supply and water demand are widening. Because it involves the most valuable and limited resource on Earth, freshwater sustainability stands out as the most significant sustainable development challenge (Figure 16.1). Risks to the economy, society, and environment rise when a community's water supplies are in short supply or in danger. Therefore, an integrated proactive management strategy is required to balance the conflicting demands for this finite resource. Therefore, an integrated proactive management strategy is required to balance the conflicting demands for this finite resource.

Water sustainability is a cooperative, neighborhood-based program that needs the active involvement of all residents. It aims to provide innovative, well-coordinated water management strategies that focus on security and value addition for all stakeholders. A community's decision on how to use its resources to accomplish its economic, social, and environmental objectives entails the development of a united vision, technology, leadership, and defined, quantifiable goals. The private sector ought to take the initiative in bringing about this change in the neighborhood. Thankfully, there are cutting-edge approaches that may assist communities in water conservation and enhance their current infrastructure in a practical manner, or alternatively, deploy new infrastructure where none is there and adopt conservation and wise management from the beginning.

16.3 The Relationships Between Water, the Local Environment, and Human Security

According to the paradigm of "human security," rather than meeting the demands of nations, security is determined by the needs of individuals. It takes into account how international security is evolving as non-state actors pose new dangers and the global economic system shifts toward multipolarity. Human security now places more emphasis on the connections between society and the environment. The establishment and maintenance of human security in connection to the environment depend heavily on water, a necessary but relatively scarce resource. In fact, according to Wolf (Wolf 1999),"the human security aspect of water scarcity seems likely to cause security threats both nationally and internationally." As was previously said, there are two methods to describe water scarcity: physically (lack of rainfall; deserts; unequal geographical distribution of water) and economically (lack of funds to purchase water; build adequate infrastructure for distribution; or purify water).

The impact of water on agriculture is one example of how the human security aspect of water scarcity manifests. It is a straightforward reciprocal relationship: when water availability declines, food production also does. Poor irrigation techniques, the most popular of which is the use of salt water and improperly treated wastewater, are caused by a lack of fresh water. In the Middle East and North Africa, the two regions with the lowest percentages of arable land and the fewest water resources in the world, soil salinity damages approximately a million hectares of arable land every year (Doble 2011). Population expansion and the reduction in food production brought on by water scarcity are strongly related.

As a result, there are worsening living circumstances and additional environmental issues, including rising water supply depletion, deforestation, and desertification. These environmental changes contribute significantly to global hunger that impacts about one billion people by causing famines, poverty, and malnutrition. The decrease in food production, arable land, and water may result in an increased reliance on foreign food aid for poor countries unable to adapt their irrigation and agriculture policies. For necessities like food and water, a stable state cannot rely on another. Food insecurity, which results from water scarcity, therefore establishes a link between state instability and insecurity as well as human security.

16.4 Water Security Evaluation

Understanding the current status and identifying problems and areas that require attention are made possible by assessing water security. Setting priorities, addressing problems, informing planning, and implementing and overseeing water security measures are all crucial steps. Therefore, understanding how to assess water security is essential. General frameworks, techniques, and instruments have been developed over the years by academic, private, governmental, and non-governmental groups with the objective of evaluating and studying water security on various scales. Water security is a broad and relatively new concept, thus numerous tools that have been created over the years to evaluate ideas like water resilience, water insecurity, water sustainability, and water governance give an evaluation for the various components covered by the concept.

16.4.1 Water Security Assessment

The Water Security Assessment understands the current situation and identifies issues and areas that require attention. This is an essential step in prioritizing and addressing problems, informing planning, and implementing and monitoring water safety measures. Therefore, it is important to know how to assess water safety.

16.4.1.1 Water Security Indicators

Water security indicators play an important role in describing the complexity of a system, dissemination of information, and translating important aspects of a complex system into an accessible format that can be understood and monitored by different stakeholders. It is a useful tool for identifying key issues and needs attention for governments and decision-makers to develop action plans and implement informed policy interventions to improve water security. The ability to compare results is an important issue in water security assessments, as it provides information and potentially allows experiences to be shared between assessed regions. Using the indicator not only provides a way to check the development of the region in terms of water security but also offers the possibility to compare the assessed cities, countries and regions. Higher or improved scores also highlight the benefits of a particular public policy and help promote its management, thus providing incentives for policymakers and politicians to improve their water management strategies and learn from their best practices.

Water security indicators, which are often used as the basis for water security assessments, are often collated in indicator systems to convey different aspects of a concept. At the national level, the Asian Development Bank's Water Safety Index (Asian Development Bank (ADB) 2016) uses an index system divided into five dimensions (household, economy, city, environment, and water hazards) to calculate the final water safety index. Among other things, the National Blueprint Framework also adopts a set of indicators to measure progress towards water-related development goals. At the city level, particularly the Urban Blueprint (Van Leeuwen et al. 2012), Sustainable Urban Water Index (Arcadis 2016), and Urban Water Resilience Index (ARUP 2019),a range of indicators are also used that allow comparison and monitoring of progress across different countries. I'm here. Many other studies in the literature employ index systems as part of proposed water safety assessment frameworks. The use of indicators and index systems is widespread, but not exclusive.

16.5 Solutions for Addressing Water Shortages

Although some human activities have made the water problem worse, technology has also been developed by humans to more efficiently get or save freshwater. Dams and reservoirs, rainwater harvesting, aqueducts, desalination, water reuse, and water conservation are all ways to deal with water shortages.

16.5.1 Dams and Reservoirs

Water can be gathered during wet periods and stored for use during dry periods in reservoirs (artificial lakes) that develop behind river dams (Figure 16.2). They can also be utilized to supply water to cities. Hydroelectricity, flood control, and recreation are additional advantages of dams and reservoirs. Evaporative water loss in arid areas and downstream river channel erosion are two downsides. Dams also slow the flow of water downstream, which may cause political unrest where rivers cross state or international borders. Another significant problem of dams is the harm they do to the ecosystem. For instance, dams alter the environment of a river by turning it into a lake and obstruct fish movement and

Figure 16.2 Nevada's Hoover Dam on the Colorado River. Lake Mead, the biggest reservoir in the country, is located behind the dam. *Source:* Photo by Dsimic.

spawning. Additionally, the temperature of the water downstream is impacted by the warming of the reservoir's surface water, which has an effect on aquatic invertebrates and fish that are acclimated to cooler water. Additionally, dams hold back sediments that would otherwise continue to flow downstream, fostering habitat and supplying nutrients.

16.5.2 Rainwater Harvesting

Figure 16.3 illustrates a sophisticated rainwater harvesting system (also known as a rainwater capture system) that has been proposed for federal buildings. However, smaller, more straightforward systems, sometimes referred to as rain barrels, can be used by individual homeowners.

A rainwater harvesting system is made up of two parts: (1) the collection system, which is made up of gutters and roof surfaces that collect rainwater and send it to (2) the input filter, a screen that captures big debris. (3) The first flush diverter clears the initial stream of rainfall of any particles not caught by the inlet filter. (4) The storage tank is made of green, food-grade polyester resin material that has been approved by the US Food and Drug Administration (FDA) and inhibits the growth of bacteria. (5) If the storage tank is full, the overflow is a drainage outlet that discharges extra water. (6) The control system keeps an eye on the filtration system and water level, while (7) the treatment system is in charge of filtering and disinfecting the water until it is fit for consumption. (9) The backflow preventer makes sure that water cannot flow backwards through the system in conditions of low water pressure (when the storage tank is low), while (8) the pump pushes water through

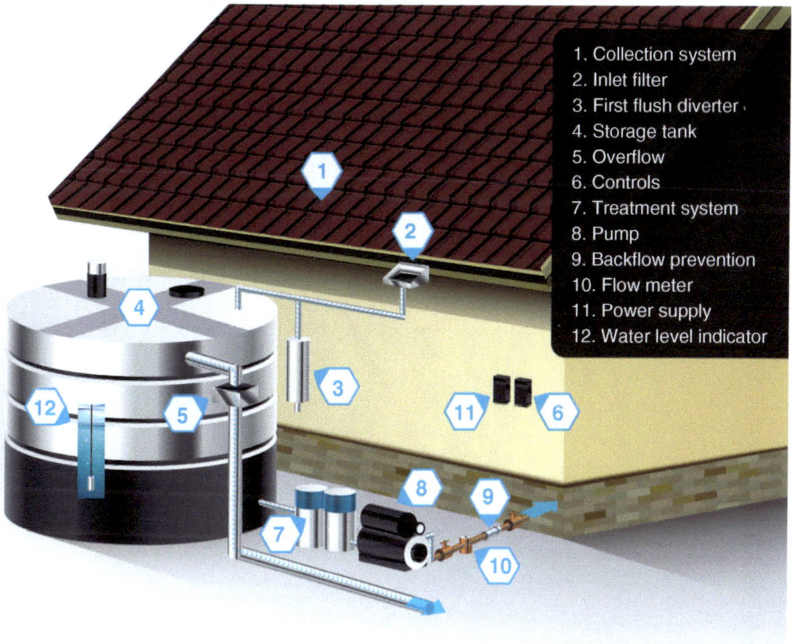

Figure 16.3 The components of a rainwater harvesting system. *Source:* Rainwater Harvesting Systems Technology Review/U.S. Department of Energy/public domain.

the system to where it will be used. (10) A flow meter that has a data logger measures the amount of water produced. (12) The water level indicator keeps track of the water level in the storage tank, and (11) the power supply can run on either conventional or solar energy.

16.5.3 Aqueducts

Water can be transported using aqueducts from a place of abundance to a place of need. Aqueducts can be divisive and politically challenging, particularly if there are significant water transfer distances. The water diversion has the potential to lead to drought in the region from where the water is drawn. For instance, after their river flow was diverted to the Los Angeles aqueduct, Owens Lake and Mono Lake in central California started to diminish (Figure 16.4). Owens Lake dried up without a water supply, becoming a significant source of particulate matter that contaminated the air during dust storms (see Air Pollution). Although Mono Lake has recovered more significantly as a result of legal action, Owens Lake is still almost entirely dry.

16.5.4 Desalination

Desalination, which includes taking dissolved salts and minerals out of seawater or saline groundwater, is one technique that can actually increase the amount of freshwater on Earth. The availability of saltwater in nearly endless quantities is a benefit of this method. Boiling, filtration, electrodialysis (using an electric current to extract the ions that make up

Figure 16.4 The Los Angeles Aqueduct diverts the flow of several eastern California rivers to the city of Los Angeles. *Source:* Los Angeles/Wikimedia Commons/CC BY-SA 3.0.

salts), and reverse osmosis are only a few methods for desalinating saltwater (Figure 16.5). The cost of producing water is substantially higher than the cost of freshwater from conventional sources because ofall these processes, which range from being moderately to extremely expensive and requiring significant energy input. A major environmental impact is also caused by the process' creation of highly salinized wastewater, which must be disposed of. The Middle East, where water is limited but energy from oil is abundant, is where desalination is most prevalent.

16.5.5 Water Reuse (Water Recycling)

Reusing water for appropriate uses, such as agriculture, municipal water supply, industrial activities, and environmental restoration, is referred to as water recycling (Figure 16.6). A single family could experience this, for instance, by adding plumbing that diverts water from the sink to flush the toilet. Large-scale water recycling is also a possibility. For instance, sewage system effluent is routinely treated to some amount, but it can be further treated to provide potable water (which is safe to drink) and then injected into depleted

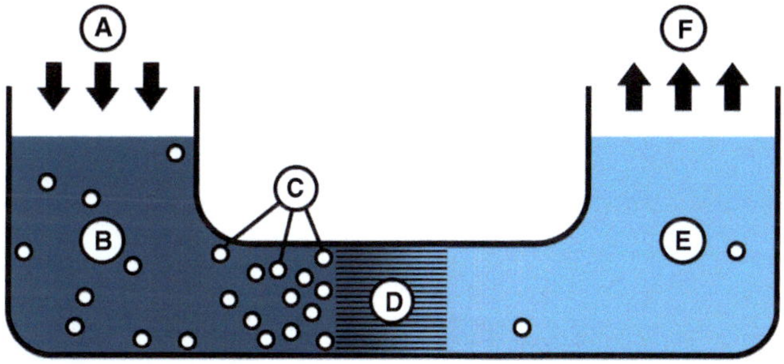

Figure 16.5 Reverse osmosis is one method of desalination. In this process, pressure (A) is applied to salt water (B), which contains ions (charged atoms or molecules) and other contaminants (C). The water is forced through a semipermeable membrane (D), through which the contaminants cannot pass. Clean freshwater that is safe to drink (E) is then distributed (F). *Source:* Colby Fisher/ Wikimedia/CC BY 3.0.

aquifers. This strategy lessens reliance on precipitation and subsequent infiltration to recharge aquifers by limiting saltwater intrusion into aquifers close to the coast. Following an educational campaign to explain the purification procedure and boost public confidence in the security of the treated wastewater, Orange County Water District in California used this method.

Drinking water treatment is performed on surface water, groundwater, and water from reservoirs. From there, it supports green infrastructure and meets home, commercial, and industrial needs. In order to make used water fit for its intended use, it is treated to that amount. Environmental restoration, groundwater recharge, agricultural reuse, surface water augmentation, portable reuse, irrigation reuse, home reuse, industrial reuse, and commercial reuse are examples of fit-for-purpose treatment and consumption. Wastewater treatment is used to process discharge from residential, commercial, and industrial structures.

16.5.6 Water Conservation

Utilizing water more effectively while using less of it is known as conservation. Water-saving technologies and lifestyle choices can both be used in the home to practice conservation. High-efficiency washing machines, low-flow showerheads, and toilets are a few examples of water-saving devices. Turning off the water while brushing your teeth is one water-saving habit, along with taking shorter showers and showers rather than baths, and replacing faulty faucets. Because a dishwasher only runs when it is full, it uses less water than hand-washing dishes. Similar to this, running fewer, bigger loads of laundry instead of more frequent, smaller loads saves water. Additionally, you can save water by picking meals with a low water footprint (like eggs) over ones with a high water footprint (like beef). There are several opportunities to conserve water when gardening. Consider planting only native, drought-tolerant plants if you live in a dry area because they require little irrigation. When you do need to irrigate your garden, make sure to do so in the morning when less water will evaporate. Drip systems help reduce evaporation by supplying only the necessary amount of water. These methods can be used on a big scale in agriculture as

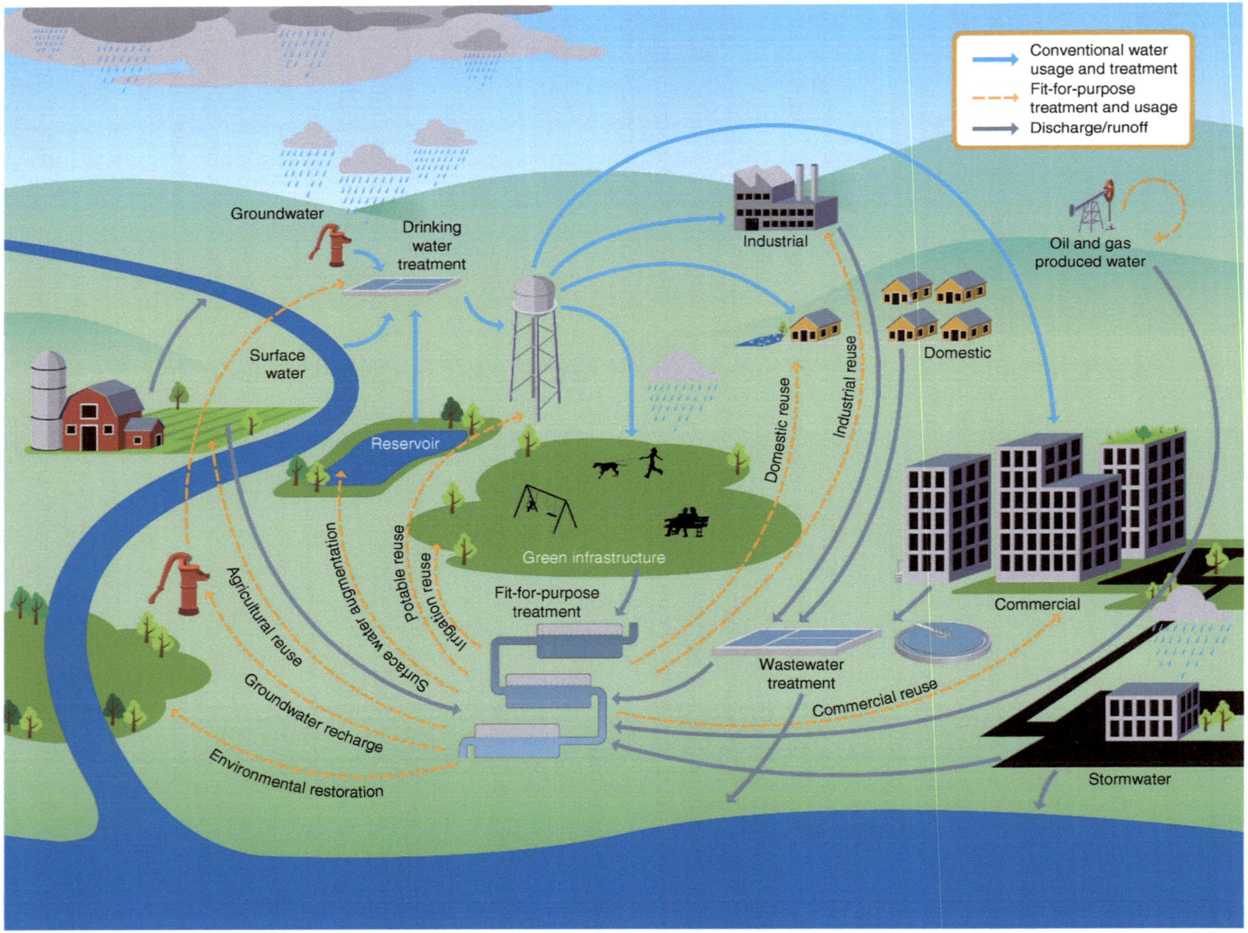

Figure 16.6 Flowchart of water reuse. *Source:* Basic Information about Water Reuse/U.S. Environmental Protection Agency/public domain.

well, which is crucial given how much more water is needed for agriculture than for urban purposes. Growing crops where natural rainfall can support them, using more effective irrigation techniques like drip irrigation, and using no-till farming, which lowers evaporative losses by covering the soil, are all examples of agricultural water conservation practices.

Bottled water is not a sustainable solution to the water crisis because it is not necessarily any safer than the US public water supply, costs on average about 700 times more than US tap water, and uses 200 billion plastic and glass bottles annually with a low recycling rate.

16.6 Conclusion

In conclusion, the critical importance of water security for both human well-being and environmental sustainability cannot be overstated. The challenges we face, from population growth to climate change, necessitate urgent action to ensure a sustainable and reliable water supply for present and future generations. Through an in-depth exploration of enabling solutions, it becomes evident that a multifaceted approach is essential to address the complex web of factors contributing to water insecurity. The amalgamation of technological innovation, robust policy frameworks, community engagement, and cross-sector collaborations offers a promising path forward. Investments in advanced water management technologies, such as smart water grids and precision irrigation systems, have the potential to optimize water usage and minimize losses. Concurrently, the development of adaptive policies that consider the interconnectedness of water resources, land use, and climate patterns is crucial. However, it is imperative to recognize that technology and policy alone cannot pave the way for sustainable water security. Engaging local communities, empowering stakeholders, and raising awareness about responsible water use are integral components of the solution. As witnessed in successful case studies, community-driven initiatives can have a profound impact on water conservation and management. The potential for change is immense. By harnessing the collective strength of governments, industries, academia, and civil society, we can usher in an era where water security is not just a distant aspiration but a tangible reality. The enabling solutions discussed in this research underscore the need for a holistic approach that leverages innovation, policy coherence, and community participation. In the journey toward water security, challenges will undoubtedly persist, but they are eclipsed by the opportunities to create a sustainable and equitable future. As we move forward, let us remain steadfast in our commitment to implementing these enabling solutions and translating them into transformative actions that secure the world's most precious resource for generations to come.

References

Arcadis (2016). *Sustainable Cities Water Index. Which Cities are Best Placed to Harness Water for Future Success?* 40. Amsterdam: Arcadis.

ARUP (2019). *City Water Resilience Assessment: Methodology*, 137. London: Arup Global Water.

Asian Development Bank (ADB) (2016). Asian water development outlook. In: *Strengthening Water Security in Asia and the Pacific*, vol. 2016, 66. Mandaluyong: Asian Development Bank. Available under a CC BY 3.0 IGO license.

Doble, J. (2011). *An Unsustainable Endeavor: Water Scarcity and Human Security in the Middle East*. The Stimeson Center. https://www.stimson.org/content/unsustainable-endeavor-water-scarcity-and-human-security-middle-east (accessed 2 February 2024).

Gerlak, A., House-Peters, L., Varady, R. et al. (2018). Water security: a review of place-based research. *Environmental Science and Policy* 82: 79–89. Retrieved from https://www.sciencedirect.com/science/article/pii/S1462901117310766.

Salim, N. (2006). Water cycle from theory to practice, derivation of a basic concept of groundwater recharge on Palestine adopting GIS techniques. PhD thesis,University of Geneva.

United Nations Educational, Scientific and Cultural Organization (UNESCO) (2013). *FreeFlow: Reaching Water Security Through Cooperation*. Paris: UNESCO and Tudor Rose.

United Nations Environment Program (UNEP) (2009). *Water Security and Ecosystem Services: The Critical Connection*. Nairobi: UNEP.

UN-Water (2013). *Water Security and the Global Water Agenda: A UN-Water Analytical Brief*. Hamilton: UN University.

Van Leeuwen, C.J., Frijns, J., Van Wezel, A., and Van De Ven, F.H.M. (2012). City blueprints: 24 indicators to assess the sustainability of the urban water cycle. *Water Resources Management* 26: 2177–2197.

Varis, O., Keskinen, M., and Kummu, M. (2017). Four dimensions of water security with a case of the indirect role of water in global food security. *Water Security* 1: 36–45. https://doi.org/10.1016/j.wasec.2017.06.002.

Wolf, A. (1999). *Water and Human Security*, 29–37. Universities Council on Water Resources. https://opensiuc.lib.siu.edu/cgi/ viewcontent.cgi?article=1171&context=jcwre (accessed 2 February 2024).

World Economic Forum (2011). *The Water-Food-Energy-Climate Nexus. Island Press, Washington, DC. (2016). The Global Risks Report 2016*, 11e. Geneva: World Economic Forum.

World Wildlife Fund (WWF) (2009). *Understanding Water Risks: A Primer on the Consequences of Water Scarcity for Government and Business*. Surrey: WWF-UK.

Index

Water Scarcity Management: Enabling Technologies, First Edition. Edited by Kanchan D. Bahukhandi, Manish Kumar, Durga P. Panday, and Tushara G.G. Chaminda.

f

g

h

i

j

k

l